Einstein

IN

BOHEMIA

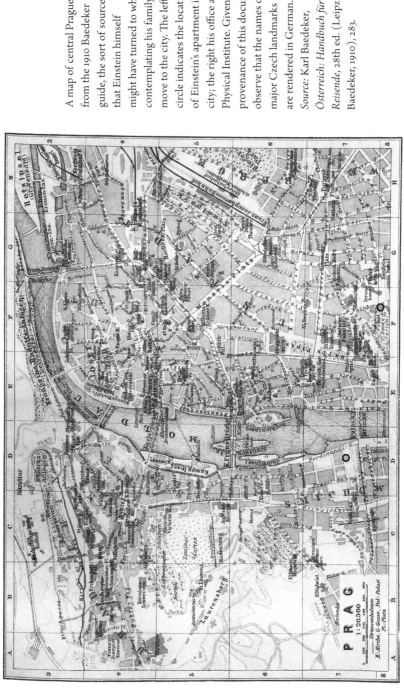

A map of central Prague from the 1910 Baedeker guide, the sort of source that Einstein himself might have turned to when contemplating his family's move to the city. The left circle indicates the location of Einstein's apartment in the city; the right his office at the Physical Institute. Given the provenance of this document, observe that the names of major Czech landmarks are rendered in German. *Source:* Karl Baedeker, *Österreich: Handbuch für Reisende,* 28th ed. (Leipzig: Baedeker, 1910), 283.

MICHAEL D. GORDIN

Einstein

IN
BOHEMIA

PRINCETON UNIVERSITY PRESS

PRINCETON *&* OXFORD

Copyright © 2020 by Princeton University Press
Published by Princeton University Press
41 William Street, Princeton, New Jersey 08540
6 Oxford Street, Woodstock, Oxfordshire OX20 1TR

press.princeton.edu

All Rights Reserved

ISBN 978-0-691-17737-3
ISBN (e-book) 978-0-691-19984-9

British Library Cataloging-in-Publication Data is available

Editorial: Eric Crahan, Pamela Weidman, and Thalia Leaf
Production Editorial: Mark Bellis
Text Design: Lorraine Doneker
Jacket Design: Chris Ferrante
Production: Jacqueline Poirier
Publicity: Sara Henning-Stout and Kate Farquhar-Thomson
Copyeditor: Sarah Vogelsong

Jacket Credit: Panoramic view of Prague, with a view of
Emmaus Monastery, the Church of the Virgin Mary, and
Palacký Bridge, ca. 1900s. Courtesy of Fotogen.
Spine Credit: Photograph of Albert Einstein by J. F. Langhans,
Prague, 1912. Courtesy of Albert Einstein Archives.

This book has been composed in Arno

Printed on acid-free paper. ∞

Printed in the United States of America

1 3 5 7 9 10 8 6 4 2

To Peter Galison,

who first taught me to understand Einstein,

and then to understand him differently.

Contents

Note to the Reader

TERMS AND PRONUNCIATION

Naming conventions present an obstacle for any book that addresses the history of Prague during the late Habsburg Empire, and those issues get more problematic when the story extends into the twentieth century. I have made every effort to use the name appropriate to the time period and the actors concerned. Thus, the region of which Prague is the dominant metropolis was Bohemia, Czechoslovakia, Czecho-Slovakia, or the Czech Republic in different periods. Almost all toponyms had a German and a Czech variant. When there is a conventional name in English—*Prague* instead of *Prag* or *Praha*—I have always used that. For other sites, I have endeavored to always present both names, but I revert to the German spelling (*Brünn* for *Brno*) when presenting the point of view of a Germanophone person. For the same reason, the *Vltava* River is sometimes the *Moldau* in these pages, *Viničná ulice* the *Weinberggasse*, and so on.

Regarding people's names, I have kept them in the form which was most customary for them in the primary context in which I discuss them. So a Czech-identified person would be *František* where a German-identified one would be *Franz*. Both names can be found in the scholarly literature. When there is a risk of confusion, I present both possibilities.

Czech orthography often presents challenges to the uninitiated. The accents represent the length of the syllable, and should not inhibit reading. Likewise, the circle over the *u* (i.e., *ů*), is the same as an accented *u*. The *haček*—what looks like an inverted circumflex over the *c*—requires some attention, as does the letter *c* when unadorned. Here is a rough guide:

c—ts as in bats
č—tch as in batch
ch—ch as in Bach
ě—insert a y sound before the e, as in Byelorussia

ň—gn as in gnocchi

ř—close to the rg sound in bourgeois

š—sh as in brash

ž—g as in Bruges

ABBREVIATIONS AND ARCHIVAL COLLECTIONS REFERENCED IN THE NOTES

AEA—Albert Einstein Archives. Hebrew University of Jerusalem, Israel.

AEDA—Albert Einstein Duplicate Archive. Department of Rare Books and Special Collections, C0701. Princeton University Library, Princeton, New Jersey.

AUK—Archive of Charles University, Ovocný trh 5, Prague 1, Czech Republic.

Cohen Papers—Robert S. Cohen Papers. Howard Gotlieb Archival Research Center, Boston University, Boston, Massachusetts.

CPAE—*The Collected Papers of Albert Einstein*, vols. 1–15 (Princeton, NJ: Princeton University Press, 1987–2018). This is an ongoing project, currently under the editorship of Diana Kormos Buchwald. So far, the volumes extend to May 1927 in Einstein's life. All volumes are available open access at https://einsteinpapers .press.princeton.edu.

HFC—Hanna Fantova Collection of Albert Einstein. Rare Books and Special Collections, C0703. Princeton University Library, Princeton, New Jersey.

LBI—Leo Baeck Institute Archives, New York City.

MÚA AV ČR—Masarykův Ústav a Archiv Akademie věd České republiky, Gabčíkova 2362/10, Prague 8, Czech Republic.

NOK—Nachlaß von Oskar Kraus. Forschungsstelle und Dokumentationszentrum für österreichische Philosophie, Universität Graz, Austria.

Struik Papers—Dirk Struik Personal Papers. Institute Archives and Special Collections, MC 418, Massachusetts Institute of Technology, Cambridge, Massachusetts.

YVS—Yad Vashem Archives, Jerusalem, Israel.

TRANSLATIONS

All unattributed translations are my own. The online version of the *CPAE* offers access to all the original texts cited, as well as an alternative translation.

INTRODUCTION

A Spacetime Interval

Einstein has become a symbol for many, a monument people
have built, a symbol that they need for their own comfort.
—*Leopold Infeld*[1]

Albert Einstein is dead. Bohemia, too, no longer exists. They have as-
cended to the realm of myths and legends, become words to conjure
with—yet they are not, in general, invoked together. Legends have their
own structure and rhythm, their own dominion over portions of our vast
cultural landscape, and these two resonate with different groups, adding
distinctly separate auras of fascination to anything they brush up against.

For 16 months, from early April 1911 to late July 1912, Albert Einstein
lived in Prague. Many people, including fans of Einstein lore or devotees
of Prague's unquestionable charm, skip over this fact. It was, after all, such
a short time, and quite early in the physicist's career too. Einstein was only
32 years old when he arrived in the city, and there was no hint of the international celebrity he would later attain. If you turn to just about any
biography of Einstein, the Prague year (and a quarter) is handled with
streamlined efficiency. How relevant could 16 months be? Historians have
dismissed it as an "interlude," a "sojourn" (sometimes a "brief" one), a
"detour," a "way station," and, most frequently, with Italianesque brio, an
"intermezzo."[2]

We should not be so hasty. For many historical icons whom we associate very specifically with the central places in their biographies, a closer
look reveals that a short period spent in an unexpected locale early in their
lives transformed their worldviews—and they in turn transformed our
world. James Joyce is almost inseparable from the Dublin he

immortalized in his fiction or the Paris where he lived and wrote in his prime, but from 1904 to 1915 he lived on and off in Trieste, then a Habsburg port city, and the impact that these periods had on him is unquestionable. Mohandas Gandhi transformed India using the political techniques he had developed as a lawyer in South Africa, yet his few years studying for the bar in London in the late 1880s profoundly structured his vision of the British Empire and his sense of India. Mary Wollstonecraft, English philosopher and apostle of women's rights, was deeply marked by her unexpected firsthand view, in the early 1790s, of the bloody Terror in Paris. The examples multiply dramatically when we come to the massive displacements caused by the Russian Revolution, the Great Depression, and the rise of Hitler's regime. Einstein would be exiled by the last of these as well; Prague was an earlier, less noticed, displacement.

That Prague would figure into a tale of European history such as Einstein's has been rarely remarked but in retrospect seems almost overdetermined. Once you start to look for it, Prague shows up as an important node in a surprisingly large number of transits across the past millennium. It was a major political center north of the Alps in the high Middle Ages, where it incubated a crucial reform movement within the Catholic Church that would continue to reverberate, often violently, through the Protestant Reformation and beyond. The Thirty Years' War (1618–1648), a brutal conflict that left Central Europe devastated, began in Prague with the tossing of two emissaries out of a window (the second of three famous "defenestrations" to occur in the city). A flashpoint of nationalist mobilization in the middle of the nineteenth century, by the dawn of the next it had become one of the most brilliant centers of literature, painting, and architecture, a rival to Paris and Vienna. Subsequently, the city's history turned much darker. The Munich negotiations that enabled Adolf Hitler to expand his territory and his war machine held the fate of Prague and Bohemia in the balance; after the carnage of the Second World War (which left the architectural heart of the city mostly intact), the Communist coup in 1948 again catapulted Prague to the world's attention. Less of a Cold War flashpoint than Berlin, Prague nonetheless grabbed the headlines twice: in 1968, when the Soviet-led invasion ended its eponymous Spring, and again in 1989, when Wenceslaus Square served as ground zero for the disintegration of European communism. This is the place that, for 16 months before World War I, was Albert Einstein's home.

Besides missing out on the intellectual interest of seeing a person simultaneously adapting to and resisting a foreign place, overlooking Einstein's time in Prague does not make much sense from the physicist's own point of view. When he moved there from Zurich in 1911, he did not know that he would decamp back to the Swiss city three semesters later. He thought he was moving his family to settle for quite some time. It is only after one knows that the Prague period was (relatively) brief that it can be dismissed as a diversion from the ostensibly "ordinary" trajectory of this extraordinary life. What if we did not read the past through the future, or through Einstein's own retrospective haze? Let us take his time in Prague the way he initially did: seriously.

This book takes as its point of departure a particular interval of spacetime—Einstein's 16 months as a professor of theoretical physics at the German University in Prague—and follows that union of place and duration both forward throughout the twentieth century and backward to the distant centuries that still reverberated in local memory. This is not how most histories are usually written or how we typically analyze lives, but my purpose here is to demonstrate that we should reconsider the customary approach. At each moment of our lives, a plethora of possibilities lies ahead of us: not just possibilities of action, but also possibilities of interpretation. Events do not possess a single meaning the moment they happen; they are refracted and reinterpreted over and over again as the future, by becoming our present, forces the past to cohere into a single, linear narrative. In the pages that follow, I aim to hold open the manifold junctures and points of departure that mark a history for as long as they resist closure, which is rather longer than you might expect. To put things in Einsteinian terms, the spacetime interval eventually becomes a defined worldline, but that does not happen immediately and is only clearly discernible in retrospect. While it is still our present, history remains open; to see how it changes, we can dive into the records of the past and hold diverse meanings up to view in our mind's eye. We can see the uncertainties implicit in Einstein's and Prague's interactions for quite some time before the narrative becomes static.

Such lines are worthwhile to trace not merely because Einstein is Einstein and Prague is Prague.

What does it actually mean to *be in a place*? We all move here and there at various times during our lives. Some of these locations, overlapping as they do with particular events or moments, assume extraordinary significance for us. We feel that we would be different people if we had not spent one summer out there or moved to that town for three years a few decades ago. We understand that places are important to us without always paying attention to the crucial role that time also plays. *When* you were *there* can matter enormously, both because of the historical moment of the place and who you happened to be just when you were there. A place can shape you—and you can shape the place—without you being aware of it. This is as true for people in the past as it is for us. We can follow someone's path through the scattered traces he or she left behind (letters, mentions in other people's memoirs, documents maintained by the state, and so on) and look at the ripples these passages propagate through their world, like those triggered by a rock tossed into a pond.

Einstein could be the rock and Prague the pond—or vice versa if you prefer. We can see the implications of that brief entanglement of place and person, neither of which registered immediately. Because, a few years after he left the city, Einstein happened to become the most famous scientist who ever lived, and because Prague has been for over a millennium a central *entrepôt* in European cultural and intellectual life, we have a trove of sources with which to reconstruct their witting and unwitting relationships. Even though he was only there for three semesters, Einstein's time in Prague, the capital of that Bohemia of yore, shaped the science, the literature, and even the politics of that city for decades to come. The same is true in reverse: for the four decades that followed his departure from Bohemia, acquaintances he had made there and ideas that he had been exposed to over a handful of months would continue to occupy him. This does not mean that Einstein's Prague period was "the year that changed everything," or even that Einstein recurred to it especially often (whether fondly or not), but rather that if we plant ourselves in 1911–1912 and foray from there across the lives of the city and the man, vast swaths of their histories can suddenly, and often surprisingly, appear connected. Neither Prague nor Einstein looks quite the same again.

You likely have a mental image of Albert Einstein that bears some resemblance to the historical individual who was born in Ulm, Germany, on 14 March 1879 and died in Princeton, New Jersey, on 18 April 1955. Depending on the context, Einstein is routinely invoked as a genius, a physicist, a pacifist, a sage, and more. Much of what is generally presented as relating to the image of Einstein is exaggerated or even apocryphal, but nonetheless a good deal can be grounded in the actions of a flesh-and-blood figure who lived through and played a significant part in some of the most dramatic, traumatic, and awe-inspiring events of the first half of the twentieth century.

Bohemia, these days, is something else entirely. The term might not recall anything other than those who have been dubbed "bohemians": unconventional artsy types, lolling around like libertines in flouncy clothes and shambolic surroundings. (Einstein, famous in his later years for his disregard of sartorial norms, haircuts, and socks, is frequently called "bohemian" in precisely this sense.)[3] "Bohemia" might be understood as the imaginary space where these folks converge. That meaning of the term was a nineteenth-century product of London and Paris and has little to do with the actual place originally known by that name.[4] The term once referred to a medieval kingdom in Central Europe that later became the far northwestern province of Austria-Hungary.[5] When the Habsburg Empire disintegrated in the wake of World War I, Bohemia began to shimmer out of existence, replaced by a designation with which it had once coexisted: *Čechy*, which is frustratingly plural. *Čechy*, together with Moravia (*Morava*), comprise the *České země*, "the Czech lands"; there is no space for "Bohemia" anymore. A delicate point of translation compounds this confusion: where German makes the distinction between "Czech" (*tschechisch*), a national identification, and "Bohemian" (*böhmisch*), a regional designation neutral with regard to nationality, in Czech the word *český* serves for both. The famous statement by Count J. M. Thun that "I am neither a Czech nor a German, but only a Bohemian" (*Ich bin weder Deutscher noch Tschech, ich bin Böhme*) is untranslatable into the Czech language.[6] Geographically, the land that used to be known as Bohemia came to constitute after World War I the western portion of what was Czechoslovakia and became in 1993 the Czech Republic (or, if you will, Czechia). Although Bohemia as a historically specific term has now vanished in the wake of global conflict, genocide, communism, and the end of the Cold War, one can occasionally find glimpses of it when strolling around its principal city: Prague.

* * * * *

Stories about Einstein and Bohemia double back, overlap, and mutually reinforce at almost every turn. A character you meet at the beginning of one narrative shows up unexpectedly years or miles later; an arcane debate about an abandoned theory of gravitation shapes ideological debates among Soviet philosophers after decades of dormancy; the medieval past proves stubbornly persistent during fin-de-siècle modernity. The diversity is so amazing, such an elaborate tapestry woven from such basic threads, that it might seem like just about everything Prague could be linked to everything Einstein, and the other way around. This is of course not the case. There are important episodes in the history of science in Prague—such as the flourishing of alchemy in the city in the sixteenth and seventeenth centuries or the important epistemological reflections of the early modern Bohemian thinker Jan Amos Komenský (Comenius)—that will not register in these pages at all. Likewise there are eminent figures who occupied Einstein's Bohemian spacetime interval who left no demonstrable trace on the man and his subsequent life, however much we might wish otherwise.

A case in point concerns another famous German-speaking Jew who lived in Prague during 1911–1912 and who was enmeshed in precisely the same social circles as Einstein: the writer Franz Kafka, author of the posthumously published novels *The Trial* (1925) and *The Castle* (1926) and the 1915 novella *The Metamorphosis* (about a man turned into a dung beetle, published nine years before Kafka's early death from tuberculosis). For much of the world today Kafka represents Prague. Because of his massive impact on global literature and thought, derived in no small part from his incisive dissection of the psychological pressures of modernity, Kafka has attracted a voluminous scholarship, some of which meticulously situates his creative energies within the city of his birth.[7] Given that both were Jewish, both stand in as symbols of important transformations of the tumultuous twentieth century, and both were in Prague at the same moment, it is overwhelmingly tempting to imagine Einstein and Kafka engaged in a meeting of minds.

The meeting happened, but the minds did not register it. On 24 May 1911, Einstein gave a talk about relativity theory in front of the local discussion circle of Bertha Fanta, an erudite, philosophically ambitious, and well-connected woman who lived above her husband's pharmacy in Old Town Square. Fanta's group had become a magnet for German-speaking intellectuals, especially those affiliated with the German University, and a pronounced complement of Jews (she and her family became increasingly Zionist over the course of the 1910s). Albert Einstein was introduced to the circle as a way to satisfy his desire for companions with whom he could play his violin. It was a matter of courtesy that the physicist, who had already earned a strong reputation for his foundational contributions to special relativity and quantum theory, would deliver a presentation about his work among them. In attendance on 24 May were his university assistant Ludwig Hopf, the prominent Prague author Max Brod, and (according to the latter's diary) Brod's friends Robert Weltsch and Franz Kafka. The discussion was apparently lively, and Hopf also held forth on the work of psychiatrists Carl Jung and Sigmund Freud. Kafka apparently did not say anything in particular.[8] The next day the group reconvened but Kafka stayed home.

Such was the fabled meeting between Einstein and Kafka: they likely shook hands and little more. There was no extended discourse on metaphysics and spacetime, no exploration of the fate of the individual in industrial civilization, no debate about aesthetics in music or literature. Neither Einstein nor Kafka ever mentioned the encounter. In itself, that might not be surprising: lots of people meet at cocktail parties and then forget their interlocutors. But these two became very prominent people in later years. In 1919, British reports of the confirmation of Einstein's theory of general relativity—work on which he began in earnest while living in Prague—catapulted the scientist to international recognition. He was front-page news around the world and in 1921 made a celebrated return to Prague to give a standing-room-only public talk. Kafka seemingly paid no attention. It is implausible that he was ignorant of either Einstein or relativity—in fact, he made a rare, and characteristically wry, mention of the latter in his diary on 10 April 1922, noting that as a child his engagement with sexual topics was "so innocent and uninterested approximately as it is today in regard to relativity theory."[9]

What about from Einstein's side? People in Berlin who knew Einstein and his second wife Elsa claimed that it was possible that Kafka had visited the physicist in his apartment on Haberlandstraße, since the two did share a mutual friend, the Hungarian-born physician Robert Klopstock. These strands are flimsy (as are the unreliable reports that Einstein returned a copy of *The Castle* to Thomas Mann in Princeton, claiming that its "perversity" kept him from finishing it).[10] Just as Kafka did not recall having met the famous Einstein after 1919, Einstein did not remember a meeting with the posthumously famous Kafka once his novels began to appear in the mid-1920s. A great deal of energy has been fruitlessly expended over the decades trying to confirm a mutual regard between these two towering figures of twentieth-century intellectual history. You can be together with someone and still be alone.

Between Einstein and the controversial Rudolf Steiner, founder of anthroposophy—an esoteric movement that argued for cultivating the mental faculties to access an objective spiritual dimension beyond the senses—a different lack of recognition took place. Welcomed across Europe, Steiner was very popular in Prague, especially among the Fanta group. (He corresponded with Bertha Fanta and visited her home on his trips to the city.) Hugo Bergmann, Fanta's son-in-law and a frequent interlocutor with Einstein during his time in Prague—and a person who we will encounter often in these pages—claimed in his diary that he "tried once, ca. 1911, to bring Einstein together with Rudolf Steiner, and also brought Einstein to Steiner's lecture, but unfortunately he had no comprehension of it."[11] A Prague-based anthroposophist named Franz Halla stated after Einstein's death that he had met the physicist at Steiner's 1911 lectures in the city on "Occult Physiology."[12] Both Bergmann and Halla were dismayed that Einstein had no appreciation for what they saw as Steiner's particular genius, especially when it came to scientific topics like non-Euclidean geometry. But their recollections depict events that could never have happened. Steiner did give lectures in Prague in 1911, but they ended in March, when Einstein had not yet left Zurich.[13] (In a bizarre coincidence, Kafka went to one of these lectures and even visited the anthroposophist in his hotel room before the latter left Bohemia.)[14] At least with Kafka and Einstein, we know they met but just did not care about the meeting. With Steiner, Einstein neither cared nor made Steiner's acquaintance.

Why do these non-events matter? They demonstrate the distinction between history and myth. The former are stories we produce in the present from the traces the past has bequeathed to us. History is fundamentally constrained by the evidence. Myths are different. They are not necessarily false, but their form is not dependent on the surviving traces of the past. They grow from our present aspirations, which enable us to find meaning in them. The interconnection of Einstein and Bohemia is a historically tractable domain, amenable to excavation in libraries and archives (in the case of the present book, those were located in Austria, the Czech Republic, Germany, Israel, and the United States). The interconnection of Einstein and Kafka is another matter entirely. It is understandable that we want to have our heroes align, to enjoy simple stories in which the good guys meet, something important happens, and everyone recognizes it. That, alas, is not always the case. The histories we *can* write are stranger and less straightforward, and therefore vastly more interesting. Throughout this book, I focus on those aspects of Einstein's time in Prague (and its aftershocks) that are trapped within our shards of evidence, and I steer away from the apocryphal or the invented, however satisfying they might feel. There is universe enough in the documents.

Situating Einstein in a place turns out to be rather common: over a dozen volumes and many more scholarly articles fall into the genre we might call "Einstein in *X*." You can easily find Einstein in Zurich, in Bern, in Switzerland, in Paris, in Berlin (several books on that), and even in Ulm, the birthplace that he left while an infant.[15] Many of these works appeared around 1979, the centenary of Einstein's birth, when celebrations around the world marked the launch of what has become a veritable Einstein industry. It is thanks to these works that the present book is possible: Einstein's archive is so comprehensive, his published collected papers (an ongoing project) so well curated, his image so vividly recorded in the memoirs of his colleagues and acquaintances that a historian can reconstruct a good deal of Einstein's life and circle to an astonishing level of resolution. The same is true for Prague in the early decades of the twentieth century, a bequest of the equally impressive Kafka industry, which has uncovered many valuable details of cultural life in the city—its art, philosophy, literature, public life—especially among its intellectual Jewish inhabitants. So, even though Einstein and Kafka never met in a significant way, we now know a vast amount about the cultural history of the city they both occupied at one moment, enough to flesh out manifold connections between person and place.

It should not, therefore, be a shock that there is a volume called *Einstein and Prague*, though it is only available in Czech. Like so many of its brethren, it appeared in 1979, under the editorship of the gravitational physicist Jiří Bičák, who is still active today at the Charles University in Prague. A slim paperback of 63 pages, it is not so much a history of the subject at hand as a collection of documents: an introduction by Bičák; a translation into Czech of a section of the biography of Einstein by the physicist and philosopher Philipp Frank; Einstein's most famous scientific paper produced in Prague, proposing that light bends by a certain amount when passing massive bodies like the sun; and Einstein's preface to the Czech translation of his popular book on relativity from 1916.[16] Bičák's book serves as a fitting Czechoslovak contribution in the context of the festivities of 1979, but it does not take a wider view of Einstein's interaction with the city. Likewise, the few shorter studies that have examined this moment have concentrated rather strictly on the scientist himself. One of the things it means to be in a place, however, is that you are usually not there by yourself. To understand Einstein in Bohemia, we need a larger cast of Bohemians (none of whom happened to be lower-case bohemians).[17]

* * * * *

A major theme of this book concerns precisely this question of belonging to a larger community, a matter that is especially tricky to nail down for Einstein, who in later years was fond of speaking of himself as a loner (despite decades of evidence of him as a social and socially engaged individual). Some forms of belonging in the modern world are thrust upon us; citizenship is a good example. It is also an instructive one for teasing out Einstein's relationship with Prague. That might seem like a non sequitur, but the conflict over citizenship has everything to do with Einstein's relationship to the category of "German," perhaps the crucial issue that shaped his every experience in the Bohemian metropolis. A brief survey of the chaotic story of the citizenship of the famously antinationalist Einstein serves to highlight many of the political and cultural assumptions that structured his world both in Prague and beyond it.

The first time we encounter documents expressing Einstein's views on citizenship, he was trying to get rid of one. He was born in Ulm, located within the German state of Württemberg, and he retained that citizenship after the family's departure for Munich, capital of the state of Bavaria, shortly afterward. As his graduation from secondary school approached, however, Einstein refused to enlist in the obligatory military service. At a month and a half shy of his seventeenth birthday (and thus still a minor) he asked his father to renounce his citizenship for him. As of 5 February 1896, when the decision was confirmed by the town of Ulm, Einstein was officially stateless, a status rather more thinkable in his day than in our own highly legalized international order.[18] He visited his family in Italy and soon settled in Switzerland to complete a year of preparatory schooling before enrolling at the Federal Institute of Technology (ETH) in Zurich. He became a Swiss citizen on 21 February 1901, almost five years after he had forsaken his previous citizenship. Despite his birth within the newly formed German *Reich*, Einstein seemed happier being

Swiss. The tension between those two forms of belonging would burst into head-on conflict two decades later.

It is hard to determine how important questions of citizenship were to Einstein. He traveled extensively internationally—at first around Europe in the 1910s and then around the world in the decade after—and he kept his passports in order. In notes to himself, however, the man who had moved from Zurich to Prague (in Austria-Hungary), back to Zurich, and finally in 1914 to Berlin, the capital city of both the state of Prussia and the German Empire, toyed with the idea that perhaps citizenship was not that important. In a document he drafted between late October and early November 1915, when the Germany he lived in was in the grips of World War I, he wrote: "The state to which I belong as a citizen plays scarcely any role in my emotional life; I consider the affiliation with a state as a business arrangement, something like the relationship to a life insurance policy."[19] So he wrote, but then he crossed it out. It was to his advantage to keep quiet about questions of citizenship. Although he was now a professor at the University of Berlin and a director of the Kaiser Wilhelm Institute for Physics (and thus a German civil servant), he insisted that he remained a Swiss citizen. The state at least tacitly concurred: unlike many of his colleagues, Einstein, as a foreign national, was not approached for war work.[20] He confirmed this status in a letter to the Berlin-Schöneberg Office of Taxation in 1920: "I am Swiss, here in Berlin since spring 1914."[21]

Einstein's cavalier attitude toward his belonging in Germany put him at odds with some of his closest friends. Unique among his colleagues living under the Central Powers, Einstein was not subject to the boycott and travel ban that the victorious powers of France, Belgium, Britain, and the United States imposed on the defeated states after the war. A well-known pacifist and opponent of the war, he was an ideal ambassador to Paris or New York (the latter of which he visited as part of a fundraising expedition for the London-based Zionist Organization) once the confirmation of general relativity hit the newsstands. He endured his celebrity in part in hopes of helping end the boycott. The eminent chemist Fritz Haber, discoverer of the eponymous process for fixing atmospheric nitrogen—and also an architect of the system of gas munitions during the Great War—vociferously protested against Einstein traveling to these (in his view) still quite hostile powers:

> If you *at this moment* journey to America, where the new president postpones the ratification of the law through which the peace between the United States and Germany ought to be proclaimed, if you *at this moment* travel with the English friends of Zionism while sanctions allow the opposition between England and us to persist with new acuity, then you are announcing publicly before the entire world that you want to be nothing other than a Swiss who by chance has his residence in Germany. I ask you to consider whether you really want to make that announcement now. Now is the moment in which adherence to Germany has a bit of martyrdom to it. . . . For the entire world you are today the most important of the German Jews. If you *at this moment* fraternize ostentatiously with the English and their friends, then people here at home will see that as a testament to the faithlessness of the Jews.[22]

(Haber's linkage of Germans and Jews as compatible but occasionally conflictual identifications broaches another theme of this book.)

Einstein's response was characteristically dismissive. "I have declined many attractive calls [i.e., professorial appointments] to Switzerland, to Holland, to Norway, and to England without even considering accepting one," he wrote. "I did that however not out of allegiance to Germany but rather to my dear German friends, of which you are one of the most excellent and wish me the best. Allegiance to the political structure of Germany would be unnatural for me as a pacifist."[23] He went on the trip to the United States, and many other places besides: Japan, China, the British Mandate of Palestine, and even Prague, all on his Swiss passport.[24] He declared in 1922 to Gilbert Murray, the Australian-born British man of letters, that "I am not an appropriate representative for German intellectuals, because I am not seen by them in their full number as *their* representative. My outspoken international attitude, Swiss citizenship, and Jewish nationality work together so that I would not be met with in a political relationship by the majority of the masses with the trust that a representative of a country must possess to be able to serve as a liaison with success."[25]

It would soon become harder for Einstein to maintain this attitude, because in 1922 he was awarded the Nobel Prize in Physics for the prior year. Neither the diplomatic imbroglio that ensued nor its resolution would be thinkable without a globalized world intimately connected by steamships and telegraphs. In the fall of 1921, Einstein had left on an invited trip to East Asia, ignoring a not-so-subtle communication from Swedish colleagues who suggested he postpone the trip because of an impending special announcement. Einstein received notification of the prize in Japan by telegram. Now there was a problem. Nobel Prizes needed to be received in person; if the laureate was not personally able to do so, his (or, disappointingly rarely, her) ambassador could stand in. Naturally, the Swiss ambassador presented himself for the honor, given that Einstein was at that very moment traveling as a citizen of the canton of Zurich. So did Rudolf Nadolny, the German ambassador to Sweden. The German version was that when Einstein assumed his Berlin position, German citizenship attached to him as a requirement of the post. Einstein had objected to this clause during negotiations in December 1913, and the Germans had not felt it necessary to resolve the matter. (A similar condition had applied when Einstein moved to Prague, and he had ignored it then as well.)[26] After 1923, Einstein relinquished his opposition to the state's narrative and settled in as a German citizen—though he still traveled on his Swiss passport at times.[27]

Einstein's newly regained German citizenship only remained settled until 25 August 1933, when Germany's National Socialist government rescinded it, transforming him into the world's most famous refugee. He was granted asylum in the United States and offered a position at the Institute for Advanced Study in Princeton, where he became a prominent advocate for those tragically displaced by Hitler's racial laws and the conflict in Europe. In 1935, the Einstein family traveled to Bermuda so that they could reenter the United States on permanent visas. On 1 October 1940, after the obligatory five-year waiting period, Einstein, his

stepdaughter Margot Einstein-Marianoff (*née* Löwenthal), and his secretary Helen Dukas traveled the short distance from Princeton to Tren-
ton, New Jersey, to take their oath of allegiance as American citizens.[28]
Even then, as visible and as prominent as he was—he was likely the most
recognizable American alive—J. Edgar Hoover's FBI explored in 1950
the possibility of stripping Einstein of his citizenship as an undesirable
alien for his open pronouncements in favor of civil rights and nuclear
disarmament.[29] That limited effort failed, and Einstein died as an
American citizen.

As one can see from the turbulent history of Einstein's citizenship, calling him a "German" is not a straightforward matter. Although he was
born in Germany, he spent much of his life trying to avoid being considered a citizen of the German state. Likewise, his confessional status as
a Jew meant that for many people in the land of his birth he was not
straightforwardly German as a matter of ethnicity. On the other hand,
it seems obtuse simply to go by his last passport and call Einstein
"American," full stop. Rather than attempting to resolve these issues by
reference to strict categories, Einstein's biography urges us to recognize
the complexity of personal and state identifications of individuals. They
are not the background to history—they are often its very substance.

The simultaneous malleability and solidity of categories was equally
true in the context of Prague. According to the census, during Einstein's
time in Prague the city's population was 93 percent "Czechs" and 7 percent
"Germans," defined by how individuals listed their "everyday language"
(*Umgangssprache*). (The proportions for Bohemia as a whole were not
so lopsided, with roughly a third of all residents set down as German.)
As is true for accounts of the many aspiring nation-states that emerged
out of multinational empires during this period, nationalism has been
the dominant framework for histories of the region that would be called
Czechoslovakia from 1918 to 1993. Sometimes, Prague is identified as a
"city of three peoples," with Czechs, Germans, and Jews living cheek-by-
jowl, sometimes discordantly but often productively.[30] (Set aside for
a moment that many of the city's residents were bilingual to some
degree and that there were Jews to be found among both linguistic

communities.[31]) More often, the history that has been told has been one of implacable hostility between Prague Germans, who despite their small numbers benefited enormously from Habsburg privileges, and nationalist Czechs, who wanted autonomy or even independence.[32]

Einstein was largely unaware of all this complexity. He interacted exclusively with people who identified as German because that was who he met at the German University. Outside of more intimate contacts, such as those he made in the Fanta circle, he only conversed with German-speakers because, knowing no Czech, he had no alternative. Nonetheless, every local he met would peg him as part of a group, as a German and sometimes as a Jew, depending on the context. Neither was a good fit. His Jewishness was, as we shall see, somewhat of a murky matter during this period, his Germanness no less so. He was neither a German citizen nor a Habsburg German; still less was he a local "Prague German." If he had to be categorized, it would be as a Swiss—a fact that almost lost him the position in Prague in the first place. The problem of *identity* is truly central to this story.

So central, in fact, that I am going to avoid using the word. Following the suggestion of the sociologist Rogers Brubaker and the historian Frederick Cooper, I opt for "identification" over "identity." As they note, doing so "invites us to specify the agents that do the identifying. And it does not presuppose that such identifying (even by powerful agents, such as the state) will necessarily result in the internal sameness, the distinctiveness, the bounded groupness that political entrepreneurs may seek to achieve. Identification—of oneself and of others—is intrinsic to social life; 'identity' in the strong sense is not."[33] Those in Prague who called themselves "Germans" or called Einstein "German" were making deliberate, highly politicized choices, just as those historians who write "German history" by confining themselves within the borders of what was after 1871 (or 1918, or 1945, or 1990) the German state without taking into account the German-speaking populations of Austria-Hungary, Switzerland, Romania, Russia, France, Belgium, Italy, Namibia, and the United States are doing.[34] For the purposes of this book, "being German" means much less than "speaking German," and thus much of what follows will be told in terms of people who were Germanophone (or Czechophone).[35]

In this way we can avoid assuming national identification and instead watch it in the process of construction.

* * * * *

The same strategy helps us navigate the kaleidoscopic history of Prague. The city was at various points the seat of the Přemyslid dynasty, the medieval capital of the Holy Roman Empire, a troublesome and rebellious counterweight to Habsburg centralism, a prosperous provincial node between the two major Germanophone imperial centers of Berlin and Vienna, the capital of a fledgling state—at first a republic, then occupied by Nazi forces, and then a communist polity—and more besides and in between. Any close acquaintance with the history of Prague in a specific period quickly teaches you that stories have a way of breaching the dams between political regimes and occasionally bending back to wash up medieval flotsam on the beaches of modernity. Although obviously physically fixed in the landscape and forced, as we all are, to move through the years in a linear fashion, when it comes to the spacetime of memory, the city demonstrates a persistent capacity to contravene expectations.

No wonder that one of the most enduring ways to speak of the place—popular among folklorists of the past and tourists of the present—is as "Magic Prague," the domain of alchemists and Golems, mystical rabbis and deranged princes, heroic mercenaries and fantastical scribblers. The shelves are lined with book after book promising a portrait of a Bohemian wonderland amid the Baroque towers and Cubist apartment blocks of the city on the banks of the Vltava (or, if one prefers the German, the Moldau).[36] Thankfully, these are not the only surveys of the history of Prague available. Instead, a series of vivid, *longue durée* narrative histories stress the harsher realities of national conflict with a decided emphasis on artistic creativity.[37] It seems that you must either choose a Prague populated by poltergeists or one abounding in aesthetes. This book offers a third option.

While the Prague one finds in these pages does have its share of poets and eerie coincidences—though you would strain to find them supernatural—the central characteristic of this Prague is as a city of knowledge, a place where science, philosophy, and rationality were serious business. Largely because of the Cold War, Prague has fallen in the popular, especially Anglophone, mind on the far side of a divide that snakes through Europe, like the erstwhile Iron Curtain, cleaving the West from the East.[38] During most of its history, up to and including Einstein's time there, that was not the case. Not only was Prague not a mystical wonderland, but it was a hub of European science, the home of Tycho Brahe and Johannes Kepler, Christian Doppler and Ernst Mach, and, yes, Albert Einstein. Bohemia as a whole was unquestionably modern. It was the seat of Austria-Hungary's large chemical industry, producing 37.7 percent of its chemicals as well as 65.5 percent of its food and 95 percent of its sugar. Prague already had an electric tram system in 1891. It hosted international conferences and was plugged into a transnational network of Germanophone universities.[39] That was, after all, how Einstein got there.

Now that we are situated with our protagonist, his city, and his time, let us turn to what kind of book this is. In each chapter, we will follow Einstein and Prague, though in some there will be rather more Einstein than Prague, and in others the reverse. There is no single way of being a person in a place at a specific time, and I hope to capture that multiplicity by guiding you in and out of the entangled pair of Einstein and Bohemia before that interaction dissipated from the memory of the public and historians alike. In order to fully grasp how Einstein understood his context and how Praguers made sense of him, sometimes I have to wander rather far from the mustachioed theorist. He will always come back. None of the stories here are arbitrary—they are all tightly connected to man, place, and time—and they are all verifiable in the documentary record. Some good stories did not make it in because they were digressions, and others were excluded because they were simply not true. Instead, we move through Einstein's life in, out, and around Prague both in real time and in later recollections of it. Einstein is a tool for narrating the history of Prague, and Prague is a tool for narrating the scientist's life.[40]

The first chapter explores the seemingly simple question of why and how Einstein settled in Prague in the first place. The next two chapters trace his life in the city, both in his scientific work to formulate a relativistic theory of gravitation and in his social life at home and with acquaintances. By the beginning of chapter 4, Einstein has left Prague, returning only once for a brief visit. Yet some of the most significant resonances of his Bohemian moment were felt in the years after he left, both by those in the city and by the physicist in Berlin and Princeton. The next three chapters explore the ripples of Einstein's Bohemia in philosophy of science, Germanophone literature, and Jewish and Zionist politics. The final chapter returns us to the unsung majority—those who identified as Czechs—and the way they fashioned their own image of Einstein at different points across the twentieth century. Historical narrative, like memory, does not always flow linearly, and sometimes the story doubles back or sneaks ahead. Likewise, the cast of characters beyond Einstein do not stay confined in cameo roles. Some, such as the philosopher and physicist Philipp Frank, the novelist and critic Max Brod, and the Zionist intellectual Hugo Bergmann, prove just as essential as our main character.

Albert Einstein is dead. Bohemia, too, no longer exists. But in the pages that follow they come together again, bringing their worlds into view once more.

CHAPTER 1

First and Second Place

I . . . was so physically and psychically spent from life in
Prague that I greeted the return to orderly and calm
circumstances as a kind of salvation. People both in Vienna
and among us in Germany had then no proper conception of
how the Germans in Prague felt.

—Carl Stumpf[1]

It can be easier to venture to a place unknown than to return to something familiar. In the new place, you do not know the stories, the scripts, the patterns that provide the loose rubric by which the locals shape their days. This helps account for that feeling of liberation when you encounter the unfamiliar: here, you have the opportunity to reinvent yourself, unconstrained by the norms and expectations that bound you in a place laden with overfamiliarity. This ease breeds two illusions: the first, that in this new place there are no barriers or taboos; the second, that those barriers that do exist do not also apply to you.

Albert Einstein knew very little about Prague before he arrived, with that little itself cobbled out of general knowledge, hearsay, perhaps a guidebook or two, and prior epistolary interactions with resident faculty and Viennese bureaucrats. His ignorance did not prevent him from becoming enveloped in a complex cultural environment, although it did hinder him from noticing how total his immersion was. History shapes you whether or not you are aware of it, and you sometimes have lines and a role allotted to you without your ken. Einstein walked into a play in Prague—one that was neither a tragedy nor a comedy, though it had elements of both—that had been running for centuries, and he would perform his part in various ways, first by virtue of his position alone, and later, as he came to appreciate some of the specifics of the story, somewhat according to his choices.

In this chapter we follow the story before the story, on several scales. For Einstein to arrive at Prague, he needed to be offered a position, and this involved its own drama, some of which Einstein knew about and some of which he did not. The minor controversies that attended his hiring as an ordinary professor of mathematical physics at the German University in Prague conditioned his reaction to the city before he accepted the position. The position itself occupied a minor role in a larger narrative, that of why there was a German University in Prague in the first place. To get a sense of the full depth of that story, one needs to appreciate the very long history of the university in Prague, founded in 1348, the oldest university in Central Europe. All of Einstein's interactions with Prague, from before he arrived until his death, were in some way mediated by this university, and its history looms large in these pages. One cannot understand why the faculty of philosophy offered a position to Einstein, what he encountered when he accepted, nor what transpired afterward without some fairly deep background about that institution.

The university in Prague was founded in 1348, and we could take that date as the creation of the German University as well—there are those who do—although more properly, the latter was created in 1882, when the preceding institution split into a German and a Czech university, separate and not quite equal. As philosopher Carl Stumpf's words at the beginning of this chapter indicate, one dominant way of viewing the history of Prague is through the lens of a constant conflict between Czechs and Germans—setting aside for the moment the Jews, who will be addressed in a later chapter—reaching all the way back to the medieval period and the rise of the city as Bohemia's metropolis. I, too, will follow this strand of tension, as it was central for many of the people in the pages that follow, especially those who identified as "German." It is, however, important not to overstate the clash. While there were periodic conflagrations, for most of the population, most of the time, residents managed to live and work side by side. Historian Tara Zahra calls this "national indifference": the deep bonds of connection across groups that are often obscured by the flames of nationalism fanned by intellectuals.[2] This is all very true, and I will flag that indifference in this chapter and the ones that follow when it appears.

National indifference was not, however, Einstein's experience of Prague, largely because he experienced the city, at least consciously, almost entirely through interactions with "Germans"—from hereon in I will dispense with the scare quotes unless necessary—and often highly politicized nationalist Germans at that. On one important level, Einstein came to understand Prague through the lens of national (and personal) conflict, and the locus of that conflict was the German University.

* * * * *

One of the first things he thought to do when it came to Prague was tell his mother. In April 1910, while still settling in after the honeymoon period of his arrival from Bern as the new extraordinary professor of physics at the University of Zurich, Einstein wrote her with exciting news, presented as an aside: "In addition there is something else interesting. I will, most probably, as an ordinary professor with significantly better salary than I now have, be called to a large university. *Where* it is I am not yet allowed to say."[3] This would have been a notable step up for the former patent clerk who had been unable to secure a job after his graduation from the Federal Institute of Technology in Zurich. (Einstein had worked at the patent office in Bern from 1902 until 1909.) He had just turned 32 years old, remarkably young for consideration for an ordinary professor post, and after just a year into a term as an extraordinary professor. (A word about terminology: in the university system of the German states, including Austria-Hungary and eastern Switzerland, an extraordinary professor occupied a status similar to an untenured assistant professor; the ordinary professorship was the highest position, "ordinary" because it was part of the statutes of the university and could not be dispensed with.[4]) Discretion—perhaps also a hint of nervous superstition?—prevented him from stating that the call might come from the German University in Prague.

As with most such openings, this one was the consequence of a retirement, that of Ferdinand Lippich, who had occupied the chair of mathematical physics since 1874 (five years before Einstein was born) and retired on 1 October 1910, three days shy of turning 72. The physicists were members of the faculty of philosophy, and the faculty made the decision to conduct a search to replace Lippich. Not that they had much choice: physics was an important field, and it seemed hard (Lippich's case notwithstanding) to hold on to talented faculty for long. In 1909 Ernst Lecher, who had succeeded the towering experimentalist and philosopher of science Ernst Mach in 1895 when the latter decamped for Vienna, himself absconded to that city after a relatively short tenure. This had become a usual affair. Prague was broadly perceived by the German professoriate as, if not quite a hardship post, certainly less desirable than positions in Berlin, Munich, or Vienna, and it was a penultimate way station in the musical chairs of Germanophone academia across both the German and Habsburg empires as talent looked for its final resting place. Lippich was a man of the old school, and the German University began seeking someone new, in multiple senses of that term, to replace him.[5]

The search committee consisted of Anton Lampa (Lecher's successor), Georg Pick (a senior mathematician who had spent his entire career at Prague), and Viktor Rothmund (a physical chemist).[6] The composition of the committee already reveals something important about the search process: it was fundamentally conservative. No institution could cover all of physics, so most chose to concentrate in a few areas in which they were already strong or had the necessary equipment to attract their top choice. The members of the committee were those already there—and often the person retiring hired his (always "his") own successor—so the choice of where to concentrate was based on these entrenched interests. As the committee empaneled in 1910 stated in its report, the choice of focus was clear: "Modern theoretical physics has experienced its most powerful stimulus from the study of electricity."[7] This would not have been the decision everywhere, as quantum theory and especially its applications to thermodynamics were hot topics (excuse the pun) at many leading faculties. But because of its members' own backgrounds and concerns, the committee decided "to recommend to the faculty only such researchers to occupy the proposed post who have assumed a position in their works on this most important problem of modern theoretical physics and thus to guarantee that our university will remain secure of a share, corresponding to its tradition, in the further development of theoretical physics."[8] Each member of the committee had his own reasons to prefer this topic. Pick had been the research assistant of Ernst Mach upon his hire at the university; Lampa was an admirer of Mach's, all of which encouraged them to look for someone who would be consistent with Mach's empiricist philosophical views; and Rothmund, the most junior of the three, was drawn to related questions of atomic physics. Like all hiring committees, the trio was to come up with a ranked list of three names, filtered by the institution's preferences.

Their choice for *primo loco*, or first position, was clear: "Einstein's research on the electrodynamics of moving bodies has made a new era."[9] This assessment came with the highest of imprimaturs, that of Max Planck, the leader of German theoretical physics, from his post in Berlin. When writing to leading colleagues from other institutions about whom they should consider, Planck—who had published all of Einstein's 1905 "miracle year" papers on the photoelectric effect, Brownian motion, and relativity theory in the *Annalen der Physik*, which he edited—lauded the work of the young Zurich professor:

> It surpasses in boldness everything that has been achieved so far in speculative natural science, indeed in philosophical epistemology, so that non-Euclidean geometry is child's play by comparison. And in fact the relativity principle, in contrast to non-Euclidean geometry—which up to now has only seriously come into use in pure mathematics— claims with full justification a real physical significance. With the extent and depth of the revolution in the domain of the physical worldview summoned by this principle one can only compare that caused by the introduction of the Copernican world system.[10]

It was somewhat unusual to propose for such a chair an extraordinary professor of such recent vintage, but the committee members were convinced that the only reason Einstein had not yet received an *ordinarius* post was his youth. Prague could get in early on a good thing. "Without a doubt Einstein has had the deepest influence on the development of modern theoretical physics," they wrote, "and there is no doubt that theoretical research in the coming years will entirely follow the path that he has blazed."[11]

It remained to fill out the other two positions on the list. The second was equally easy: Gustav Jaumann, ordinary professor of physics at the polytechnic in Brno (known in German as Brünn), who had been an extraordinary professor at the German University in Prague before being promoted to the chair in Moravia. Jaumann was thus more familiar to the locals, but for that reason also a bit riskier, as the committee noted: "Jaumann's position in contemporary physics is an isolated one; this is connected on the one hand with the fact that his entire mode of thinking diverges from the mode of thinking of the dominant theory, on the other hand it is however also not characteristic of his mode of presentation to attract those who think differently to a closer engagement with his theory."[12] Although they provided a list of honors for Jaumann—which they had not done for Einstein—and noted that he was more senior in age, there was a distinct lack of enthusiasm for this familiar face. Hence, *secondo loco*. If Einstein turned them down, they would settle for Jaumann.

The committee named Emil Kohl to the third position on the list. There was little to say about Kohl, and the committee included all of it in their report. He was from Vienna, where he had received his doctorate in 1890—15 years before Einstein had earned his—and had then taught in secondary school while working on his habilitation, the second doctorate required for a professorship at a German university. He received this in 1903 also from Vienna and had been named as *tertio loco* for a post in Czernowitz (in the far eastern reaches of the Habsburg Empire) only recently. Needless to say, the job went to someone higher on the list, and he remained as an instructor (*Privatdozent*) in Vienna, working on fluid dynamics and waves in matter, including seismic phenomena.[13] This was not someone working at the cutting edge of electromagnetism, nor someone who had strong connections to Mach, as did Jaumann and—as we shall see in a later chapter—Einstein.

Had he been granted the opportunity to read the report when it was sent off to the Ministry of Education, Planck would have been pleased, as would any of the leading theoretical physicists in German space. This was a list designed to hire one and only one outstanding candidate: Albert Einstein. The ministry, however, had other plans. In July 1910, as a postscript to a letter to Arnold Sommerfeld, professor of theoretical physics in Munich, Einstein related the bad news: "I am not going to Prague. The Ministry has—as I learn from Prague—made difficulties."[14] The minister of education, Karl von Stürgkh, as was his prerogative, had inverted the order of the first and second positions. Jaumann received the call.

To make sense of this decision, it is essential to understand both why there was a *German* University in Prague and the difficult status of this institution within an empire run out of Vienna. This means stepping back—way back—to the long history of the university in Prague. Einstein almost certainly knew almost none of what follows, but that does not mean it did not matter to him every day he spent in the city. We start in the fourteenth century.

* * * * *

For some German nationalists both in Einstein's time and since, there had *always* been a German University in Prague, and it had been founded in 1348 by Holy Roman Emperor Charles IV, who made the Bohemian capital the center of his empire.[15] Arguments about continuity that last for centuries are hard to pull off, especially in Central Europe, where dynastic tumult and catastrophic warfare have more often been the rule than the exception, and so it proves here. Much of the confusion stems from semantic ambiguity about what precisely one means when calling a university "German."

The Holy Roman Empire, which Charles IV helmed, is often called, in German, the first German Reich—in reference to which the Nazis would, in time, name their dominion the third—but Charles was also the king of Bohemia. Born in Prague to a mother of the Přemyslid dynasty, he had been raised in France and Italy, represented the House of Luxembourg, and spoke Czech as well as German (but preferred Spanish). Is "German" supposed to be an ethnic category? If so, it is devilishly difficult to figure out what the students of the university Charles founded might have defined themselves as, even if one were to irresponsibly set aside decades of debate about what "ethnicity" might have meant in medieval Europe. Therefore people who want to make this argument often turn to language: if the university denizens spoke German, they were Germans. But from its foundation, the university in Prague, like *all* universities in Central and Western Europe until centuries later, was run in Latin. Insistence on a national identification of the university is pointless when referring to the historical origins of the university and only became meaningful, and pointed, in the nineteenth century.

What we can say for certain is that a Slavic people who called themselves Czechs had established a dynasty (the Přemyslid) centuries before the establishment of a university, and also that before the twelfth century there were already people who spoke German living and working among the Czechs.[16] Over time Prague grew, and grew prosperous, and Charles IV wanted to see his capital become a major metropolis. He executed a number of building projects and reforms to elevate the city, such as the founding of the New Town in March 1348 across the Vltava (Moldau) River from the Hradčany castle that still towers over the city. Creating a university fit directly into these ambitions. The pope approved Charles's request to establish a *studium generale* on 26 January 1347, and the emperor acted by issuing an imperial and royal charter for the institution on 7 April 1348. This meant that theology could now be taught in the heart of Europe, and faculties of arts and medicine were also created at this time. (The law faculty was organized somewhat separately in 1372.)[17] Very quickly the institution began to attract students from both the region and farther afield, and this very attractiveness contained the seed of later problems.

The medieval university needed to organize its burgeoning student and faculty population somehow, and the statutes of 1360 borrowed a technique from the arts faculty at the University of Paris—which otherwise had a quite different governance structure—to separate the population into four *nationes* based on land of origin and then make decisions, such as who was to serve as rector, with each group casting a vote.[18] The term *nationes* is one of those pernicious false cognates that might make one think that it had something to do with "nations," and perhaps even "nationalism" in our modern sense. (Nationalist interpretations of the university appropriating it in precisely this manner don't help.) But the *nationes* were quite explicitly geographic rather than ethnic, and each comprised speakers of many languages—all of whom were at the very least bilingual, since they did all university business in Latin. The Bohemian *natio* included students and faculty from Bohemia, Moravia, Hungary, and neighboring territories. The Bavarian *natio* comprised those from Bavaria, Austria, Tyrol, Frankia, Swabia, Switzerland, Hessen, the Rhineland, the Netherlands, Westphalia, and part of Hannover—many of the contemporary residents of these areas would be rather shocked to be regarded as "Bavarian." (Einstein himself would have fallen in this *natio*, either as a Swabian or as a Swiss.) The two remaining *nationes* were the Saxon (Brandenburg, Braunschweig, the Hanseatic cities, Mecklenburg, Lower Saxony, Norway, Denmark, Finland, and the Baltics) and the Polish (Poland, Lithuania, Prussia, Schleswig, Thuringia, Saxony, and others). None of these populations, it should be obvious, was homogeneous, and each accounted for a different proportion of the student body across the first half-century of the university. Over this period the Bohemian *natio* grew from 12 to 22 percent of the university's graduates, while the Polish grew even more, from 16 to 30 percent, and the Bavarian started at 30 percent before dropping to 20 percent.[19] Of course, there were conflicts—what university system does not have them?—but when they occurred they were more likely to be theological and confessional than anything else.

It was just such a confessional dispute that led to the first of three major divisions of the university in Prague, that of 1409. The problem had started about 25 years earlier, when the papal schism of 1378—during which one pontiff decamped to Avignon and the other held fast in Rome, both decrying the illegitimacy of the other—sent shockwaves across Latin Christendom that finally hit Bohemia in 1384. The theological masters and doctors of the university debated which side to support, but the question was not only, or even not primarily, a doctrinal one. For Wenceslaus IV, Charles's son and successor, it was a political dilemma of significant proportions, and he wanted backing from the university to be neutral in the dispute. (He also wanted to set up a third pope in the German states, but never mind that now.) The Bohemian *natio* was willing to support him, but the other three were for Rome. To get the desired result, Wenceslaus did what short-sighted rulers have done repeatedly before and since: he amended the franchise. He granted the Bohemians three votes to the others' single vote in a decree issued at Kutná Hora (Kuttenberg) in 1409.[20]

That solved one problem and created a whole slew of others. The Polish, Saxon, and Bavarian *nationes*—both students and masters— preferred to leave rather than put up with second-class status. Between 500 and 800 members of the university community pulled up stakes, some of them founding another great school at Leipzig, others heading to recent establishments in Heidelberg, Erfurt, Vienna, or Cologne.[21] (The Polish *natio* particularly preferred nearby Leipzig.) These numbers were sizable, given that the city as a whole had a population on the order of 40,000 at this time. For later nationalists of both Czech and German persuasion, this was the first split of the university, and whether it was interpreted as a triumph of national reclamation of a local institution from foreigners or as a catastrophe born of xenophobia depended on one's slant. Indisputable across the political spectrum is that the Decree of Kutná Hora diminished the stature of Prague from that of a European city to that of a regional one. As even R. W. Seton-Watson, British historian, political activist, and devoted advocate of Czechoslovak nationhood, put it evenhandedly in 1943: "That Prague had been the intellectual centre of the whole Reich under Charles IV is true; that it was no longer so after the lapse of a generation is equally certain."[22] That same year, Nazi historian Wolfgang Wolfram von Wolmar could only concur: "The work of the German emperor Charles IV, the first university of the empire, was degraded by his incompetent son into a provincial school and was stripped of its European reputation."[23] It is unlikely that these two would agree on anything else.

The decline of the influence of Prague's university in the fifteenth century was probably overdetermined, however, as Bohemia was during this period engulfed in the traumatic religious conflict known as the Hussite Wars. Jan Hus, after whom this strife is named, was not himself a violent man. A product of the University of Prague, he was ordained as a priest in 1400, at the high point of the city's intellectual life. Two years later he began advocating a reform of certain church practices along some of the lines proposed by the influential Oxonian John Wyclif. Symbolically most salient to Hus's followers (though not to Hus himself) was the demand that at Holy Communion the parishioners receive both the wafer and the chalice—that is, "of both kinds," *sub utraque specie* in Latin. Those who later marched under the banner of Hus's martyrdom eventually adopted the term "Utraquist" to designate their interpretation of Catholic doctrine, and for over two centuries they functioned in Bohemia as a semi-independent entity, often conducting services in Czech (as Hus also had).[24] The price of that quasi-autonomy was enormous bloodshed, which broke out after the fiery execution of Hus as a heretic at the Council of Constance in 1415. The violence swept across the region, eventually endowing Hus and the later Hussites with the retrospective reputation as forerunners of Martin Luther's Reformation and the ensuing wars of religion.[25]

Throughout this period, the university continued to exist, functioning very much as a regional university. In 1417, it declared itself officially Utraquist, and within two years the Catholic masters had left Prague. They remained excluded from the university until 1462, though even after that date limited Catholic influence lingered. A century later, in 1562, the Catholic Counter-Reformation sent the recently established Jesuit order to Prague to set up a separate institution of higher education that would exist in parallel with the Utraquist university: a "Clementinum" to the preexisting "Carolinum." And so the university in Prague was again divided, although in this instance the split was initiated by the establishment of a new, unintegrated school. Once again, the issue was not language or ethnicity—though it would later be interpreted as such by nationalists of both persuasions—but a confessional standoff between Catholics and Utraquists.[26]

Thus Prague had two distinct universities in 1618, when two imperial officials and a secretary were thrown out of a window at Prague Castle onto a dung heap and set off what would later come to be known as the Thirty Years' War. This conflict lasted a much shorter time in Prague. In November 1620, on a hill just east of the city (today, inside its perimeter) named White Mountain (Bílá Hora), Utraquist forces were defeated by the Habsburg armies of Holy Roman Emperor Ferdinand II, aided by the Catholic League, and the Counter-Reformation arrived in Prague in earnest. The symbolism of the defeat at White Mountain—interpreted later as the attempted extirpation of the Czech nation, but at the time seen as the beginning of the end of the Utraquist Church—still resonates within Czech political mythology.[27]

The re-Catholicization of Bohemia reached into all aspects of life in the region, and the university was no exception.[28] The split within the institution came to an end with the abolition of the Utraquist Carolinum two days after White Mountain, though its remnants were united with the Clementinum in 1654 as the now-renamed "Charles-Ferdinand University." The fusion was done faculty by faculty, with the Clementinum controlling the religiously sensitive theological and philosophical faculties (which included what would later become the natural sciences), and the Carolinum holding onto medicine, law, and the ceremonial functions of the university. Administratively, the Jesuits were displaced from their control of the institution (though they were still very prominent in teaching), and it was set directly under the control of the monarch.[29] The institution remained Habsburg, Catholic, and Latin-speaking until the end of the eighteenth century, when a series of reforms from Rome and then Vienna began a new chapter.

* * * * *

On 21 July 1773, Pope Clement XV dissolved the Jesuit order. Given the prominence of Jesuits within the halls of the Charles-Ferdinand University in Prague, this had enormous implications. The university became no less Catholic, but it did become significantly less Latin. With the Jesuits gone—the order finally left Bohemia in 1781—the pressure to conduct courses in that language began to slacken. German began to creep in as a language of instruction.[30] This was the default replacement for a number of reasons: First, correspondence between Bohemia and Vienna was conducted in German, while Czech, associated with Utraquism and rebellion, had been suppressed in the elite circles that fed the university.

Second, Emperor Joseph II had proclaimed German the language of administration for the multilingual Habsburg Empire, a move made for efficiency of administration more than any other reason, and in 1781 he issued his Edict of Toleration (*Toleranzpatent*), which removed certain restrictions on non-Catholic Christians (though it also contained a series of strictures designed to assimilate and Germanize Jews). Both reforms ended up Germanizing the character of the university.[31] Lutherans and Calvinists could now attend the institution. The close temporal juxtaposition of these various changes focused attention on the university as a politically sensitive space, and a *linguistic* one at that.

German became the official language of instruction in Prague in July 1784, a little over a year before it replaced Latin at the university in Vienna, though its implementation was complex and by no means universally opposed to the spread of Czech as a scholarly language. As part of Joseph's reforms, he made six years of basic school attendance compulsory for children across the empire, preferentially in the students' native language, which had the effect of increasing Czech literacy. Even earlier, in 1774, Joseph's mother Maria Theresa had created a chair in Czech at the university in Vienna; in 1791 a similar chair, first held by František Martin Pelcl, was established in Prague. Czech was also widely introduced into instruction after 1784 in courses on pastoral theology as well as midwifery, professions that demanded contact with the Czechophone peasantry in the countryside. Yet Latin held on tenaciously. An 1804 imperial decree switched the language used by the philosophical faculty, especially in the fields of philosophy, pure mathematics, and physics, back to Latin, which was still the language of most of the texts. (German was reintroduced in those fields in 1824.)[32]

There was no mistaking the fact that the university in Prague had shifted from a Latin, Catholic institution to a German-dominated (but still Catholic) one. As Czech literacy spread and intellectuals educated within the university chose to identify themselves with the Czech language and construct a Slavic nationalist vision of Bohemia's future—a process known as the "revival" (*obrození*) and its protagonists as "awakeners" (*buditelé*)—academic journals in both languages proliferated, producing a segmentation of scholarship by language.[33] As often happens with bilingualism in contexts of power asymmetry, the Czechs read the German work, but fewer Germans were even cognizant of the extent of their counterparts' accomplishments.

"In the year 1848," after the failed European revolutions, which were particularly violent in Prague, noted the *fin-de-siècle* Czech historian Jaroslav Goll, "the Prague university ceased to be German and became utraquist, which lasted until 1882."[34] Goll's choice of term is fascinating, reappropriating as it does the Hussite term meaning "of both wafer and wine" to mean "of both German and Czech language."[35] The halls of the university now resounded with lectures on numerous subjects that were offered in either language depending on the capacities and inclinations of the professoriate. In the philosophy faculty, there were Czech lectures in philosophy, aesthetics, zoology, botany, mineralogy, geology, mathematics, general and Austrian history, classical and Slavic philology, geography, chemistry, and physics (both mathematical and experimental).[36] As it more than doubled in size, the student body shifted to individuals who identified themselves as primarily speakers of Czech (though data on this front are notoriously soft, especially before the end of the nineteenth century). Self-described "Czechs" grew from 30 percent of the medical faculty to 56 percent, and from 53 percent of the law faculty to 62 percent, while in the philosophical faculty they approached 73 percent and in theology reached as high as 83 percent, meaning that by 1882 the university had shifted from approximate equality in 1855 to roughly two-thirds Czech and one-third German.[37] Yet even as the university became more linguistically and demographically diverse internally, the aftershocks of 1848 also pushed its more thorough integration into the broader network of German universities, encompassing those of the German states, Austria, and Switzerland. This was the network that we already saw activated in the search for a physicist that yielded Albert Einstein's name.

The pressure to reconsider the official policy—German dominant, Czech and other regional languages tolerated on an ad hoc basis—grew across the Habsburg Empire. The university in Budapest had already shifted from German to Hungarian in 1860, and in the 1870s the universities at Kraków and Lviv changed to Polish. Would the university in

Prague become Czech as well? There was a comparison closer to hand: the Polytechnic in Prague, the oldest institution of its kind in Austria-Hungary (founded in 1807, eight years before the Vienna Polytechnic). In 1869, the government decided to split the Polytechnic, which had been utraquist since 1863, into two separate institutions, one German and one Czech—a significant transformation of the educational landscape accomplished with remarkably little fuss.[38] Even though there were now two separate institutions, which creates the impression of two separate peoples, one should not forget that national indifference was still at play, however asymmetrically. Czechs still regularly went to the German Polytechnic (comprising over 10 percent of the student body in 1900, but declining thereafter), while no Germans ever attended the Czech.[39] The Prague Polytechnics offered an alternative model for the university's future to either continued utraquism or wholesale transformation of an institution into a Czechophone one.[40]

The question for Prague's university was: Should it stay united, or would perhaps a split make for better policy? The answer entirely depended on what one thought a preferable outcome to be. Matters reached the breaking point in the early 1880s as the city of Prague grew in both size and prosperity, which intensified a demographic transition within the Bohemian capital that now became visible. In 1880 the Austrian census took explicit note of the "everyday language" (*Umgangssprache*). Instead of framing inquiries about the population in terms of potentially volatile ethnographic criteria, Viennese officials introduced this option in order to facilitate international comparisons.[41] It had the effect of channeling perceived ethnic anxieties into a linguistic frame.[42]

It is difficult to evaluate these anxieties by a quantitative standard, as the pre- and post-1880 censuses simply collected different data and therefore are not comparable, even when taking into account the increasing density of the city and its significant geographic expansion through its incorporation of new urban regions beyond the traditional four of Old Town, New Town, Lesser Town (Malá Strana), and Hradčany. This inhibits historians' ability to get a sharp picture of the relative sizes of the populations that chose to identify themselves as either Czech or German.

In 1848, counts of the four historical districts tallied 66,046 Germans (66 percent of the total population), overwhelming the remaining Czech population and the 6,400 identified as Jews. The numbers in decades immediately following prove hard to reconcile (sometimes including suburbs later incorporated into Prague's municipality, sometimes registering Jews, and sometimes doing neither). What is clear is that by 1890 German-speakers did not constitute more than 19 percent of the population of the Old Town, and there were even fewer elsewhere in the city. The newer districts that were integrated into the city from 1883 to 1901 were overwhelmingly Czech-speaking and dropped the proportion further.[43]

The German population declined for three reasons. Cities across Europe in the nineteenth century ballooned due to migration of rural populations to urban industrial jobs, and this was no less true in Austria-Hungary. There were, however, significant differences in these migrants' destinations: Czech-identified peasants in the countryside tended to aim for Prague above all; Germans, however, might opt for Berlin or Vienna. As of 1900, for example, 60 percent of Prague's residents had been born outside the city, and most of these had come from Czechophone villages.[44] Together with a low birthrate among German families in the Bohemian capital, these two factors combined to produce the perception of a stark reversal of fortune for Germans. That in turn led to the third reason: the tendency of Prague's Jews over time to increasingly answer "Czech" on the census when asked about everyday language, a transition that factored into much of the nationalist politicking around Jews in the city.[45] Yet despite this, in 1912—during Einstein's time in the city—when estimates pegged Germans at roughly 7 percent of the population or less, there were still 196 German associations in the city and 36 distinct student groups to service a population of 32,800.[46] Feeling beleaguered, German activists abandoned the liberal vision of Germanness as a voluntary identity of progress and moved to a narrower, ethnicized understanding that was increasingly exclusionary.[47] The dynamic would bring forth bitter fruit in the 1930s.

Reading back from the ethnic-baiting disasters that led to World War II—an essential component of Einstein's relationship with the city in later years—it is understandable that a significant historical literature has emerged that chronicles how the boundaries between two once-neighborly communities developed into an unbridgeable gulf of antagonism, sometimes with the goal of blaming either the Czechs or the Germans for starting it.[48] It is easy to take this antagonistic line too far, with the effect of naturalizing the nationalist split rather than explaining it. There remained widespread bilingualism throughout this era, calling into question precisely the simplistic binary answers that the census offered.[49] While it is important to remember this, it is also important to note that Einstein did not interact with this fluid, nationally indifferent Prague, because he himself counted as monolingual in this context (his passable French and rudimentary English not being relevant in these surroundings). He was understood, quite simply, as a "German," for the straightforward reason that his job was, after all, as a professor at the *German* University. This explicitly tagged his post to the language, and language had by this time become identity. But at our point in the story, not only had Einstein not received the offer thanks to ministerial interference, but there was not even a German University yet.

That would come in 1882. Its reluctant midwife would be Ernst Mach, the physicist who had established the reputation of that science in the Bohemian capital, inspiring the later local enthusiasm for Einstein, and who happened to be during the academic year of 1879–1880 the rector of Prague University, one of the highest posts a professor could attain in Habsburg Bohemia.[50] Mach had originally hoped to keep the university together, maintaining its utraquist status but holding onto German leadership, but both political pressures and enrollment realities indicated that if the university stayed united, it would eventually become entirely Czech. To preserve a German institution, the link to tradition had to be sacrificed.[51] A petition to divide the university on linguistic lines had in fact been submitted to the emperor as long ago as 12 August 1872, but it took a protracted course through the various houses of representatives, regional and central, before being approved by Emperor Franz Joseph on 11 April 1881 and promulgated in law on 23 February 1882. By that fall, there were two universities in Prague where there had previously been one. (The theology faculty stayed intact and caused Mach no end of problems in his second term as rector—this time only of the German

University—in 1883–1884, and he resigned in frustration. It would also divide in 1891.) Both the Czech and German universities were taken to be equal heirs of the original institution promulgated by Charles IV five centuries earlier, but they were both in a more immediate sense equally young, products of the age of mass politics. The university had been split for the third time, and it would be granted a half-century's reprieve from further trauma.

Administratively, the division could have been nightmarish chaos, but its parts were sorted out relatively quickly, a testament to how much the post-1848 utraquist *modus vivendi* had already evolved into a divided university in all but name. Individual professors selected which university they wished to join, and research institutes allotted to individual fields followed their directors. The proliferation of institutes was a recent development: between 1878 and 1880, institutes opened for chemistry, zoology, mineralogy and geology, and mathematical and experimental physics.[52] Given the realities of who occupied the senior chairs in 1882, that meant most institutes—including the last, where Einstein would work—went to the German University. Obviously, new buildings could not be conjured in an instant, and both philosophical faculties stayed in the Clementinum and both law faculties in the Carolinum, with the two universities sharing the grand hall (*Aula*), the Germans on even days and the Czechs on odd days, as decided by lot. There was a debate in the medical school as to who would get the cadavers on which days. Students opted to use different doors depending on nationality, and professors in the different universities grew further apart until interaction virtually ceased altogether, with rare exceptions.[53]

The Czech University emerged from the split with almost no institutes and persistently underfunded, but triumphant. Goll would recall it 20 years later as the end of an extended conflict, of which Bohemia had seen so many: "We strode through this period more beside one another than with one another, without love but also without hate—*neque odio, neque amore*—but still as neighbors that live and let live. It was like an armistice."[54] The student population grew enormously at the Czech University, an outcome that could have easily been predicted, as Czech-speakers had only one university to attend while German-speakers had dozens. By 1910 the Czech University had 4,128 students, while the

German had 1,733. Nonetheless, Vienna offered 1,763,000 crowns to the Czech University, only 151,000 more than the German was allotted. The disparity was most evident at the medical school, where the German University could disburse 769,300 crowns for its 344 students, while the Czech had to make do with 724,400 for its 631.[55] Student numbers kept dropping on the German side and rising on the Czech, while the faculty lines stayed equivalent at both institutions.[56] If this was a peace treaty, it is unclear who won the war.

Faculty at the German University jealously fought to hold onto the privileges they had even as their status slipped, further encouraging younger faculty to seek more prestigious positions elsewhere. Without the willingness of some Czech students to attend science courses at the German University the situation would have been even worse; although many Czechs boycotted the twin institution, there was still crossover in mathematics, physics, and medicine. It did not work the other way: while in 1900 there were 77 native-speaking Czechs at the German University, and in 1910 there were 45, there were zero native-speaking Germans at the Czech University in either of those years.[57] In 1909 the city council issued a glossy brochure entitled *Prague as a German City of Higher Education* to entice German-speaking students from abroad to come study there, despite what they might have heard about the Czech character of the city.[58] The challenge of attracting German students was analogous to the problem of attracting German professors, and Albert Einstein was precisely the kind of young, exciting person who could help boost the visibility of the institution *vis-à-vis* its burgeoning Czech counterpart.[59] That is, if the faculty were ever permitted to make him an offer. Now that we have seen some of the deep context that structured the tragicomedy of the quest to bring Einstein to Prague, it is time to return to where we left him: in limbo.

* * * * *

Minister von Stürgkh would not accept the initial list sent him by the Prague physicists. It just would not do. He informed Lampa, Pick, and Rothumnd that he would invert the order of the top two candidates on the list and make the offer instead to Gustav Jaumann. (There was no solace for Emil Kohl, whose progress had once again stalled at third place.) There was nothing especially unusual about this, from the point of view of either ministerial meddling with rank ordering or the choice of Jaumann.

If anything, Lampa should have expected it. When the position he currently occupied had opened up at the German University, a committee of the faculty (including Lippich and Pick, as well as three others) had contacted professors at other institutions to come up with a rank-order list and in June 1909 had ranked Anton Lampa second. In the first position they had placed Johannes Stark, then at the technical university of Aachen, who would later discover the Stark effect (the splitting of an atom's spectral lines in an electric field) and win the Nobel Prize in Physics in 1919. He would also famously become an early supporter of Adolf Hitler, a leader in the anti-Semitic campaign for an "Aryan Physics" against the so-called "Jewish Physics" of relativity and quantum theory, and an important scientific administrator in the Nazi state in the 1930s. But that was all in the future; at this moment he was a prominent experimental physicist at the cutting edge of the science. He was also a Bavarian and a citizen of the German Reich. Lampa was closer to home. Born in Budapest in 1868, he had grown up in various towns in Bohemia, studied at the University of Vienna, and habilitated there in 1897, experimenting with Hertz electromagnetic waves. When von Stürkgh, already minister of education, received the rank-order list in 1909, he opted for Lampa because he was Austrian, full stop. (The physicist who came in third, Stefan Meyer, was a colleague of Lampa's from Vienna.) Lampa had gladly taken the job, but now he was frustrated that the same ministerial trick was working against him.[60] (To add a further irony to this game of musical chairs, when Jaumann had been hired at Brno, Lampa had been in third position.)[61]

Bureaucratic fiat was, naturally, irritating, but Jaumann was not at all an absurd person to be ranked ahead of Einstein. Given that every reader of these lines knew quite a bit about Einstein before picking up this book yet has likely never heard of Jaumann, that statement requires a bit of explanation. Gustav Jaumann was born on 18 April 1863 in Karánsebes in the Kingdom of Hungary (now Caransebeş in southwestern

Romania), and his family moved to Prague when he was a child. He studied at the universities in both Vienna and Prague, beginning his doctorate at the latter under Ernst Mach in 1884. Mach immediately hired the talented student as an assistant, and Jaumann implemented important improvements in the electrometer for the lab. He earned his Ph.D. in April 1890 and habilitated an astonishing six months later. In 1893 he jointly authored a textbook with Mach and was hired as an extraordinary professor at the German University. He left for the polytechnic in Brno in 1903.[62] Jaumann was, in short, a local and familiar candidate in 1910. Yet he was placed behind Einstein, which had something to do with not only the latter's astonishing recent achievements, but also some of the aversion that comes with familiarity.

Jaumann was a student of Ernst Mach in more than just physics. Philosophically, he took one of Mach's most controversial positions—a skepticism toward atoms—much further than his teacher. When Jaumann had been a student, this had been a minority view, but still defensible. By 1917, Philipp Frank—himself a strong admirer of Mach's philosophy of science and Einstein's successor to the chair in mathematical physics at Prague (and later still his most influential biographer)—considered Jaumann's slavish following of Mach's misguided views pernicious: "One cannot deny that Gustav Jaumann has undertaken this task with strong constructive force in numerous works. I however do not believe that success in the real spirit of Mach's doctrines has been achieved."[63] When Jaumann died in 1924, he was in the midst of composing a 2,000-page textbook filled with continuous differential equations to demonstrate a physics with unified forces and without any atomic particles—the position of a borderline heretic. When the committee was deliberating in 1910 Jaumann seemed even then behind the times. And as much as Jaumann clearly admired Mach, by the time he left for Brno in 1903, the feeling was not mutual.

Perhaps the emergent conflict started in 1893 with their jointly authored textbook for *gymnasium* students, which came out in two nearly identical versions, one for teachers and one for students. Mach cut the two chapters Jaumann wrote from the student edition, deeming them too hard for the task at hand.[64] Relations deteriorated after that. In Mach's so-called "Prague Testament," composed over an extended period and detailing his wishes for the physics institute after he left for Vienna in 1895, the old master singled out Jaumann for particular abuse, demanding that he not be appointed his successor:

> Jaumann worked here for ten years, so to speak, on his own behalf, and received a salary for it. I demanded only very few tasks as an assistant from him in the first years and later none at all. I was on the contrary in a constant state of self-defense against his penchant for extravagance and disorder. His work as an assistant, preparation for lectures, restoration of apparatus, free manufacture of new devices, management of materials, correspondence, were seen to in the last four years exclusively by my son Ludwig alone, with the aim of supporting me. He established with great effort exemplary order in Jaumann's administration.[65]

Needless to say, Jaumann, who was then at the German University, did not get the call to Prague at that time. But it seems he never learned of his fall from favor. In 1906, after Mach himself had retired, Jaumann wrote him asking for support in his application for a position that had been vacated by the suicide of Mach's Viennese colleague Ludwig Boltzmann.[66] He did not get that post either, and his next opportunity to leave Brno came with von Stürgkh's intervention.

Von Stürgkh performed such inversions all the time, and he considered it good policy. When the case of Lippich's successor came before him, he pondered the particulars and then chose the Austrian ranked number two. In a memorandum to the emperor on 16 December 1910, he explained:

> Although the professoriate nominated Professor Dr. *Einstein* to the first position because it placed especial value on his brilliant achievements in the area of modern theoretical physics, I nonetheless believed I should first begin negotiations with Professor *Jaumann* from Brünn, named to the second position, since he also has published entirely

outstanding works and might open a broad and fruitful field of his original research activity through assumption of a theoretical-physical chair. In addition this could mean at least that one of the numerous young physicists of the Vienna School could take *Jaumann*'s vacated chair at the German higher technical school in Brünn.[67]

Einstein was not privy to this reasoning, and he did not interpret the machinations as a response to either his citizenship or even the fact that his wife was a Serbian, and thus a member of an ethnic group that conservative Austrian officials viewed with great suspicion. For Einstein, there was only one reason for the inversion: "I was not called to Prague. I was only put forward by the faculty; the Ministry however has not accepted the proposal because of my Semitic descent."[68] This was, as the documents show, not the real reason, but Einstein believed it until his death, as did many locals in both Prague and Vienna, an urban legend that would prove important in Einstein's complex self-identification as a Jew.

Maybe it would not turn out so bad for Einstein in the end. After all, sometimes the person in second position got the job. When Ernst Mach, to date the most successful physicist in the university's history, had been hired in Prague in 1867, he had also been in second position behind Adalbert von Wahltenhofen yet got the post when the latter declined the offer.[69] The same had actually already happened to Einstein in his current position at the University of Zurich. When the extraordinary professorship had opened up in 1909, Einstein had been *secondo loco*, behind his former classmate Friedrich Adler. When Adler had learned that Einstein, whom he considered more talented, had been ranked behind him, he had ceded the position and devoted himself to revolutionary politics instead. A few years later, in 1916, in opposition to the brutal atrocities of the Great War, the former physicist would walk up to von Stürgkh, now prime minister, as the latter was dining at a Viennese hotel and shoot him dead. Einstein would successfully lobby to have the death penalty waived for Adler.[70] By then, Einstein had become world famous. But at this point in our narrative, he was still stuck in Zurich and in second position on Prague's list.

Einstein eventually got the position, a gift from none other than Jaumann, who opted to stay at Brno and declined the post. There is a popular story about why he made that decision, one that fits what we know about Jaumann's abrasive personality. In his biography of Einstein, Philipp Frank reports Jaumann's answer to the call, upon learning that originally he had been in the second position, as a direct quotation: "If Einstein has been proposed as first choice because of the belief that he has greater achievements to his credit, then I will have nothing to do with a university that chases after modernity and does not appreciate true merit."[71] Here, as throughout his 1947 biography, Frank provides no references to sources, even for direct quotations such as this one. It's a great story, one that is especially ironic considering the titanic stature Einstein later acquired; but it is probably not true, and certainly not in the verbatim form in which we have received it. Archival evidence indicates that Jaumann considered the offer but rejected it because of breakdowns in salary negotiations, and also because he was quite familiar with the conditions in Prague.[72] If he were to move, it would be to a place like Vienna; otherwise, he was happy where he was. Nonetheless, this has not stopped most Einstein biographers from quoting Frank's apocryphal account—certainly derived from rumors circulating around Prague—as direct testimony. This would not be the only time that Frank's cavalier attitude to citation pushed the mythology of Einstein down false tracks.

With the candidate for first position withdrawn, the road was cleared for Einstein, and he was accordingly called to Prague. The salary offer was very good, especially compared to what he was receiving in Zurich: 6,400 crowns a year, with a raise of 800 crowns after both the fifth and tenth years, and 1,000 after the fifteenth and twentieth years, followed by a pension vested in 10 years.[73] (The ministry expected Einstein to last in Prague that long, which he did not; it also expected the Habsburg Empire to last, and in eight years it was gone as well.) Von Stürgkh pointed to an important condition attached to Einstein's assumption of the post: "Finally I note that the adoption of Austrian citizenship is tied with your appointment; Your Honor wants thus to immediately take steps to effect the release from your current citizenship."[74] Despite extensive searching, I have been able to find no evidence that Einstein ever became a subject of the Habsburg emperor; it is absolutely certain that he never relinquished his Swiss citizenship. Given the absence of any traces in this most bureaucratic of domains, it seems likely that Einstein just ignored this comment. As we have seen, he would ignore a similar demand from Berlin three years later.[75]

In Switzerland, there was little understanding of this drama of first and second positions, but rumors got out that Einstein was being courted by Prague, and it made his local supporters anxious. Students at Zurich petitioned the university to do everything it could to retain the young physicist: "The undersigned students of the lectures of Prof. Einstein wish to hereby ask you to do the utmost possible to retain this outstanding researcher & teacher at our university."[76] The counteroffers were insufficient—an extraordinary professor salary, even a generous one, could not compete with the promotion that Prague offered, and on 20 January 1911, Einstein formally resigned from Zurich.[77] The family was moving to Prague.

* * * * *

"It is mildly puzzling to me why Einstein made this move," wrote Abraham Pais, a physicist and one of Einstein's many biographers. "He liked Zürich. Mileva liked Zürich. He had colleagues to talk to and friends to play music with. He had been given a raise."[78] Perhaps no one was more distraught about it than Mileva Einstein, *née* Marić. She wrote to her close childhood friend, Helene Savić, in January 1911, two months before the move, that she was filled with apprehension:

My husband, after mature reflection and because of material advantages, has accepted the invitation to go to Prague, and in March we shall leave Zurich to move over there. I cannot put it otherwise than to say that I am not going there gladly and that I expect very little pleasure from life there. I am a little afraid for the children, who have to grow up in those unpleasant circumstances. My only comfort is that I might be imagining it all in dark colors, or that this invitation might be followed by another that might take us away from there. Do you know Prague at all?[79]

That, indeed, was the question: What did any of them know about this place? This is a difficult question to answer, as the decision was made so quickly that no friends seem to have been consulted about it by post, thus leaving us no surviving trails in the correspondence. Had he turned to the Germanophone periodicals emanating from Prague for guidance, Einstein might have paused. In March 1911, the conservative periodical *Deutsche Arbeit* published a nostalgic piece about how wonderful Prague had been in 1811, when it had been (ostensibly) a fully German city.[80] That, like Stumpf's reminiscence at the start of this chapter, would have colored the situation more darkly. More likely, Einstein did what any German-speaking traveler around Central Europe of the time would have done: consulted the Baedeker guide to the city. The most recent version was the 1910 edition for Austria—Prague was, indeed, one of the major metropolises of that country—and it presented a series of impressive sights to be visited, as was fitting for a travel guide. It mentioned Tycho Brahe and Johannes Kepler as scientific luminaries and noted that the city was home to four institutions of learning, two polytechnics and two universities divided by language. It also dubbed the German University "the oldest institution of higher learning in the German Empire."[81] There was a sense in which that was true, but it papered over a good deal of history that would matter to the young family of four—the Einsteins' sons, Hans Albert and Eduard, were six years and eight months old, respectively—when they slowly began to get to know the city.

They took their time on the trip. They left Zurich on 30 March, stopping in Munich, where Einstein visited the Dutch physicist Peter Debye and the field's leading light, Arnold Sommerfeld. Then they headed on to Bohemia, arriving at the Hotel Viktoria at 18 Jungmannstraße (Jungamannova třída) on 3 April 1911. Einstein's appointment began the following day. They stayed at the hotel for a few days before settling into their apartment on the third floor of a new building on Třebízkého ulice 1215, in the rapidly expanding neighborhood of Smíchov on the western bank of the Moldau. (The building still stands, but the street name and house number have changed to Lesnická ulice 7.)[82] Einstein had moved from first position on the list to second and back again, oscillating between centrality and marginality. And so it was to be in Prague, a city that was both a metropole and a province, depending on how you looked at it. It was that dual status that would enable Einstein to carry out his most intensive physics research in the city, while at the same time casting him in a bit part in a play that had been unfolding since the Middle Ages.

CHAPTER 2

The Speed of Light

Thus the Einsteinian relativity theory of 1905 sinks into the dust. Will a new, more general relativity principle arise like the phoenix from the ashes? Or will one return to absolute space? And summon back the much-maligned ether so that it can carry, besides the electromagnetic field, also the gravitational one?

—Max Abraham[1]

On 9 June 1911, Einstein received a letter in Prague from an unexpected source: the Belgian soap magnate Ernst Solvay.[2] Einstein had met Solvay once before, on the occasion of accepting his first honorary doctorate in July 1909 at the University of Geneva, a few short years after receiving his original doctorate. The Geneva ceremonies marked another of the dramatic changes in fortune that had greeted the former patent clerk who had just that same year been appointed extraordinary professor at Zurich. He was being honored on the same dais as such luminaries as physicist Marie Curie from Paris and chemist Wilhelm Ostwald from Leipzig, the latter long a hero of Einstein's (his father had tried to get his wayward son a job with Ostwald immediately after graduation) and on the eve of being awarded the Nobel Prize in Chemistry. Joining the three scientists were Ernst Zahn, a restaurant keeper from Göschnen in the canton of Uri who was widely acclaimed for his authentic, earthy folk prose, and Solvay, who stood at the apex of an empire based on his innovations in the manufacture of soda ash. To be sure, Einstein remembered Solvay, but this was not who Einstein expected to contact him at his post at the German University in Prague.[3]

Solvay had written to invite Einstein to a conference to be held in Brussels at the end of that October on the topic of "Radiation and the Quanta." Solvay was a passionate follower of physics, and the chemist Walther Nernst, a pioneer in thermodynamics and professor at the university in Berlin, had persuaded him to bankroll a gathering of the leading minds working on topics related to the new quantum theory. Nernst had drawn up the list of invitees for the meeting, Einstein's name prominently among them. It had been an obvious choice, given Einstein's 1905 paper explaining the puzzling photoelectric effect—whereby electromagnetic radiation of high frequency (e.g., blue or ultraviolet light) ejects electrons from metallic surfaces, now the basis of solar panels—through the postulation of light particles ("quanta"), a hypothesis that contravened the orthodox interpretation of light as a wave. This insight, which solidified the hypothesis of discrete energy packets in thermal radiation that had been proposed by Nernst's colleague Max Planck in 1900, would be cited by the Swedish Academy when it awarded Einstein the Nobel Prize in Physics for 1921. On the other hand, Einstein was rather young and had only just received his first full-time position at a university. He was an indispensable invitee, but also noticeably junior to the rest of the attendees (see figure 1).

The Solvay Conference of 1911 was a milestone in the history of physics, and it has been credited with solidifying the very notion that "modern physics" had displaced what came to be understood as "classical physics."[4] The agenda of the meeting was clearly focused on the decade-old quantum theory, which had already generated an overwhelming number of puzzles related to thermodynamics, electromagnetic radiation, atomic spectra, and more. It was *the* hot topic of contemporary physics, and Einstein had been at the center of it from the outset. He delivered the final of the four keynotes, on the quantum theory of specific heats, and it was a tremendous success.[5] He cut a fine figure, impressing the pantheon of theoretical physicists, many of whom he was meeting for the first time, including Hendrik A. Lorentz, the Dutch physicist whose work on the electrodynamics of moving bodies had been crucial to Einstein's special theory of relativity.

FIGURE 1. The participants in the Solvay Conference of 1911. Albert Einstein is the figure
standing second from the right. Marie Curie (the only woman invitee) is sitting at the table in
discussion with French physicist Henri Poincaré. Sitting at the head of the table are Ernst
Solvay and, to his left, Hendrik Lorentz, who chaired the meeting. Seated at the far left is
Walther Nernst, whose brainchild this meeting was. Finally, standing second from the left at
the blackboard is Max Planck, next to his famous equation, the intellectual impetus for the
gathering. *Source:* Benjamin Couprie, 1911, https://commons.wikimedia.org/wiki
/File:1911_Solvay_conference.jpg.

Given that it represented such a salient mark of his own lofty status
in the discipline and the opportunity to debate vigorously with his most
respected colleagues, one might think Einstein would have been enthusiastic about what his Swiss friend Michele Besso had prospectively
called "the Brussels witches' sabbath."[6] Despite its social charms, however, Einstein seemed rather lukewarm about the conference's intellectual content. "Nothing came out of Brussels," he wrote to his former
assistant Ludwig Hopf, "but it was overall a delightful spectacle."[7] To his

close friend Heinrich Zangger back in Zurich, he joked that "the entire affair would have been a *delicium* for diabolical Jesuit fathers."[8] Despite the later mythology that Einstein spurned all social conventions, he clearly had enough social graces to write to Solvay that the "Congress will always form one of the most beautiful memories of my life."[9] (But not enough to spell the industrialist's name correctly: he had it as "Solvey.")

Einstein's light-hearted nonchalance about the meeting stemmed from the misalignment of his current research with the agenda of the conference, and this had everything to do with where he was then based geographically. If you were unaware of the details of Einstein's biography and the mutable borders of European history, the list of attendees printed in the original French proceedings of the Solvay Conference might strike you as perplexing. The participants were separated by country, a fairly typical connection in this age of nation-based competitive internationalism (similar to the recently inaugurated Olympic Games), and down the list one comes across this line: "For Austria, professors A. Einstein of Prague; F. Hasenöhrl of Vienna."[10] This reminds us of two things: Einstein was coming to Brussels from Prague, and Prague was at this moment an Austrian city. Both facts transport us back to the physicist's local context of 1911.

Although early in his time in Prague Einstein continued to pursue some questions related to quantum topics and occasionally fielded correspondence about them, he had sharply changed his research area when he arrived in Bohemia.[11] Or, more precisely, he had changed it *back*. Einstein published almost a dozen papers while in Prague, and with very few exceptions they all concentrated on one topic: how to expand his special theory of relativity so it could encompass accelerated frames of reference, something that he had understood since 1907 would necessarily entail a new theory of gravity. Moving to Prague was a caesura in his personal life. It broke off relationships with friends and implanted him and his family in an alien city where they had no established ties. But it was also importantly an *intellectual* rupture, in this case with the quantum theory. Einstein took advantage of the lack of established local expectations, as well as what he interpreted as an impoverished intellectual community, to turn inward to explore the general theory of relativity.

Today, when we know that it is possible to formulate a self-consistent general relativity and that this theory provides a powerful interpretation of gravity—in fact, that it remains, over a century later, the centerpiece of gravitational physics and cosmology largely in the form Einstein endowed it—the choice seems a bold but reasonable one. It did not strike his peers that way. Around 1910, there were several unsolved anomalies in the dominant Newtonian theory of gravity that had been recognized for two centuries, but these were not seen as pressing problems, certainly not in the face of the fascinating vistas opened up by quantum physics. The physicists drawn to gravity were marginal players who often found it difficult to maintain a steady job while working in this backwater of physics. There was definitely nobody of the stature of an Einstein (even factoring in his relative youth) to measure himself against, yet he engaged in detailed and often blistering polemics during his three semesters in Prague as he struggled to develop what came to be known as the "static theory."

He failed. His attempt to generalize special relativity, the most distinctive effort in physics of his time in Prague, foundered on a series of internal inconsistencies, and Einstein abandoned the project by the summer of 1912, when he left Prague and returned to Zurich for a new position at the Federal Institute of Technology. He then teamed up with his college classmate and ETH professor of mathematics Marcel Grossmann to develop what came to be called the "outline" (*Entwurf*) theory of general relativity, a problematic though promising approach that three years later would evolve into the gravitational field equations still in use today. Historians have reconstructed Einstein's chaotic path from the *Entwurf* theory to general relativity in minute detail, scouring his famous "Zurich notebook" for clues to the course of the physicist's negotiation of different puzzles posed by his new tensor formalism, which necessitated a curved spacetime.[12] But this was not the theory he had crafted in Prague: that one worked with "flat" (i.e., Euclidean) spacetime, and the only lasting residue of it was a prediction of the bending of starlight around the sun's gravitational field during a solar eclipse, an important conjecture that would play an outsized role in making Einstein's reputation.

Had Einstein been killed by a runaway horse cart in July of 1912 as he was arranging his affairs to move back to Switzerland, we would not think of him today as the architect of general relativity. (The scenario is not fanciful; Marie Curie's husband Pierre had died in precisely such a manner on 19 April 1906.) Einstein would be seen as a trailblazer for his insights in 1907 but ultimately as a failure in bringing the theory to fruition. That he would eventually solve the problem of general relativity was not obvious in 1912, yet implicitly historians and physicists have tended to assume this by concentrating on Einstein's path once he arrived in Zurich—that is, when he was on a course that would eventually manifest as "the right track"—thus relegating the Prague theory to the shadows as an unsuccessful distraction. If what you are interested in is understanding the theory Einstein unveiled in Berlin in November 1915, this approach makes sense. If, on the other hand, we want to appreciate Einstein as a physicist like many others, wandering into mistaken cul-de-sacs and admitting his mistakes, and also to understand how the path toward one of the most successful physical theories of all time was deeply rooted in the opportunities and constraints imposed by Prague, we must treat the static theory as Einstein did: very seriously.

* * * * *

No idea has ever emerged full-blown from the mind of its creator, an Athena bursting forth from Zeus's skull. All theories, in physics no less than in art, have contexts that shape and condition their birth. In the case of Einstein's shift from quantum theory to gravity, that context was the German University in Prague, where he arrived in early April 1911 to take up a post as ordinary professor of mathematical physics. He was flush with a salary of 8,672 crowns and the resident of a brand-new apartment across the river from the new institute of theoretical physics that he would helm, located at Weinberggasse, or Vinična ulice 3 (later renumbered 7). The summer semester began on 20 April, and Einstein geared up to teach his courses and integrate into the community of the city on the Moldau.

The semester lasted until the end of July, and he spent many hours a week teaching three different courses: "The Mechanics of Discrete Mass

Points," which he taught to 13 students for three hours a week; "Thermo-dynamics," to 12 students for two hours a week; and a weekly seminar to 6 additional students. His winter semester resembled the summer one. From October 1911 until Palm Sunday 1912, he taught the same courses, with doubled enrollments in "Thermodynamics," at the Clementinum, right at the base of the Charles Bridge in the historic halls of the sixteenth-century Jesuit college, while the seminar was conducted back at the institute in the New Town. His final semester—by which point his departure had become public knowledge—comprised courses on "Mechanics of Continua," "The Molecular Theory of Heat," and another seminar, held on Friday evenings, in which seven students forsook the pleasures of the town to discuss theoretical physics in detail.[13] There was nothing extraordinary about this teaching load: these were basic, foundational courses, disconnected from his current research, and he garnered the usual enrollments one would expect in the increasingly strapped German University. The only significant change across these three semesters was that the number of auditors in Einstein's courses started high with eight and then fell to two for the following semesters, as his novelty wore off.

Three of those auditors came from the Czech University, which was not especially unusual in technical subjects like theoretical physics. Perhaps more remarkable is that Einstein did not seem to recognize that there was a nationalist valence to these students' choice to audit his classes. Here, again, Einstein's ignorance of the local context did not mean that his auditors did not experience some tension, or relief that Einstein did not make an issue of it. Since student reminiscences of Einstein's teaching from this period are rare, and these particular ones are largely unknown outside Czech scholarship, it is worth quoting one of them at length. This comes from Miroslav Hrabák, himself a resident of Einstein's neighborhood of Smíchov and the son of a metal lathe operator:

Einstein was calm, of small stature, with flowing hair. We met him always with a Virginia cigar, even during lectures, when he constantly walked around with the index finger of his right hand on his nose and his thumb on his chin. His presentation of the material was very demanding, he presupposed a high level of knowledge of the students. He wrote the formulas and examples messily on the board; even a colleague, confused by a disorganized calculation, once complained about something. Usually placid, he made a fuss—he retorted that he is not a sergeant (*feldvébl*), but then he wrote it in order. He in general looked down on the Austro-Hungarian order which reigned in the school. I took two examinations with him: I remember that for one of them he examined me from eight in the morning until noon, when the famous midday firing of the cannon alerted him that time was up. He interrogated me about all of physics. Then he put on his coat and we went down together, across the Palacký bridge to Smíchov, where I lived.[14]

In the wake of the tumults of the twentieth century, the interviewer had found Hrabák working at the Slavonic Library in Prague. This report, and the few others like it, engage in the posthumous romanticization of Einstein that seems unavoidable among those who had encountered him decades earlier. Nonetheless, we obtain a picture of students exhibiting the respect for an instructor typical of that time and place.

The respect was not necessarily mutual. Einstein walked into his first class with relatively low expectations. "My colleague Lampa tells me that the scientific interest of the students here is very feeble," he wrote to Zangger three days after his arrival in the city, and two weeks before his first lecture. "But I believe one's voice will also ring through this forest according to the manner in which one calls. I will not let my illusions be devastated so easily."[15] His lecture notes that survive show that his courses derived from the similar ones he had taught in Zurich, but now he evidently gave less attention to his pedagogy.[16] Presumably, this had something to do with his focus on his own research. According to another later reminiscence from an auditor who had graduated in philosophy a few years earlier, Hugo Bergmann, Einstein tried to put a positive spin on what must have seemed like mindless drudgery. "Actually I am glad to be allowed to lecture on these elementary matters, for in my own research it often happens that I pursue and explore some thought," Bergmann remembered him saying, "only to realize in the end that I have been wandering in a maze; all those weeks of hunting a phantom would have been lost, had I not been giving my lectures and thus done something useful during all that time."[17] It seems, as we saw earlier, that the students did not notice his abstraction, and a later professor at the German University who had arrived for graduate study just after Einstein had left recalled that he had a reputation for being accessible to students.[18]

If stimulation were not going to come from his students, Einstein might have hoped he could engage his colleagues on the details of his scientific research, but this was also largely a mirage. His research assistant during his final semester in Prague, Otto Stern, related in an interview preserved on a magnetic tape stored in the archives of the ETH that "Einstein was completely isolated in Prague (1912), even though there were four colleges there: a German university, a Czech university, a German polytechnic, and a Czech polytechnic. At none of them was there a single person with whom Einstein could talk over the matters which truly interested him. He did so *nolens volens* with me. . . . The only really intelligent man there was a mathematician named Pick."[19]

Georg Pick was the only one Einstein claimed he would miss upon his return to Zurich.[20] Pick had been born in August 1859 in Vienna and had studied physics at the university there from 1875 to 1879, the year of Einstein's birth. The differences in their ages and career experiences marked them as different sorts of scholars. Pick earned his doctorate at Königsberg with legendary mathematician Karl Weierstrass in 1880, and then moved to Prague to serve as assistant to Ernst Mach, then professor of experimental physics. He filed his habilitation there in 1882 (the year of the split); it was entitled "On the Integration of Hyperelliptical Differentials Through Logarithms." He stayed at the German University for the rest of his career, during which he published 67 works mostly in complex analysis and differential geometry, but with individual studies in a wide array of other specializations. When Anton Puchta, a professor of mathematics, resigned his post in 1888, Pick stepped in and was appointed ordinary professor four years later. He taught for 46 years in the faculty of philosophy, even serving as dean during the 1900–1901 academic year, and upon retirement moved to Vienna, only returning to Prague with the *Anschluss* between Austria and Nazi Germany, 79 years old and in poor health. He would eventually die in 1942 shortly after his deportation to the concentration camp at Theresienstadt, a victim of Hitler's Protectorate of Bohemia and Moravia.[21] That, of course, lay in the future when Einstein met him. The physicist admired this senior mathematician, and the two discussed his ongoing research into gravitation and relativity theory. This was his only significant interchange with local faculty about these topics.

For all his intellectual isolation from both students and colleagues, Einstein seemed to work happily at Prague. He was most pleased with his professorial accommodations. "I have a magnificent institute here in which I work very contentedly," he bubbled to Marcel Grossmann a few weeks after his arrival in Bohemia. However, he could not resist immediately following this statement with a few complaints about the city surrounding the Institute for Theoretical Physics: "Otherwise it is less homey (the Czech language, bedbugs, miserable water, etc.). The Czechs are, by the way, much more harmless than one thinks. I scarcely know my colleagues yet. The administration is very bureaucratic. Infinite quantities of paper-pushing for the most insignificant nonsense."[22] Einstein mostly appreciated the rich library on Weinberggasse, where Ernst Mach had established the new physics institute in 1879, moving it from the center of the Old Town on Obstmarkt (Ovocný trh), just before the division of the university. As was the rule, the institute followed its director to the German University.[23] Einstein had an office on an upper floor, and he essentially cabined himself in there with notebooks and references. "This summer I will not leave Prague because I urgently need the vacation for work," he told Michele Besso. "My position and my institute here make me very happy. Only the people are so alien to me."[24] Intellectually alone, he settled into what he came to Prague for: thinking about gravitation. He had been waiting a long time for this opportunity.

* * * * *

The year 1905 has long been known as Einstein's *annus mirabilis*, or "miracle year," and for good reason. In a rapid burst of productivity in the winter and spring, he sent off articles detailing his new theory of

Brownian motion, the photoelectric effect, and of course the special theory of relativity. Each of these could (and eventually did) transform separate domains of theoretical physics: atomic theory, statistical mechanics, and electromagnetism. The attention of theoretical physicists homed in on the patent clerk from Bern, whose theories were rapidly assimilated into the mainstream of the science. As noted earlier, the implications of Einstein's light quanta for almost every area of physics became a dominant, though disputed, topic for more than a decade to come among his peers, but that was only because the tremendous consequences of relativity theory had in large part been widely accepted. It would take some time yet for Einstein's bold fiat excommunicating from existence the luminiferous ether—the hypothesized and, until Einstein, universally accepted medium upon which light waves and other forms of electromagnetic radiation were presumed to travel—to become commonplace, ushered in by the development of a four-dimensional mathematical interpretation of spacetime created by Einstein's former mathematics teacher at the ETH, Hermann Minkowski (now at Göttingen), in 1907.[25]

That same year, Johannes Stark invited Einstein to write a review essay on relativity theory for the *Jahrbuch der Radioaktivität und Elektronik,* an opportunity that allowed him to summarize and tie together the various refinements that had been made to special relativity during the prior two years—including the derivation of the mass-energy relation encoded in his most famous equation, $E = mc^2$—and to lay out some directions for future research. In the final section, he considered what would be required to generalize the special theory of relativity. The latter was "special" because it described the relationship between space, time, and the speed of light solely for inertial frames of reference, that is, those moving at a constant speed with respect to each other. What would happen, Einstein wondered, if one tried to apply the theory to *all* frames of reference, including those that were accelerating?

To understand the challenge of the problem, it helps to review the basic features of special relativity. Imagine—to take one of Einstein's classic examples—a train hurtling along at an exceedingly high, but constant, velocity (think science-fictional speeds, such as a significant

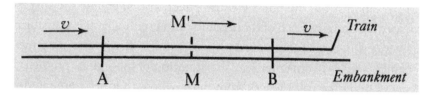

FIGURE 2. Einstein's schematic 1916 illustration of the thought experiment about lightning
bolts hitting the front and back of a train. *M'* is the observer sitting on the train, *M* is the
woman at the station, and *A* and *B* represent the points where the lightning bolts hit the
ground. In this rendition, the velocity *v* represents the train speeding to the right at a constant
rate. It is, however, just as legitimate—indeed, essential for the scenario—that *M'* perceives
himself as stationary and *M* and her embankment as zipping off to the left at a constant
velocity of –*v*. The speed of light, *c*, which is constant for all observers, is not depicted.
Source: Albert Einstein, *Relativity: The Special and the General Theory,* tr. Robert W. Lawson
(New York: Three Rivers Press, 1961 [1916]), 29.

fraction of the speed of light), passing by a stationhouse on the embankment (see figure 2). There is a gentleman sitting on the roof of the train
at its exact midpoint, with two mirrors angled to each other so that he can
view the front and the back of the train at the same time. In the middle
of the station platform, another person is standing; she also has a similar
contraption that lets her observe both ends of the train. Just as the two
individuals are opposite each other, the woman notices a freak of nature: lightning bolts have streaked out of the heavens and struck the tracks
just at the front and back of the train! Holy smokes, she thinks, I'll have
to speak to the rooftop rider later about this event. When the two get
together, however, they find they have different accounts of the event.
Where the woman saw both lightning bolts strike simultaneously, the
man observed a lightning bolt strike the front of the train, and then a few
moments later a *second* bolt hit the back of the train. Who is correct? As
special relativity shows, they both are.

Here's how to resolve the difference of perception. From the frame of
reference of the station, the two lightning bolts are simultaneous, and the
reason that the man sees them staggered is that he is rushing toward the
light beam heading from the front of the train, encountering it first, and
the light beam from the rear takes some extra instants to catch up to him.
But—and this is the crucial point—since the motion here is constant,
it is just as correct for the man on the train to believe that *he* is stationary

and the platform is moving in the opposite direction away from him at *Star Trek* speed. In his frame of reference, the lightning bolt at the front of the train happened first, the one at the rear second, but the platform was rushing toward the rear of the train so fast that the woman measured the beam from the back just as the one from the front finally reached her. The notion of simultaneity depends on one's frame of reference. From this basic qualitative scenario, the seemingly paradoxical properties of which stem entirely from the assumptions that the frames of reference are equivalent and that the speed of light moves at a constant velocity in every frame, one can derive the astonishing consequences of special relativity.

If one is accelerating, though, things work differently. Imagine now that you are riding in a windowless elevator being accelerated upward. You feel your weight on the floor quite firmly as it pushes against your feet. If a light beam now were to penetrate the side of the elevator, it would seem to follow an arc downward, as the accelerating elevator zoomed upward. That's odd: light follows a curved path. But here's the wrinkle Einstein noticed: There is no way to tell from within the elevator whether you are being accelerated upward by an incredibly powerful winch attached to a cable on the roof or whether you are simply standing at rest in a gravitational field. In both situations, the light's path would bend, and you would feel pressure from the floor. If one followed the consequences out to their conclusion, this would mean that any attempt to generalize special relativity to accelerated frames would be indistinguishable from a theory of gravitation. This was Einstein's speculation in 1907 in his article for the *Jahrbuch*.

This insight relied on an assumption that Einstein understood to be universal, what he called the "equivalence principle." In classical physics, there were two different understandings of mass, our measure of the quantity of matter in any given body: gravitational mass, or the quantity of matter that exerted a gravitational force according to Newton's law of universal gravitation; and inertial mass, or the quantity of matter that resisted changes in motion, as in Newton's second law that force equals mass times acceleration ($F = ma$). The explanation for motion was entirely different depending on whether one had inertial or gravitational mass in mind; the odd thing is that the physical results were always the same. As Einstein noted in 1907, "This proportionality between inertial and gravitational mass holds however without exception for all bodies to the level of precision heretofore attained, so that we must until there is justification of the contrary consider it generally applicable."[26] If the equivalence principle always held true, as the detailed measurements by Hungarian physicist Roland von Eötvös indicated they did, then perhaps there was only one kind of mass, and one kind of physics.[27] A generalization of special relativity would necessarily be a theory of gravity.

It was an audacious proposal. "I am extremely excited about your new conception of gravitation," physicist Max von Laue wrote to Einstein upon reading the *Jahrbuch* essay.[28] Einstein was clearly itching to see whether his hunch would bear out. The following year he wrote to Arnold Sommerfeld in Munich to underscore the potential of the idea: "The fact that in a gravitational field all bodies are accelerated equally inclines one precisely very much to the idea that an accelerated coordinate system and an acceleration-free coordinate system with a homogeneous gravitational field appear as completely equivalent things. One arrives on the grounds of this assumption at very plausible consequences."[29] Not everyone was equally enthusiastic. In 1911, in the first major book on relativity theory, von Laue thought that the probability that Einstein's intuition would be validated was relatively low. He ended his summary of the consequences of the equivalence principle with a series of questions about the causes of gravity, the results of gravitational momentum, and its effect on light beams. "No one can presently answer all this," he concluded, "and precisely the circumstance that Newton's law, victorious so far, fully satisfies the astronomers, places the hope for an answer in the foreseeable future to a minimum."[30]

Despite some initial notice taken of Einstein's speculations at the end of his 1907 paper, by the time the physicist arrived in Prague in spring 1911, most established theoretical physicists would have agreed with von Laue's pessimistic verdict: it was a nice notion, but not a likely one. This was not so much instinctive conservatism—among the architects of quantum theory, conservatism in physical speculation was not especially pronounced—as an assessment of opportunity costs and an induction about the unillustrious examples presented by past attempts to reform

Newtonian gravity. The costs were straightforward enough: although there were, to be sure, some anomalies presented by Newtonian universal gravitation (such as the well-known mismatch between the theory and the observed advance of the perihelion of Mercury), surely the challenges in statistical mechanics and atomic physics were more pressing and more likely to yield empirical verification. As for the gallery of discarded explanations for the cause of gravity, there was little reason for optimism.[31] Even a relativistic theory of gravity developed around 1911 by as gifted a theorist as Henri Poincaré, who had formulated his own special theory of relativity alongside Einstein's, was largely ignored, given that it was not rooted in the dominant tradition of field theory.[32]

This climate of aloofness helps explain what has sometimes been posited as Einstein's "delay" in exploiting his 1907 intuitions about the generalization of relativity. Looking back in 1949 on the four years between the *Jahrbuch* essay and his arrival at the Institute of Theoretical Physics on the Weinberggasse in Prague, Einstein had this to say: "Why were another seven years required for the construction of the general theory of relativity? The main reason lies in the fact that it is not so easy to free oneself from the idea that co-ordinates must have an immediate metrical meaning."[33] Notice the three additional years Einstein added to this account: he was not measuring the distance between his early notes and Prague, but between them and the formulation of full-blown general relativity in Berlin. Einstein had not only forgotten the distractions of quantum theory in postponing his work on general relativity, but reduced everything to the technical challenges that beset him after he began his collaboration with Grossmann in Zurich. Almost four decades later, Einstein had even forgotten that Prague had provided him the setting and the respite to devote himself to this question. He had also forgotten the static theory that he had left in tatters on his office floor at the German University. It is time to bring them back to light.

* * * * *

Light was at the core of the static theory, in two senses: the constancy of its speed and its potential to bend in a gravitational field. The speed of light dominated special relativity, making the novel conception of spacetime thinkable. Indeed, one of the first mentions of Einstein's new turn toward full-time research on general relativity after his move to Prague placed particular emphasis on the issue of light's speed. "The relativistic treatment of gravitation creates serious difficulties," he wrote to Jakob Laub. "I consider it probable that the principle of the constancy of the velocity of light in its customary formulation only holds for spaces of constant gravitational potential."[34] Einstein's willingness to relax his former insistence on the constancy of the speed of light in all frames of reference dominated all of his publications on the topic in Prague.[35]

This is especially noteworthy given the role that this stricture played in his 1905 work on the electrodynamics of moving bodies. The classic paper introducing special relativity begins with an elementary thought experiment concerning electromagnetic induction with a magnet and a coil of wire, and then immediately introduces two principles that ushered physics into a post-Newtonian age. First, Einstein expanded the classic Galilean principle of relativity, now insisting that the laws of physics—including those of electromagnetism—must be precisely the same in all inertial frames of reference, that is, all frames that are moving at a constant velocity with respect to each other. The goal of general relativity was also to incorporate accelerated frames of reference. (Think about being on an airplane: you know when you are taking off or landing because the acceleration thrusts you back into your seat or forward against your seatbelt; when you are flying at a constant velocity, however, as far as the physics of tossing a ball is concerned, you might as well be standing still.) The second principle is more counterintuitive: "Light in empty space always propagates with speed V, independent of the state of motion of the emitting body."[36] (Einstein would quickly adopt the convention of using c to label the speed of light.)

In the summer of 1911, Einstein had an idea that there was a connection between the speed of light and the gravitation potential. "I, by contrast, have on the basis of some rumination—already somewhat daring but that still has much going for it—arrived at the view," he wrote to the Dutch physicist Willem Julius that August, "that the difference in gravitational potential could be a cause of the shifting of the [spectral] lines.

From these considerations follows also a bending of light rays by gravitational fields."[37] We have already seen that the intuition that light rays would bend in a gravitational field followed from the thought experiment about the elevator hurtling through space. By June, Einstein had figured out how to calculate the degree of bending by modifying the gravitational potential, the factor that determined the strength of the gravitational field at each point of spacetime. As he then put it slightly more technically in the most influential of his Prague physics papers published in the *Annalen der Physik* on 1 September of that year:

> We must for time measurement at a place that has a gravitational potential Φ relative to the origin use a clock that—set at the coordinate origin—runs $(1+\Phi/c^2)$ times slower than that clock with which time is measured at the origin. We call the speed of light at the origin point c_0, so the speed of light c in a place with gravitational potential Φ is given through the equation
>
> $$c = c_0(1+\Phi/c^2).$$
>
> The principle of the constancy of the speed of light according to this theory does not hold in the same manner as it was preserved as an axiom in conventional relativity theory.[38]

Indeed it does not. The point of this equation is to argue that instead of using the classic Newtonian gravitational potential formula (known as the Poisson equation), one could instead track the changes of the gravitational potential Φ by introducing a speed of light that changes at every single point in spacetime. (It would be, however, constant for any two observers at the same spacetime point or experiencing the same gravitational potential.) One way of thinking about this is to return to Einstein's classic equation of the relation of mass to energy, $E = mc^2$. In a gravitational field, the energy would fluctuate, and he could capture that either by changing mass or changing the speed of light. He opted for the latter.

Why is this important? The title of this first Prague article on what would later be called the static theory, "On the Influence of Gravitation on the Propagation of Light," makes the stakes fairly clear. "I have already sought an answer to the question of whether the propagation of light would be influenced through gravity in an article which appeared three years ago. I return to the topic because my presentation of the situation then did not satisfy me, all the more however because I now realize belatedly that one of the most important consequences of that observation is amenable to experimental testing," he wrote in the opening paragraph. "That is, namely, that light beams which pass in the vicinity of the sun will according to the proposed theory experience a deflection by its gravitational field so that one would find a sensible enlargement of almost a second of arc of the angular distance of a fixed star appearing near the sun."[39] That would be large enough to be measured, but it would require a solar eclipse to be able to see starlight streaming past the sun. Jupiter, the most massive planet in our solar system, would have made for a more convenient observational test, but unfortunately he calculated that the effect would be 100 times weaker there than around the sun, too small for contemporary instruments to detect.[40]

If this paper had been all Einstein had published on the static theory, his Prague theory of gravitation would have gone down in history as a success. This prediction of the bending of starlight around massive bodies like the sun later became the most publicly ballyhooed empirical result of general relativity, and its renown transformed Einstein into the tremendously famous physicist we know today. That event, however, would not occur until spring 1919, when the British astronomer Arthur Stanley Eddington organized two expeditions to measure the potential bending of starlight during a total eclipse of the sun. Eddington's own goals were not just physical: a devout Quaker, he was concerned with repairing the shattered relations among European nations due to the Great War, and a British test of a German pacifist's theory of the universe seemed to fit the bill. The positive results of this test, touted in newspapers worldwide, were the impetus for Einstein's ascension to international fame. Given the drama of Eddington's adventure and its epic consequences, it is understandable that many think of the bending of starlight in the context of Anglo-German relations and then global notoriety, as Americans in California continued to probe the results (eventually confirming them), and reactionary anti-Einsteiners touted a "German" (i.e., not Jewish) precursor whom they argued should receive the credit. This focus on the transnational Einstein is very much valid.[41] There is also, however, a complementary local story in Prague.

Einstein's newfound interest in astronomical observations brought him into contact with Leo Wenzel Pollak, a demonstrator at the Institute for Cosmic Physics at the German University. Pollak found the claim of starlight bending around the sun to be intriguing, and he wrote about it to a colleague in Berlin named Erwin Finlay Freundlich in late August 1911—after Einstein had submitted his paper to the *Annalen* but before it had been published.[42] Freundlich was excited by the idea, and he and Einstein entered into discussions about how it might be tested. Almost everything about the static theory was new at this point, and Einstein was not at all confident that there would be a positive result. "If no such deflection exists then the assumptions of the theory are not correct. One must keep in view precisely that these assumptions, as they already suggest themselves, are indeed really bold," he wrote to Freundlich in September. "If only we had a much larger planet than Jupiter! But nature has not seen fit to concern itself with making it comfortable for us to figure out its laws."[43] The only feasible test would require a solar eclipse, and it would not necessarily be easy. Total solar eclipses are not that common, and the test would need one that reached totality over a region where observing conditions would be favorable and where instruments and photographic plates could be on hand to measure a very slight deflection. Also, it couldn't be cloudy.[44]

Einstein continued to encourage Freundlich's interest in conducting measurements to test the theory in early 1912: "It concerns a question of entirely fundamental significance. From the theoretical standpoint there seems to be a real probability that the effect actually exists."[45] Others were skeptical. Max von Laue, one of the earliest supporters of special relativity, fretted to Einstein: "Whether the astronomical test suggested by you is easily implemented, I don't know; I fear however, that if the deflection is observed, it could always be attributed to the variation of the index of refraction of the solar atmosphere, over the composition of which there are also still many possible hypotheses."[46] Nevertheless, Freundlich seemed determined, and so Einstein wrote to the Viennese philosopher and physicist Ernst Mach—former doyen of Prague physics—in 1913 from his post at the ETH in Zurich that there would be an eclipse test of the new gravitational theory the following year.[47]

The plan was to go to the Crimea. Freundlich laid all the arguments out in a 1914 article and then headed in the summer to the Black Sea peninsula to set up his equipment.[48] With eclipses, it's all in the timing, and the timing was bad for a German astronomer to be traipsing around a strategically important peninsula in the Russian Empire. Freundlich and his team were interned when World War I broke out, and they were thus unable to make the measurements of the star field. (It turned out to be cloudy anyway, so even had they been at liberty it would not have made a difference.) The glory was left for the future, and for Eddington. Still, Einstein never forgot his earliest astronomical supporter, and in his preface to the third edition of Freundlich's 1920 book on general relativity—the first to be published after the spectacular news of confirmation coming out of Britain—Einstein tipped his hat to Freundlich's entrepreneurial intuition: "Mr. Freundlich is not only a renowned presenter of the matter as an expert of the present area of knowledge; he was also the first of his colleagues who had earnestly occupied himself with the testing of the theory."[49]

Einstein and Freundlich remained in touch. The famous physicist—now in Berlin—helped find work for the perpetually underemployed astronomer, including having him design the solar telescope for the art nouveau "Einstein Tower" observatory that was under construction in nearby Potsdam. The ruptures following Adolf Hitler's rise to power in Germany in 1933 dispersed both of them, Einstein first to Caltech (where the definitive confirmations of his gravity theory had been conducted) and then Princeton, and Freundlich to the winds. He ended up in Istanbul teaching until 1937, whereupon he moved to Prague, that city that had first brought the two of them together, as a professor of astronomy. He was not there for long, Central Europe being a dangerous neighborhood at the time, but did manage to have an important influence on Zdeněk Kopal, a Czech student who later occupied a distinguished professorship in astronomy at Manchester for three decades. In January 1939—after the amputation of the Sudetenland from Czechoslovakia but before the Nazi occupation of what would become the Protectorate of Bohemia and Moravia, Freundlich decamped for the University of St. Andrews in Scotland (on Eddington's recommendation). He taught there until 1959.[50] Upon retirement, he returned to Wiesbaden, his hometown, as a professor at the nearby University of Mainz. He died five years later.

* * * * *

Back in early 1912, in Prague, Einstein was hard at work pushing beyond his initial insights about the equivalence principle and the potential bending of starlight around massive bodies to produce a full-blown general theory of relativity. His starting point was already set: he would work with a modified Poisson equation—$\Delta c = kc\rho$, where k is a constant, ρ is the mass density, the c on the right is the speed of light in a vacuum, and the Δ is an old-fashioned rendition of the Laplacian operator (which measures how rapidly a function varies across space)—to model the gravitational potential with spatially variable speed of light against a flat (Euclidean) spacetime. From here, he needed to work out a consistent theory. It was grueling labor. He wrote to his former assistant Hopf that he could be excused for his dilatoriness as a correspondent "because I am working *like a horse*, even if the cart does not always move very far from the spot."[51] Einstein liked the equine simile, which he repeated a week later to Heinrich Zangger: "But I have worked and been driven like a horse. The main subject was a relativity-theoretic work about gravitation. It was finished a few days ago."[52] Well, he had hoped it was. In March 1912 he was forced to confess again to Zangger that "I am up to my ears in the problem of gravitation, so that I cannot summon the energy to write a letter."[53]

Nonetheless, despite the downs, there were definitely also ups, and Einstein repeatedly thought he had cracked the problem. "The research into the statics of gravitation (point mechanics[,] electromagnetics[,] gravitational statics) is finished and pleases me very much," he wrote his friend and colleague Paul Ehrenfest in February 1912. "I truly believe that I have found a piece of truth. Now I am thinking about the dynamic case, once again moving from the more special to the more general."[54] He was confident enough to write enthusiastically to Lorentz, whose physical intuition he admired perhaps more than anyone else's: "The second matter concerns the relationship of gravitational field—acceleration field—speed of light. Simple and beautiful things emerge entirely inevitably from this. The speed of light c is variable. It determines the force of gravity."[55]

He wrote an article outlining what he believed was a complete static theory and sent it off to Wilhelm Wien, editor of the *Annalen der Physik*. "I am sending you a paper for the *Annalen*. A great deal of sweat went into it, but I now have total confidence in the matter," he wrote in his cover letter. "Now I am looking for the dynamics of gravitation. That however won't come quickly!"[56] He was hoping that this theory would finally put to rest the criticisms of one of his rivals, a German physicist based in Turin named Max Abraham. His confidence was misplaced. Two weeks later, Einstein suddenly realized he had made a mistake and wrote Wien asking for him to refrain from publishing the paper. Then he had second thoughts about his second thoughts, sending a further letter on the same day: "This morning I asked you to send me back my manuscript and now I ask that you hold on to it. Certainly not everything that is in the work is tenable. But I believe the matter should be left as is so that those who are interested in the problem can see how I came to the formulas."[57] The problem was how to move from the static to the dynamic case, and it was a tough nut to crack.

Nine days later, he sent another article to Wien, further elaborating the static theory: "I send you here the continuation of the research about the static gravitational field for the *Annalen*. I have tormented myself dreadfully over this matter, but believe now I have found the correct sense."[58] He sent the good news to Hopf as well: "I have now rigorously derived the theory of gravitation for the *static* field. The result is wonderfully beautiful and astonishingly simple. Abraham's theory is entirely false. I will surely have a duel of pens with him."[59]

On this, at least, Einstein was most definitely correct. Abraham was not one to take dismissal lying down. Born in 1875 in the then Prussian city of Danzig (now Gdańsk, Poland), Abraham was an astonishingly versatile physicist of Einstein's own generation. Trained at the University of Berlin under Max Planck, he graduated and assumed a *Privatdozent* position at Göttingen, which he occupied for nine years. In this role he was situated better, academically speaking, than Einstein was at the Bern Patent Office. In every area of his research he came butting up against Einstein. In 1902 he published a theory of the electron that competed with Lorentz's model, only to have Einstein's own modification of Lorentz's theory, special relativity, take over. (Abraham never relinquished his own theory.) He continued to look for professional advancement, taking a job at the University of Illinois in the United States in 1909 but abandoning it after a few months to assume a teaching post at the polytechnic in Milan. Still on Einstein's heels, he sent an explication of his first theory of gravity to the *Rendiconti della R. Accademia dei Lincei* in December 1911.[60] Einstein read it in early 1912 and thought it atrocious.

The feeling was mutual. Einstein, as we have seen, felt he had to give up the constancy of the speed of light in order to have a functional theory. "I have now worked out a theory of the statics of gravitation that steps out from the schema of relativity theory in that it violates the principle of the constancy of the speed of light," he wrote to the Polish physicist Marian von Smoluchowski on 24 March. "I deem it rather certain that this principle only applies insofar as the gravitational potential can be considered as constant."[61] Abraham considered this a contradiction in terms. Given his steadfast opposition to Einstein's theory of special relativity and his embrace of his own electron theory, he crowed that Einstein "has given up the postulate of the constancy of the speed of light, essential for his earlier theory; in a recently published work he allows the demand of invariance of the equations of motion under Lorentz transformations to lapse, and with that deals relativity theory its death blow."[62]

This sparring back and forth formed a constant subtext in Einstein's on-again, off-again submissions to Wien. In the February paper (which appeared only on 23 May), he noted that "Abraham's system of equations cannot be reconciled with the equivalence hypothesis, and this conception of time and space cannot be considered correct already from a purely formal mathematical standpoint."[63] The equivalence principle lay at the root of their disagreement: for Einstein it was the entire justification for assuming that relativity theory would have any implications for gravity; for Abraham, it was nonsense. As Einstein himself would come to realize by late spring 1912, largely from the force of Abraham's criticisms, it was not possible to maintain the equivalence principle as an absolute rule.[64] Consider our elevator rider. She could imagine that there was a gravitational field emerging from the floor of the elevator car, but it would be a strange sort of gravitational field that emerged entirely perpendicular to the floor. On Earth, the planet's spherical shape produces a radial gravitational field that cannot be imitated by any elevator. An equivalence principle could only hold infinitesimally. Abraham preened at what he saw as Einstein's retreat.[65]

For the physicist in Prague, however, it was no such thing; it was simply a gradual revision of the theory in view of conceptual difficulties (such as another problem in accounting for momentum conservation). By the beginning of summer 1912, on the eve of moving back to Zurich, Einstein thought any further dialogue would be fruitless. He had begun his research on gravitation already with some reservations, tinged with respect, about this familiar sparring partner. "He must be a curious, immoderate person. Even his quite sharp judgment seems to be influenced by his passion," he wrote in 1910. "I gladly concede to him that he calculates better than I do, that he knows more books than I do, and whatever else he wants."[66] These comments encapsulate the central element of his objection to Abraham, that he "operates formally without thinking physically!"[67] The equivalence principle was a perfect battleground for precisely this conflict: Einstein considered it a wonderful heuristic guiding his physical intuition about how matter should behave in accelerated frames of reference, while all Abraham could focus on were its (undoubted) calculational flaws. For Einstein, this was simply not good physics, and his letters from this period are peppered with sarcastic and dismissive barbs that he tossed—behind Abraham's back—to his friends about the man from Turin: "Abraham's theory is created off the top of his head, i.e., created purely from considerations of mathematical beauty, and is completely untenable. I cannot at all comprehend how an intelligent man can let himself be enchanted by such superficiality."[68] In that same letter, he confessed that he had been hoodwinked by the theory for a whole fortnight before coming to his senses. Abraham's theory was not invariant under Lorentz transformations yet relied on Minkowski's space-time geometry, which required such invariance. Fatal internal contradictions emerged in consequence.[69]

Einstein's tendency to grumble about Abraham in his letters to friends presents another contrast between the two scientists—who were, it bears underscoring, two of the very few theorists working actively on developing a new theory of gravitation. Einstein may have had serious problems with the kind of physics Abraham was doing (and also his temperament), but he did not lower the tone in public. Although he considered Abraham's "recently published theory of gravitation to be a severe blunder," he made efforts in April 1912 to try to obtain for Abraham a better job in Zurich (of all places). "It is just incomprehensible that this truly significant man, because of some high-spirited sarcastic comments which he made several years ago, is to be shunned like a leper," he wrote.[70]

Publicly, however, the exchange soon began to descend into polemic—pushed by Abraham's tone and rhetorical style—until Einstein no longer felt dialogue was productive. On 4 July 1912 his final published response to Abraham concerning the static theory of gravitation was received at the *Annalen der Physik*. In it he stated that he was amazed that "Abraham writes that I have given the death blow to relativity theory through the denial of the postulate of the constancy of the speed of light and through it the corresponding renunciation of the invariance of systems of equations through Lorentz transformations," even though he had always insisted that the two postulates of special relativity were "entirely independent of each other."[71] Relaxing the constraints on light in no way invalidated the more important first principle of relativity in all inertial frames of reference. "We do not have the slightest reason to doubt the general validity of the relativity principle," he continued. "By contrast I am of the view that the principle of the constancy of the speed of light can be considered to be correct insofar that one constrains oneself to spacetime regions of constant gravitational potential. In my opinion here lies the limit of the applicability not of the relativity principle indeed but of the principle of the constancy of the speed of light and thus our current relativity theory."[72] He was ready to concede some of Abraham's points about the validity of the equivalence principle only in infinitesimally local contexts. "But I see here no reason to forego the equivalence principle also for the infinitely small; nobody will be able to deny that this principle is a natural extrapolation of one of the most general experiential statements of physics."[73]

That was his last explicit engagement with Abraham's ideas. When the latter jumped in for yet another continuation of the fight, Einstein preferred to keep his own counsel. In late August 1912, he sent a final statement to the *Annalen*: "Since each of us has expressed his point of view with the necessary detail, I consider it unnecessary to answer Abraham's preceding note. I would like here for the time being only to request that the reader not interpret my silence as agreement."[74] By this time he was already back in Zurich and ready to start a new approach to general relativity relying on tensor mathematics and curved spacetime. He discarded both Abraham and the static theory in a single move.

* * * * *

Einstein may have publicly abandoned Abraham, but Abraham had not abandoned him, predicting the eventual capsizing of relativity theory against the shoals of his continued reliance on the equivalence principle.[75] In Einstein's letters to friends discussing, among other things, the development with Marcel Grossmann of the *Entwurf* theory of general relativity, he still could vent his spleen in acerbic asides: "Abraham's new theory is, so far as I can see, logically correct, yet it is still an embarrassing stillbirth," he wrote to Arnold Sommerfeld in October 1912, a year after the Solvay Conference.[76]

Einstein could ignore Abraham as he continued to work on gravitation, but Abraham could not symmetrically ignore him. Einstein was a major player in theoretical physics, despite gravity not being a major playing field of that discipline. Any mainstream engagement with gravitation had to pay attention to the publications coming out of the ETH and, after 1914, from Berlin. Abraham continued to pursue his agenda, but he was getting less and less attention, even as he drew novel implications from his theories, such as gravitational waves.[77] In 1914, Abraham announced the end of relativity:

> Any relativity theory, both the special one of 1905 and the general one of 1913, will thus founder on the shoals of gravity. The relativistic ideas are obviously not broad enough to serve as the framework for a complete world picture.
>
> Yet relativity theory remains a historic achievement in the criticism of the concepts of space and time. It has taught us that these concepts depend upon their representations, that we form them from the behavior of measurement of length and time intervals using yardsticks and clocks, and they are subject to the vicissitudes of those. This secures for relativity theory an honorable burial.[78]

From above the fray, Einstein continued to grudgingly admire Abraham's persistence: "He complains powerfully in *Scienza* against everything relativity, but with understanding."[79] Einstein even wrote yet another letter in support of Abraham's hiring, this time in Aachen in 1921.[80] The irascible antagonist died just one year later.

Once the two ceased jousting, the field of interlocutors in gravitational physics narrowed considerably. The two main alternatives to the line Einstein was pursuing after his abandonment of the static theory were those of Gustav Mie, then at the small northern German university in Greifswald, and Gunnar Nordström, a talented theorist from Helsinki, at the time a city within the Russian Empire. Einstein did not have much patience for Mie, but he followed Nordström's work carefully, regarding it as the most significant alternative to his own line of research, and one that had the potential to be correct.[81] Educated at the universities in both Helsinki and Göttingen (he received his doctorate from the former in 1910), Nordström was only two years younger than Einstein and in precisely the same generation.[82] He was also, as his geographic origins indicate, marginal to the mainstream of Germanophone physics, as was every researcher on gravitation in this period with the exception of Einstein.

Nordström's work was published in late 1912—thus after Einstein had already left Prague—but his impetus for it stemmed from the static theory Einstein had developed in Bohemia. He was impressed with

Einstein's results that showed the connection between the generalization of relativity to accelerated frames of reference and a new theory of gravitation, but he wanted to develop it without the problematic equivalence principle. He also insisted on using flat, Euclidean spacetime. In the appendix to one of his important early articles, Nordström acknowledged his fruitful constructive correspondence with Einstein, a compliment that Einstein returned in print in 1913 when he maintained that Nordström's static theory was the only alternative to his own that he considered viable.[83] In many respects, the theories were equivalent; where they differed most substantially was in the prediction of whether light would bend in strong gravitational fields: Einstein's theory said it would; Nordström's said it would not. That question would be spectacularly resolved in 1919. In this moment before the Great War, however, Einstein continued to encourage Nordström, and during the war he helped him cross heavily militarized borders by issuing him bona fides.[84] The relationship remained cordial until the Finn's premature death in 1923. In 1915, in fact, Einstein was involved in placing Nordström's name in second position on the hiring committee list for a post in Berlin.[85]

Speaking of those in *secondo loco* who were associated with Einstein, there was another marginal player in the pursuit of gravity in 1912 who was, however, never acknowledged by him: Gustav Jaumann, the man who had initially been offered the Prague position—after some ministerial legerdemain—instead of Einstein. Among theoretical physicists in this period, Jaumann had a reputation for maintaining a rigid and unwavering commitment to continuum physics. Following his mentor Ernst Mach's epistemological skepticism about the ontological reality of atoms, Jaumann had built an entire physics out of continuous fields and substances, a technical feat as mathematically challenging as it was out of step with the mainstream. In his treatment, gravity was just like temperature, a potential built out of the Poisson equation, and he claimed that his great achievement was to explain the long-term stability of gravitational systems—as well as the anomalous perihelion advance of Mercury, which later became one of the three classic tests of Einstein's theory of general relativity.[86]

As far as I have been able to detect, Einstein and Jaumann never met, and the former seems never to have mentioned the latter. The converse is not true. In 1913, Gustav Mie, needling Einstein as was his wont, defended the priority of Max Abraham's gravitational theory over the *Entwurf*. (Einstein objected, noting that the theories were different, so priority was not really at issue.) A few months later, Jaumann wrote in to a journal that his own theory had priority even over that of Abraham. "The fundamental idea of my gravity theory is that the Poisson differential equation is the rudiment of an effective physical equation and must be completed through *introduction of the fluxion of some physical variables*," he stated, "more precisely the total fluxion of the first order *of the gravitational potential itself* to an effective physical law, which thus also represents the temporal propagation of gravitational effects."[87] There was no reaction. When reviewing the literature on this topic in 1914, Abraham remarked that Jaumann's work was so far afield that he was going to omit it in his analysis.[88] The only credence the scientist from Brno's priority received was in a laudatory obituary upon his death in 1924.[89]

* * * * *

Einstein left Prague as he had come: without a workable theory of general relativity. But the city proved essential to the transformation of his research profile by giving him the conceptual (and literal) space to detach himself from the pressures of the ever-growing quantum theory and enabling him to follow a hunch from several years earlier about the equivalence principle and gravitation. The theory he produced between the fall of 1911 and the spring of 1912—the static theory—ended up foundering on its own assumptions, but not without teaching Einstein a few valuable points. The first was that some implications of any theory of general relativity, such as the bending of starlight around a gravitational body such as the sun, would be amenable to empirical verification, a realization that led him to use his contacts in Prague to connect with Erwin Freundlich in Berlin and initiate an eclipse expedition. Although the understanding of the precise magnitude of the bending of this light would change as the final theory of general relativity took form, the centrality of this phenomenon would not—indeed, it would form the catalyst of Einstein's global celebrity.

Second, the constant back-and-forth in letters to colleagues and in printed duels with Abraham considerably refined his understanding of the equivalence principle and its limitations. The initial insight about the need to modify special relativity's insistence on the constancy of the speed of light when faced with a changing gravitational potential, however, did not. The *Entwurf* theory carried over this central finding: "In earlier works I have shown that the equivalence principle implies that in a static gravitational field the speed of light c depends on the gravitational potential. [. . .] c is not to be understood as a constant, but as a function of the spatial coordinates that represents a measure for the gravitational potential."[90] In the spring of 1912, Einstein realized—almost certainly in conversations with Georg Pick—that he might have to relinquish flat spacetime in favor of a Riemannian geometry, an idea he pursued with vigor with Grossmann in the fall.[91]

The very success of general relativity in the form in which it was presented in Berlin in 1915–1916 has served to occlude the significance of the physicist's time in Prague, the moment when he first began to work in earnest on gravitation. As a collaboration of historians over the past 15 years has shown, Einstein pursued in Zurich and then in Berlin two parallel routes toward general relativity: one physical and one mathematical. His eventual triumph encouraged him to emphasize the mathematical route—the quest for general covariance—over the physical intuitions that had guided him in his first ventures in 1907 and then again in Prague. This was how Einstein related the history of the theory that made him unimaginably famous, and an effect of this presentation has been to diminish the apparent significance of the months he spent in Prague.[92] It is important not to overstate the contrary point: if Einstein had died in that freak horse cart accident in the wake of his fight with Abraham, we might associate gravitation more with a modified Nordström theory than with Einstein—the theory as of summer 1912 was not an embryonic form of general relativity. On the other hand, it is hard to see how he could have reoriented himself so single-mindedly on this path without the experience of splendid isolation that he had enjoyed in Prague.

That isolation had benefits but also costs. Both family and professional disappointments induced Einstein to start exploring ways out of Bohemia a semester after he arrived. In late August 1911, the university in Utrecht wrote inquiring whether he would be interested in a position there. It only took Einstein four days to respond decisively in the negative: "I have been here in Prague only 4 months and am now happy to have adapted somewhat to the foreign conditions. Thus I much regret that I must ask you to consider another colleague for the vacant position, and I give you my heartfelt thanks that you wanted to place me in your milieu."[93] His Dutch contact, Willem Julius, tried again in late September, and Einstein once again turned the proposal down. He wrote his Zurich friend Heinrich Zangger "that I however right away declined for the reason that I could not so soon again decide to change my residence." (Although, in this same letter, he also told Zangger that his refusal had nothing to do with a newly acquired fondness for the city: "But that I would abandon half-barbaric Prague with a light conscience, that is certain.")[94] Or, as he wrote Julius at greater length, after commenting on his enjoyable interactions with Dutch colleagues: "These pleasant personal experiences have made my decision to remain here truly difficult; but I am now resolved to do so. Think of yourself in my present situation! Here I have a spacious institute with a magnificent library and I don't have to struggle with the difficulties of the language, which with my awful ponderousness in learning languages comes heavily into consideration for me!"[95] It is quite telling that Einstein, a professor in a bilingual city where his own tongue was in the overwhelming minority, thought of Prague as essentially Germanophone.

Utrecht persisted. In November, Einstein somewhat modified the answer he had been consistently offering. "Before I left my Zurich homeland [*Heimat*] for Prague," he wrote Julius, "I promised *privatim* in Zurich to communicate with them before I took any other position so that the administration of the Polytechnic could be allowed to make an offer in the event they found that a good idea."[96] He asked Zangger and Grossmann to make inquiries back in Switzerland—*not* at the University of

Zurich that he had left, but at his alma mater, the ETH.[97] After a lengthy exchange of letters with administrators in Zurich, including a debate about whether they would come to Prague to negotiate with Einstein or he would make the trip to Zurich (in the end, Einstein traveled), he received an official call in late January 1912.[98]

What had changed from his earlier determination to remain in Prague? The most likely cause was his experience in Brussels at the Solvay Conference. Whatever his privately dismissive comments about the intellectual results of the gathering, it seems that the experience of engaging in conversation with the luminaries of his field registered with Einstein. He made an impression on several of them as well. Much as the German University in Prague had earlier asked Max Planck for his endorsement of this former patent clerk, the ETH bureaucracy contacted Marie Curie in Paris. "In Brussels, where I attended a scientific congress in which Mr. Einstein took part, I was able to appreciate the clarity of his mind, the comprehension of his writings, and the depth of his knowledge," she responded. "If one considers that Mr. Einstein is still very young, one is correct to base on him the greatest hopes and to see in him one of the premier theoreticians of the future."[99] Einstein now looked around Prague and, instead of seeing quiet where he could concentrate on gravity, lamented the absence of colleagues against whose intellect he could sharpen his pathbreaking ideas. What had just a few months earlier appeared as an exciting new adventure in this Germanic outpost among the Habsburg Slavs had come to seem provincial, even backward. Once he saw Prague this way, he could never unsee it, and Zurich's invitation proved impossible to resist.

CHAPTER 3

Anti-Prague

Since then I must always, when I want to write something about a foreign region, think: "Better to forgo the local color, you will only disgrace yourself." Yes, it turns out as though the residents formally have a right for the exclusive representation of their homeland.

Almost all good travelers have also respected that and cannot stress often enough that their comments are only meant subjectively; no dependable exact account, only a mistaken attempt—and an interesting one precisely through its mistakes. Precisely because of the fact that these observers refuse in advance any ethnographic and scholarly whitewashing, they obtain the courage to present directly their coincidental experiences unaffectedly, without responsibility and, where they seem to establish laws, one softly hears the skeptical, almost ironic undertone.

—Max Brod[1]

From Albert Einstein's point of view, especially in later years, the most significant thing he took away from Prague was something he had left behind: the static theory of gravitation. It had not been a success in terms of leading directly to a full-blown theory of general relativity, but it had provided a passage out of the quantum rut and set him on a course to one of the most remarkable developments in the history of physics. In this way of looking at things, Prague was simply a backdrop, a city in which he happened to live while working on a particular set of equations, but he might as well have been in Bonn or Königsberg or Vienna—any university town where he could function seamlessly in German, shuttling back and forth from his apartment to his offices at the local physics institute. Yet this was not Einstein's actual experience of the city, which impinged on his thoughts and shaped his life in countless ways—even if he wasn't conscious of them.

Take that daily walk to the institute, which was highlighted by all observers as the metronomic backbone of the physicist's life in Bohemia. Even today, with much greater traffic—there is a busy tram station along the way that holds up pedestrians—the journey lasts about a quarter of an hour if one walks briskly, as Einstein did in those days. The heart of Prague did not suffer too many architectural atrocities across the violent twentieth century, and many of the buildings and landmarks that Einstein would have seen we can still observe today. He left no surviving trace of his constant traverses of what he would certainly have called the Moldau River (known as the Vltava in Czech both then and now), and it is possible that he did not notice much of it, lost in daydreams about the equivalence principle or the numbing blindness of the habitual commute. But we can reveal what he did not deem worth mentioning, perhaps did not even notice, and in doing so we can encounter the real Prague of 1911 and 1912, the Prague in which Einstein lived.

He would have begun this daily walk by descending the stairs of what was then Třebízkého ulice 1215, emerging out of the portico of the new art deco building on the south side of this short residential street in the neighborhood of Smíchov. He would have turned right and walked toward the Moldau, and when he hit it made another right along the banks, heading for the Palackého Most, or Palacký Bridge, which he would take across the river. The bridge was built from 1876 to 1878, according to the designs of Bedřich Münzberger (whose name is a characteristic blend of Czech and German monikers), to solve a pressing problem for the industrial firms located deeper in Smíchov. Factories like the Ringhoffer Machine Works needed a way to move heavy material into the heart of the city, and for that a stone bridge was essential. Chain suspension affairs like the Francis I Bridge would not do, and the ornate towers of the iconic Charles Bridge farther north—today a statue-studded pedestrian thoroughfare bearing throngs of tourists ambling from the castle to the Old Town Square—made it inconvenient for the movement of machinery. The Palacký Bridge was a modern bridge for a modern city.[2]

As Einstein turned left onto the bridge to cross the river, he would have seen something that is no longer there. In those years the Palacký Bridge was framed by four sculptural groups based on themes from Czech folklore, including the beautiful soothsaying mother of the people, Libuše, and her chosen husband and mythic founder of the first dynasty in Bohemia, Přemysl. They would have greeted Einstein on the Smíchov side of the bridge as he walked to work. The statues had been designed by the canonical Bohemian sculptor Václav Myslbek and installed during the 10 years following 1887. They were thus a later nationalist adornment to what had come to be an increasingly symbolic bridge. (The statues were damaged during a U.S. air raid on the city on Valentine's Day 1945; after being restored, they were moved from the bridge to the gardens at Vyšehrad, down the eastern bank of the river.) As the city's chief stone bridge, the fourteenth-century Charles Bridge, became a symbol of Catholicism and Habsburg power—a symbol of the "Germans"—the Palacký Bridge would become the opposite. The stone was subtly blue, white, and red, the national colors of the Crown of Bohemia. Einstein almost certainly did not know who the figures were or what the colors meant; for him, this was simply the easiest way to get to work.

For the entirety of his time in Prague, he would have encountered a bit of trouble as he alit from the bridge to enter the New Town. There was a large construction site there—so large that it is really astonishing that Einstein never commented upon it in letters to distant friends. It must have been an impressive sight, as dozens of workers labored to erect a gigantic monument to František Palacký (1798–1876), historian and central nationalist hero of 1848.[3] Palacký remained committed to what eventually became Austria-Hungary—he felt it was the only way to protect the Czechs amid the surges of German nation-building to the west and north and Russian pressure from the east—but he was a constant thorn in its side, leading the Old Czech political party to demand ever greater autonomy within the framework of the Habsburg Empire. After his death, as Czech politics radicalized into increasingly confrontational postures, Palacký was transformed into a convenient emblem of oppositionist politics. His monument would heighten the symbolism of his bridge.

The foundation stone for the monument was laid on the centenary of his birth in June 1898, in a ceremony commemorated by Slavic personalities near and far, such as Grand Duke Konstantin, president of the Academy of Sciences in St. Petersburg (of which Palacký was a member), who sent a sympathetic greeting.[4] Stanislav Sucharda, the leading student of Myslbek—designer of the mythic figures on the bridge, and also the statue of Wenceslas in his eponymous square, the most internationally recognizable sculpture in Prague—composed an enormous montage of figures (figure 3). In it Palacký himself, seated in the center and gazing across the river, is overshadowed by his granite backdrop, itself surmounted by the gendered symbolism of angels and victory. The bronze groups offer an allegorical retelling of Czech history in a nationalistically tendentious variant: they are designated with titles like

FIGURE 3. The Palacký monument, unveiled in 1912, facing the Palacký Bridge across the Vltava/Moldau toward Einstein's neighborhood of Smíchov. *Source:* Rémi Diligent, 2006, https://commons.wikimedia.org/wiki/File:Frantisek_Palacky_monument_(global).jpg.

"Oppression," "History Tells the Story," and "The Awakener." The whole ensemble took six years to design and months to erect.[5] It must have been noisy. Einstein could not have failed to notice it as he passed along the north edge of the structure down a side street on his way to the Physics Institute on Weinberggasse, a few blocks away.

Public officials, both Habsburg and Czech nationalist, freighted public monuments with enormous significance in the decades surrounding the turn of the century, and the Palacký memorial was no different.[6] During Einstein's time in Prague, memorials meant to convey the growing political strength of Czech activists popped up in the most visible surroundings. A plaster cast of the giant memorial to Bohemian religious reformer Jan Hus was presented in the Old Town Square in March 1911, there for Einstein to see when he took in the obligatory sights upon his arrival. On 11 April, a few days after his first entry to the city, the Prague City Council affirmed the original 1899 ordinance that had approved the memorial—a rebuke to the Catholic Habsburg symbol that graced the square, the Marian Column (which would be torn down in November 1918 a week after the declaration of Czechoslovak independence by a nationalist mob returning from a gathering at White Mountain). Palacký himself had been instrumental in elevating the image of Hus as a touchstone of Czech nationalism, despite the overwhelming Catholicism of the Bohemian population.[7] In 1912, Myslbek's bronze Wenceslas statue was unveiled, replacing a stone monument, removed in 1879, that had been there since 1680 and had been damaged in the 1848 uprisings.[8] So salient were these recent additions to the public image of the city that the 1913 edition of the Baedeker guide made sure to insert a reference to the new Palacký tableau.[9]

It would have been hard for Einstein not to notice any of this, given how central each was to the landscape of his own path through the city. Even the unveiling must have caused a ruckus, disrupting his path to work in his final months at the university. The Palacký monument was ceremonially unveiled in late June 1912, timed to coincide with a huge gathering (*Slet*) of the Czech nationalist gymnastics league, the Sokol. Athletics in this period were central to Czech nationalists, who participated in the 1912 Olympics as a separate grouping from Austria named

"Bohemia"—though not, as in 1908, under an entirely separate flag and anthem.[10] Crowds for the Slet began to flood into the city in May for public demonstrations of physical prowess and the touting of symbols of ethnic autonomy; the events continued into the first week of July. As many as 300,000 people converged on the city, with one day alone garnering a high of 120,000.[11] It must have been quite a crowd at the base of the Palacký Bridge.

None of this seems to have registered in any tangible way with Einstein, at least not enough to leave a trace in a letter, a notebook, or a reminiscence of a friend. All this fervor both gymnastic and granite invigorated the city's overwhelming Czech majority yet passed the physics professor at the German University by. He surely observed it, and just as surely put it out of his mind. This chapter is about the Prague Einstein did see and the Prague he did not. Rather than comparing a "German Prague" to a "Czech Prague" in a fashion that would resonate with the boosters behind the Palacký project, Einstein was involved in a different set of binaries. Prague was his new home. He began to assemble impressions of the city, understandably drawing them by contrast to Zurich, the Swiss metropole he had left. It was not a flattering comparison for Bohemia: Zurich was for Einstein an anti-Prague, and Prague became an anti-Zurich. All of his comments on his new environs bear the implicit or explicit mark of its failure to live up to his standards. Einstein's understanding of himself as a displaced Swiss structured his contradictory feelings toward the city, while his social milieu, including some of the leading Germanophone intellectuals of the day, simply registered him as a German joining their island in the Slavic sea. He may have forgotten it all after he left, but while there his interactions with the city were passionate, heavily gendered, and ultimately superficial—though no less strongly felt for their lack of depth.

* * * * *

Einstein would later harbor a fondness for Czechoslovakia, the state that would emerge out of the ashes of the Habsburg Empire in 1918, with its capital in Prague. This affection clearly colored the stories he would tell his friends, family, and early biographers about his time in the city, generating the impression that he had spent his three semesters at the German University most contentedly. In a 1930 biography, Einstein's stepson-in-law Rudolf Kayser—writing as Anton Reiser, partly as a concession to the physicist, who objected to the whole project and forbade its publication in Germany (he acceded to its appearance abroad)—painted a poetic vision of Prague that bears some traces of the oral exchanges from which it must have been drawn:

> He was in provincial Austria, a land of nationalist struggles; in an old romantic city of Catholic tradition. The old buildings, the dark streets, the magic of ruin, evoked an atmosphere of a decidedly artistic sort. The gravestones of the old Jewish cemetery told a thousand years' story of his race. Einstein's artist's nature was stirred by all this.
>
> He had gone with fear and misgiving to his new home. He soon felt adapted to his conditions; the city filled him with enthusiasm. In contrast to his first experiences at Zurich, his position pleased him.[12]

Einstein's most important biographer and successor as professor of physics at the German University, Philipp Frank, likewise wrote that Einstein "often remembered his stay in Prague, where he learned to value the Austrian people, with pleasure. He had never shared the Prussians' suspicious views about the Austrians."[13] Even Hans Albert, Einstein's elder son, who had been old enough to retain his own memories of the city, corrected a Czechoslovak journalist in the 1970s about misapprehensions that Einstein had not liked the city: "If they say that my father was not enthusiastic about Prague, they are greatly misled. He loved Prague and he spoke about it with enthusiasm even many years later."[14]

Hans Albert and the others aside, Einstein's correspondence of the time presents a very different picture, one that started with superficial exoticization and ended in grumpy complaining. The process began with his first letter back to Zurich on 5 April 1911, as he looked on the new adventure with anticipation: "We arrived here after a wearying journey & right away have already found an apartment. But there are countless difficulties to overcome when you arrive to such different and unfamiliar conditions."[15] Two days later, after moving out of a hotel and into the apartment on Třebízkého ulice, he sent a letter to his close Zurich friend, the physician Heinrich Zangger, which framed the city as a land of picturesque sites and seething nationalist tensions:

> Prague is a beautiful city to look at. The people are, each according their own destiny, arrogant, shabby-genteel, obsequious. They are masters in cooking. A certain grace is present in many of them. Houses & surroundings are somewhat dirty & gone to seed. The animosity between Germans & Czechs appears significant. Example. I ask our institute's porter where one can find wool blankets. My predecessor—Herr Lippich—learns that he had recommended to us a store owned by a Czech. He immediately sends his maid to us to ask me to buy the blankets in a "German" shop.[16]

Virulent nationalism was the exception rather than the rule among the residents of Bohemia, but Einstein had no way of knowing that. His interaction with his predecessor shaped his approach to the city. While he would never adopt the nationalist rhetoric of Lippich or his colleague Anton Lampa, he would not hesitate by November to tell Zangger in a letter that Bohemians were intrinsically dishonest.[17]

Indeed, Einstein continually oscillated between feelings of tremendous satisfaction in his new position and acute frustration, the latter often directed at the residents of Prague—understood implicitly to be the Czech majority. In a letter to Michele Besso, Einstein began with praise of his wonderful institute and his hopes to get a lot of work done over the summer before slipping with virtually no transition into a complaint that "the people are alien to me" in Prague. "There are absolutely no people with natural feeling: unemotional and a peculiar mix of snobbish and servile, without any sort of goodwill toward their fellows. Ostentatious luxury next to creeping misery on the streets. A bleakness of thought without faith."[18]

Visitors brought back the familiar, often making the situation worse by reminding him of how pleasant life was elsewhere. After almost a year in the city, he included in a chatty missive to family friends, Alfred and Clara Stern, an extended diatribe about Prague:

In winter I had a visit from my mother and from a school friend & thus became properly acquainted with the architectural beauties of Prague. A short stay in Prague—only for pleasure—is extraordinarily worth it. But here life has its dubious dark side. No drinkable water[.] Much misery alongside boasting and arrogance. Class prejudices. Little true cultivation [*Bildung*]. Everything is byzantine and priest-ridden. My older bear-cub must go to Catholic religious instruction and—*horribile dictu*—to church. The ink-shitting at the office is endless—everything, so it seems, to give the retinue of scribes in the state chancelleries an apparent justification for existence (That sentence is not bad; if it were printed it would perhaps be memorialized in your pretty collection). When I come to the institute, a servile person smelling of alcohol bows and says "most obedient servant." Something like individuality is uncommon here, is also only rarely found among the students. Surely the Imperial *gymnasium* professors have already corrupted a good deal. But there are nonetheless a few splendid chaps among my students.[19]

It goes without saying that this was an unfair representation of the city and its residents, but it is a revealing one. Notice that he had lived in the city for months without taking the opportunity to acquaint himself with the major sights—it was only his mother's visit that had roused him to become a tourist. This ignorance of the town's attractions had not prevented him after only a month in residence from trying to entice Besso to visit in the same letter quoted earlier. Writing to Zangger a little later, one sees the surface images already tainted by dissatisfaction with people: "Thus I hope to get you to spill the beans and at the same time show you the wonderfully beautiful city, the city of these barbarians. These people truly have a backwards culture. I have so far not discovered true scientific interests among my colleagues, only a certain haughtiness."[20] (Both friends eventually came for visits.)[21]

Einstein's views of life in Prague were based on surface impressions, generalizations of a few encounters to entire classes of people. Without spending years in a place, indeed, it is hard to understand how else one would build up a picture of a city. Such superficiality had also marked his arrival in Bern, which he explored before assuming his post at the patent office. "It is charming here in Bern. An antiquated, extremely cozy city in which [one] can live exactly as in Zurich," he had written to his fian-cée, Mileva Marić. "On both sides of the streets stretch really old arcade passages, so that even in the heaviest rain one can go from one end of the city to the other without becoming noticeably wet. In the houses it is uncommonly clean, I saw this everywhere yesterday when I looked for my room."[22]

Prague did not generate such happy associations. One senses that his frustration with his colleagues, with the quality of the students, and with the stalling of his quest for general relativity all had something to do with his exasperation. "As good as I have it here superficially, I cannot shake the feeling of being in a kind of exile," he wrote to Carl Schröter, profes-sor of botany at the ETH, in early February 1912. "One who has spent his formative years in a democratic society cannot properly get used to such a caste system as they have here. One never stops finding the nobil-ity and dignity nonsense comical. Since I've been here, I now properly know how to value the simple and healthy habits and customs of the Swiss. Here there is also none of that lively scientific interest among the students as in Zurich."[23] The capital of Bohemia could never live up to the image of his former city on the lake.

Even before he had begun to entertain the idea of returning to Zurich, Einstein started to position it as the anti-Prague in almost every reference to his environment. Where Zurich was imagined clean, he rendered Prague dirty; where Zurich was imagined modern, he rendered Prague backward. Where Zurich was deemed "German," Prague was identified as Slav. "I like it very much, even if Prague is no Zurich," he wrote within three weeks of arrival. "I have an institute with a rather good library and few official duties. I already work the whole day in the institute, at pres-ent about molecular motion in solid bodies. Also new research on ra-diation theory is planned. Truly otherwise it is different here than in Zu-rich. The air is full of soot, the water is life-threatening, the people are superficial, shallow, and uncouth, if also, as it seems, in general good-hearted."[24] But there was no need to worry, he affirmed two months later; they would get by: "Things go well for us here, even though life here is not so entirely pleasant as in Switzerland, quite besides the fact that we are strangers here. There is no water here that you can drink before boiling it. The population for the most part cannot speak German and acts hostile toward Germans. Even the students are less intelligent and industrious as in Switzerland. But I have a beautiful institute with a rich library."[25]

Einstein's reaction to the Bohemians and to his new circumstances was knee-jerk, prejudiced, and highly stereotypical—down to the complaint about potable water, a staple of entitled travelers' accounts then and now—but it was also completely human. He and his young family were largely alone. They were without friends and social ties in a city where doing one's shopping often posed a linguistic challenge. It was natural to compare Prague to Zurich, the only city in his peripatetic life where he (and his wife, as we shall see) had fully felt at home. Yet Einstein's continuous presentation of Prague as backward significantly distorts certain facts. First, with a population of over half a million, Prague was vastly larger than Zurich. Einstein, who had not lived in such a large city since his teenage years in Munich, was rather taken aback by its size. Prague was also modern in terms of infrastructure. When the Einsteins had moved to Bern, they had had to use oil lamps to light their home; in Zurich, they had used gas; in Prague, they had electricity (and a live-in maid).[26] One might think that the son of a man who had made his living wiring cities for power would have had a greater appreciation for Prague's sophistication in this regard.

Interestingly, this personal contrast of Bohemia with Switzerland had a persistent cultural resonance. Not only the Einsteins, it seems, contemplated Prague and Zurich together. When reading through contemporary sources from the decades before and after Einstein's tenure in Prague, one comes across a host of references contemplating Switzerland as a model for Bohemia. These comparisons all had two main characteristics: they were broadly optimistic, and they were narrowly political.

To compare Bohemia with Switzerland was to juxtapose a subordinate region of a sprawling land empire to a sovereign nation, and that was precisely the reason why Czech nationalist activists did so. By focusing on the obvious similarities between the two regions—landlocked, economically productive, and, most important of all, containing a multilingual populace—these writers used Switzerland as a metaphorical tool after the revolutions of 1848, when it was invoked by Augustin Smetana to render thinkable the notion of an independent Bohemia.[27] (Or perhaps a federated Habsburg domain in which Bohemia was granted significant regional autonomy; the Swiss model was flexible depending on the scale on which it was applied.)[28] Before the emergence of the Helvetic analogy, perhaps the most dominant activist comparison for the Bohemian situation was with Daniel O'Connell's Irish Repeal movement, which stressed the notion of Bohemia as colonized by Austrian Germans; this proved substantially less palatable to the international community, and its use decreased in frequency across the nineteenth century.[29]

The heyday of the Swiss/Bohemian twinning was the Paris Peace Conference at the end of World War I, when Edvard Beneš, the foreign minister of the new state of Czechoslovakia, repeatedly invoked Switzerland as a way of assuring the great powers that his country, broken off from the now-extinct multiethnic and multilingual Habsburg Empire, would be able to protect its own multiethnic and multilingual minority groups, which included a sizable proportion (approaching one-third of the population) of German-identified citizens. Although this argument did not gain much traction with the ambassadors present, Beneš returned to it again and again.[30] The comparison remained a talking point in the 1920s, even as the majority of the Czechoslovak "German" political parties pursued a policy of noncooperation with the state. Prague philosopher Emanuel Rádl, for example, sought to reassure a Germanophone readership that the minorities had nothing to fear, since their situation "is also the case in Switzerland between the German- and the French-speaking population and, as I believe, also in Canada between the French and the English."[31]

All of these developments lay in the future during Albert Einstein's time in Prague, however, when he was seemingly unaware of the occasional musing about the vitality of Switzerland as an exemplar for the more politically attuned and creative of his contemporaries. The political version of the Swiss analogy functioned at the scale of Bohemia as a whole, considering the Czechs and Germans counterparts of the dominant Germans and French in the Alpine confederation. (Other nationalities like Hungarians and Jews were left out of the former framework, much as the Italians were left out of the latter.) Einstein's anti-Prague of Zurich worked on a much smaller scale. His mental Switzerland was not a multilingual country; it was a monolingual, Germanophone city. As in Zurich, his entire social circle in Prague consisted of people who identified themselves as "Germans," with the important exception of his wife. What follows is an exploration of the contours of this space, working inward to his domestic sphere from the public world that an ordinary professor at the German University was expected to inhabit.

* * * * *

The first important social event for anyone in that position was the inaugural lecture. On 23 May 1911, about a month and a half after Einstein's arrival in the city and the beginning of his first semester of teaching, the German-language newspaper *Bohemia* carried an advertisement that the following day, at 7 p.m., those who arrived at the lecture hall of the physics institute of the university (they meant the German University, naturally) would have the chance "to meet our university's new theoretical physicist, Professor Einstein, who will speak about a topic from his research specialization, the *relativity principle*, in which he has achieved outstanding things." In order to induce the cultivated strata of the German elite to attend, the advertisement added an additional blandishment: "In conclusion the Viennese representative of the Zeiss firm of Jena will demonstrate some beautiful experiments with a new filter apparatus for ultraviolet light."[32]

We do not have a transcript of the talk that Einstein gave, which we know covered the special theory of relativity, but by all accounts it was a success. Gerhard Kowalewski, a newly arrived professor of mathematics at the German Polytechnic in Prague and over the course of a year a member of Einstein's relatively small social circle, was blown away. "The impression which his inaugural lecture made on me is unforgettable," he wrote in his memoirs. "The entire Prague intelligentsia had gathered and filled the largest lecture hall which could be found in the natural-scientific institutes. . . . Many listeners were astonished that relativity theory was something so simple. Einstein described his theory in a masterful manner for the broader circles of the educated."[33] Einstein was also impressed with Kowalewski and put him on a list of three nominees when a position opened up for a mathematician at the German University, next to Viennese scholar Hans Hahn, one of the initial drivers behind the philosophical movement that would flourish into the Vienna Circle.[34] In 1912, shortly after Einstein left the university, Kowalewski joined its faculty.

Kowalewski was one of the many acquaintances from Einstein's 16 months in Prague who reached out to him in later years in hopes of rekindling former connections. "Do you still remember me, Professor G. Kowalewski, from your regrettably so brief time in Prague?" he wrote to the physicist in 1922. "I am now professor of pure mathematics at the Dresden Polytechnic"—he had moved in 1920, shortly after Czechoslovak independence. "My wife really wanted to leave Prague, and in Dresden there was such a good and honorable opportunity. . . . Only in the hope that you still have a little interest in me from your time in Prague do I dare to burden your time with such things."[35] Einstein responded to the scientific content of the rest of the letter but did not mention Prague, nor did he refer to it in a further epistolary exchange with Kowalewski in 1926.[36]

Noteworthy here is not that a former colleague wished to approach someone who was now a global celebrity—that is understandable enough—but the frequency of such missives in his later years, which raises a deeper point. Einstein did not simply arrive in Prague, work on general relativity, and then leave. On the contrary, he participated quite actively in the various forms of bourgeois sociability that were typical of that time and place. He was not a professor at the German University in title alone; he performed that post's cultural role just as fully. Prague was not a way station for Einstein, at least upon his arrival there in spring 1911. He set about building a network of friends and was especially keen to find a group of acquaintances with whom he could play violin. His mathematician colleague Georg Pick had a quartet in which Einstein participated (as Kowalewski tells us), but the physicist was always looking for more places where he could socialize musically.[37] It was in this capacity that he first became acquainted with Bertha Fanta and her salon.

Fanta held regular gatherings at her home above her husband Max's successful pharmacy, the Unicorn, directly on the Old Town Square. A circle of Germanophone intellectuals, many Jewish but also some (like Kowalewski) not, convened there to discuss philosophy and literature, gossip, and play music. Kowalewski tells us that "Einstein also sometimes appeared at these evenings."[38] The philosopher, Zionist, and university librarian Hugo Bergmann—who was married to Fanta's daughter Else—attended some of Einstein's courses as an auditor and claimed that he was the one who first brought Einstein over to the Fantas', where Bergmann regarded him as "a frequent visitor."[39] Regardless of how often Einstein came, we know that he appeared enough to make an impression. It was at this house that the possible sole meeting between Franz Kafka and Einstein took place, as the young writer was also an occasional attendee, although he became increasingly reluctant to visit, complaining to his close friend Max Brod in 1914: "Tomorrow I'm not coming to Fanta's; I don't like going there."[40]

Kafka may not have enjoyed the Fanta circle, but many others did. Bertha Fanta was widely regarded as a remarkable woman. Born into the Sohr family in Libochowitz (now Libochovice) in 1866, she moved early to Prague and spent most of her life there. We know little about her childhood, how she met Max, or the early years of their marriage, but she did leave an extended set of autobiographical notes about her journey into the philosophy of Franz Brentano, a rare first-person testimony from a woman of this era.[41] In the early years of the twentieth century, she was the center of a salon that met at the Louvre Café at the edge of the Old Town and was mostly devoted to adoring discussions of Brentano's philosophy. (Max Brod was ostensibly expelled from the circle for a story he had written in which a womanizing civil servant was presented as a Brentano adherent.)[42] Kafka, who had attended the Louvre Café gatherings even before Brod, left with his friend in sympathy—though, to be frank, he had also grown tired of it, much as he would of Fanta's subsequent domestic salon.[43]

As her new circle developed, Fanta, heavily under the influence of Bergmann, shifted her emphasis toward literature and also Zionism. (She was also very fond of the anthroposophy of Rudolf Steiner, who visited her house whenever he was in the city.)[44] Bergmann remained devoted to his mother-in-law, dedicating his 1913 book on the philosophy of math to her and lamenting her 1918 death even 15 years later.[45] Max Brod, too, was deeply moved by Fanta's sudden death, presenting one of the eulogies at her funeral. As he wrote about the loss to Kafka: "Now a heavy blow has struck me: the sudden death of Mrs. Bertha Fanta, that you surely know about from today's newspaper. I really liked this woman. She was a completely pure person and she conducted a passionate war against her small defects. She always held so firmly to life, feared death, philosophized about immortality."[46] He added that toward the end of her life she had been learning Hebrew and intended to emigrate to Palestine with her daughter and Bergmann.

The discussions at Bertha Fanta's provided Einstein's introduction to Zionism. Her salon may also have been where he was first exposed to the then quite popular ideas of Sigmund Freud, the radical psychiatrist from the Habsburg metropole.[47] If so, the vector would have been Ludwig Hopf, Einstein's assistant who had accompanied him from Zurich to Prague and was quite interested in psychoanalysis. In fact, Else Bergmann believed it was Hopf who first brought Einstein around.[48] Several participants remembered collective discussions of Immanuel Kant's *Critique of Pure Reason* in which Einstein came across as a strict Kantian, a recollection that would likely have amused him. Yet another candidate for the person who first introduced Einstein to the Fanta circle was his colleague, the philosopher Christian von Ehrenfels, a lapsed Brentanist who at that very moment was enjoying quite a *succès de scandale* for his eugenic theories of sexual polygyny.[49]

Einstein relished a good debate about Kant as much as the next educated Germanophone intellectual of his generation, but he kept coming back for the music, and it was the music that he himself would recall once the rest of the Fanta circle's doings receded from his memory. On one occasion his playing was accompanied by Brod, but the most important musical connection he made through the circle was surely Ottilie Nagel. A gifted pianist, Nagel was the sister-in-law of Moriz Winternitz, a professor of Sanskrit at the German University, with whom Einstein developed a close friendship. But while the chance to play his violin is what Einstein remembered from his time at the Fantas', the reminiscences of the other attendees barely commented on it. Instead, they recalled a talk Einstein delivered to the assembled intellectuals on special relativity. He did make a presentation of the sort, but our best source for the event, Brod's diary, recalls the date as 24 May, which happens to have been the day of Einstein's inaugural lecture—a reasonable conflation, but not one that helps us learn what he said. Brod also notes that his friends Robert Weltsch and Kafka were in attendance, as was Hopf, with whom they fell into a conversation about Freud and Jung. The discussion continued the following day, but without Kafka, who declined to return.[50] Whatever Einstein said seems to have slipped away without comment. Also absent from these recollections of Fanta's circle is any mention of the physicist's wife. She had a very different experience of Prague, and it is time that we turned to her.

* * * * *

It will be best if I call her Einsteinová. In Czech, women's last names are adjectival, formed by adding an -ová suffix onto the husband's or father's last name: Kundera to Kunderová, Havel to Havlová, Navrátil to Navrátilová. True, most of the Germanophone milieu she would have interacted with in Prague would have referred to her as "Frau Einstein," but the live-in maid, the merchants at the shops, and the porters would have called her Einsteinová. I adopt this name because all the alternatives usually deployed in histories about the Einstein family seem to me unsatisfactory. One cannot just write "Einstein," because there would be confusion with her husband. Simply calling her by her first name, Mileva, is diminishing: everyone else in the story gets a last name. Likewise, calling her Marić (or, in its Hungarian version, Marity), her maiden name—which she assumed again, sometimes hyphenated as

"Einstein-Marić," after her divorce—negates the central fact of her life in Prague: she was not simply in Bohemia as "Marić"; she was there as Einstein's *wife*. That is precisely the literal meaning of the Czech naming convention, so despite the artificiality it is what I will use.

Einsteinová had not wanted to move to Prague. She had written her friends about the relocation with apprehension, seeking information about this foreign, exotic metropolis. In some ways, it was not that foreign to her. She had been born into a Serbian family on the Hungarian side of the Austro-Hungarian Empire, and so she had been raised in a Habsburg milieu (her father had been a minor civil servant). Habsburg territory meant Habsburg rules, including the prohibition on girls' attending *gymnasia*, so her father had sent her to Serbia, where she could get an education. Raised speaking German—the language of social mobility—at home, she had gone off to the ETH in Zurich to pursue higher education in physics, a choice again in response to the ban on female entry into the universities for credit in her own country. (The ETH began allowing women to study there in 1876, and the first graduation of a woman from the physics section was in 1894.) There she had met and begun a relationship with fellow student Albert Einstein, and they had married in 1903. (A child born the year previously had been given up for adoption in Novi Sad, Serbia.) She had followed Einstein to Bern and then back to Zurich, the city where she had felt most comfortable.[51]

Prague was decidedly uncomfortable for Einsteinová. Some of this was for the same reasons cited by Einstein: bedbugs and poor water. Some was the consequence of the utterly conventional gendered family dynamics of the couple. She had to take care of the household, manage the servants and the shopping, ensure there was enough boiled water to drink, and take care of the children. The extent of the domestic obligations also grew with the move to Prague, since the apartment was bigger and her husband's elevated status demanded a more elaborate lifestyle.[52] The family was also larger. Hans Albert was already in school, but the infant Eduard (called Tete or Teddy at home) was a difficult child, often sickly and upset. Einsteinová was also *always* in Prague. Einstein traveled a good deal: to Germany for talks, to Zurich to negotiate a return to the ETH, to Brussels for the Solvay Conference. Einsteinová, though she was entitled to a Swiss passport, never got one and always traveled with Einstein under his.[53] When he was gone, she was confined to the city.

Confined seems to be the right word. In the few letters written by Einsteinová from Prague that have survived, she eagerly looked forward to some diversion from the daily domestic duties. In early October 1911, she leaped at the idea of going to Switzerland to see her husband, who was on one of his many travels. "It is now already an eternity since we have seen each other; will you still recognize me?" she wrote. "Should I really come to Zurich? The weather is wonderful here. Glorious autumn weather, wonderful, do we want to do something?"[54] Einstein was increasingly absent even when he was in the city, beyond attending to his courses or working in his office on general relativity. He spent a good deal of time, sometimes entire Sundays, with the Winternitz family. One of that couple's several children—Einstein was good with children—who recorded his memories of this period recalled that Einsteinová and the physicist's children never came, not once.[55] She had to fend for herself (and them) on those leisurely weekends. Hans Albert, eight years old by the time he left Prague, remembered the city vividly, attributing his lifelong career in hydraulics to his childhood fascination with the flowing Moldau. His wife's biography of him points to darker memories as well: "Hans Albert's recollection [was] that when he was eight years old, he began to sense tension and discord between his father and mother."[56]

There were several reasons for the marital friction, none of which would have been intelligible to young Hans Albert. The most significant was connected to Einstein's travels. On a trip to Berlin to talk to physicists in April 1912—his first to that city, as it happens—he connected with a cousin of his (on both his mother's and father's sides) named Elsa Löwenthal, née Einstein and recently divorced from Max Löwenthal, whose last name she had assumed. An affair ensued almost immediately.[57] Einstein took pains to conceal it from Einsteinová. In his first love letter to Löwenthal, he begged her to write to him at his office address: "Sci. Institute Weinberggasse. Prague. If you have the chance, write me again, but only if it makes you happy. I will always destroy the letters, as you wished. The first I have already destroyed."[58] Later, after the Einsteins returned to Zurich, the distance between the pair was farther but the correspondence continued, Einstein taking care to find another address away from Einsteinová's eyes.[59] Even from Zurich, traces of Prague tinged the affair. In April 1913, at Löwenthal's request, he sent her a photograph of himself taken right off Wenceslas Square.[60]

An incident that scandalized the physics community just after the Solvay Conference ended in early November 1911 gives us a window into Einstein's casual views about marital fidelity. It concerned Marie Curie, whom Einstein had encountered at the conference; herself a Slav (Polish-born), she had been widowed for five years since her husband Pierre's early death. Beginning in the summer of 1911, rumors had begun to swirl about a possible affair with 38-year-old physicist, and her married colleague, Paul Langevin, who was also present in Brussels. While the meeting was underway, the Parisian press published intimate letters about their relationship. Both Curie's and Langevin's names were dragged through the mud, and Einstein was horrified. He wrote to his friend Zanger in Zurich:

> The tabloid horror story in the newspapers is nonsense. That Langevin wants to get divorced has long been known. If he loves Mrs. Curie & she him—then they do not need to run off together because they both have enough opportunities in Paris to see each other. I did not at all get the impression that there was anything special between them, but rather that all three met in harmless pleasantness. I also do not believe that Mrs. Curie is hungry for power or for anything else. She is an unpretentious, honest person whose duties and burdens almost overwhelm her. She has a sparkling intelligence, but despite her passionate nature is not attractive enough to become a danger to anyone.[61]

He continued to offer her moral support from Prague: "I feel compelled to tell you how I have learned to marvel at your spirit, your energy, and your honesty, and that I consider myself fortunate to have made your personal acquaintance in Brussels. Whoever is not among the reptiles will as always be glad that we have such individuals as you and also Langevin among us, real people one feels happy to be in contact with."[62] At this point, Einstein himself was not yet engaged in his own extramarital affair; this was purely a matter of decency. It is also noticeable that he did not seem concerned about the impact of these events on Madame Langevin.

Other reasons for the growing chasm between Einstein and Einsteinová were more nebulous, perhaps not even clearly grasped by either party. There were many things that proved asymmetric in the Einsteins' unpleasant reactions to Prague, but one important one was that it clearly mattered that Einsteinová was not just a woman, but a *Slavic* woman in a city where the relationship between Germans and Slavs—and specifically German men and Slavic women—was especially fraught.[63] Philipp Frank, who as Einstein's successor at the German University was likely privy to gossip that circulated about the Einsteins after their departure, included a passage in his canonical biography of Einsteinová's husband stating that "she was as a Slav not very inclined to fit herself into the circle of professors' wives. Because in their conversation the inferiority of the Slavic peoples and the greatness of the Germans played a central role. Even if it was not entirely sincerely meant, it still resulted in not exactly a pleasant atmosphere for a Slav."[64] There were probably several reasons why the Einsteins did not socialize with the other professors as much as was the norm—Einstein's own dislike of such occasions and absorption in his work high among them—but the drawing room snubs must have played a role. Many later biographers make a point of Slavicizing Einsteinová, exoticizing this highly Germanized citizen of Switzerland in stereotypical ways.[65] Even Hans Albert did so, calling his mother "a typical Slav, capable of strong negative feelings, and once hurt she could not forgive."[66]

Even more than being a Slav, it mattered what kind of Slav she was *not*. Einsteinová was no Czech, and thus she was blocked from access to alternative social circles that would have been widely available in Prague. Einstein was entirely oblivious to this distinction; in fact, just before his arrival in Bohemia, he invited a colleague from Croatian Zagreb (also in the Austro-Hungarian Empire) to visit him in Prague, juxtaposing the invitation with a comment about the nearness of Zagreb to Serbia, as if all Slavic regions were neighbors, or at least mentally associated with each other.[67] Although Serbian and Czech are not mutually intelligible, someone with fluency in reading the former could make her way through written Czech, and thus it is entirely possible that Einsteinová was also aware of the anti-German hostility to a much greater degree than Einstein, particularly as Czech merchants and city residents would understandably have related to her in no other guise than as a German professor's *German* wife. Despite the very noticeable presence of pan-Slavic and neo-Slavic political and cultural movements—such as the Sokol Slet celebrated in late spring 1912 and the 1908 Slavic congress in Prague on the sixtieth anniversary of the first such gathering in the same city—the implied hierarchy that ranked the more Westernized, "civilized" Slavs (Poles and Czechs) above the politically powerful Slavs (Russians) and then everyone else was pronounced.[68]

As to the kind of Slav Einsteinová was, that had additional dimensions. Bergmann mentioned once meeting Einstein's "first (Yugoslav) wife."[69] Formally independent since 1878, Serbia's earlier conflicts had been with the Ottoman Empire it bordered to the south. Political focus soon shifted to the Habsburg Empire to the north, where Serbians resented Austro-Hungarian interference in the Balkans and complained of the treatment they received from other southern Slavs, especially Bosnians. One of Einsteinová's biographers considered Vienna's hostility to Serbians a major source of discrimination toward her in Einstein's hiring process, as well as a factor in how the other professors (and professors' wives) of the German University treated her.[70] The deteriorating situation between Austria-Hungary and Serbia—which would in 1914 provide the context for the assassination of Archduke Franz Ferdinand that launched the Great War—was a topic of much discussion in Prague, which had become "a real headquarters for Southern Slavs" at the end of the first decade of the twentieth century.[71] These Slavic gatherings provided no haven for Einsteinová, who did not identify herself with them and thus was frozen out on all sides. Einsteinová's status, in some people's eyes, as primarily a "Serbian" does provide the context, however, for one of Einstein's only political statements before his pacifist awakening in World War I: "If only the Austrians can keep calm; a conflict with Austria would be bad for the Serbs, even in the case of victory. I believe, however, that the saber-rattling means little," he wrote to Einsteinová's friend Helene Savić in Novi Sad.[72]

Aside from the geopolitics inflecting Einsteinová's position, the very local, Prague-specific cultural resonances of the Einsteins' marriage—"German" man, "Slavic" woman—influenced how others related to them, even if the couple themselves did not entirely comprehend all or even most of these tensions. Here we can highlight one short literary work, *A Czech Serving Girl* (*Ein tschechisches Dienstmädchen*), which its author, the same Max Brod who graced the Fanta circle that Einsteinová apparently never attended, published in 1909, two years before the Einsteins' arrival in Prague. The first edition of 124 pages highlighted the city's local color: the cover featured an illustration by Lucian Bernhard of the Malá Strana and the Hradschin castle, seen through the tower of the Charles Bridge. It sold out, and a second edition was issued before the end of the year.[73] It is by no means a great piece of literature, but the work and its reception provide important insights into the specifically gendered aspects of the German man–Slavic woman dyad that was echoed in the Einsteins' marriage.

The story is told in the first person by William Schurhaft, an "indifferent" young Viennese similar to the protagonists of Brod's early fiction, in which he was heavily under the influence of Arthur Schopenhauer's philosophy, emphasizing apathy and detachment from the everyday world. At least, that is Schurhaft when the novel opens. He has been sent by his businessman father to Prague in a paternal attempt to knock him out of his torpor and engage him in life: the father imagines that the Baroque scenery, the history of the city, and the omnipresent tensions of the Czech–German conflict will rouse the lad. The extracts of his father's letters sound uncannily like the correspondence Einstein would send back to his friends in Zurich a few years later. At first, Schurhaft's boredom continues: "But I do not like Prague at all. I get no impression from its curiosities, about which people have told me so much."[74] He does eventually come to life, though not in the way his father had hoped. A young woman—specifically highlighted in the title as "Czech" rather than the more neutral-sounding "Bohemian" (*böhmisch*)—named Pepí Vlková fills in for another servant in his apartment for one day, and Schurhaft becomes obsessed with her. (Pepí is also the name of a serving girl in Franz Kafka's much later novel *The Castle*; many of Kafka's female characters have Czech names.) He scours the city for her, learning some of the language and interacting with Czech locals and city landmarks (like the astronomical clock in the Old Town Square) as he woos and finally beds her. Once again she vanishes, and Schurhaft, self-obsessed as always, is distraught. He finally finds out that she has committed suicide, throwing herself into the river after a dispute with her abusive husband—about whom Schurhaft, who had spent so much energy exoticizing Pepí as an avatar of her culture rather than considering her as a person, knew nothing. Her corpse washes up at the base of the Palacký Bridge, where a few years after the novel's publication there would be a monument by which a German University professor of physics would walk to work every day.

The book was hotly criticized when it appeared, much to Brod's astonishment. Here he had thought he was *praising* Czech culture by extolling its beauty and admiring the mellifluous sound of the language. Indeed, throughout his long career in the city Brod often functioned as an interpreter of Czech culture (for example, the works of the writer Jaroslav Hašek and the composer Leoš Janáček) to Germanophone audiences.[75] He spoke the language well and perceived himself to be a Bohemian, committed to Prague in both its German and its Czech guises. But that is not how Czech critics understood him. Even when they thought the book relatively harmless, they saw it as indicative of how Germans perceived Czechs:

> Max Brod's just-published small novel, *Ein tschechisches Dienstmädchen*, is a characteristic endeavor to endow artistic elements to the trivial. The author doesn't always succeed, there are many forced combinations here, but some pages nonetheless yield beauty. The book's plot is on the whole unimportant, the mood is mostly confined to how the author looks at Prague. The author's impartiality is surprising—there are no attacks on Czechs, no insinuations à la [Karl Hans] Strobl, however here and there a rather ironic "*strammdeutsch*" [i.e., uptight German] viewpoint.[76]

The harshest review came from the writer Božena Benešová, a noted feminist in a society that was increasingly mobilized around women's rights, who saw the book as treating Czech women as sex objects and the Czech people as the passive recipients of German condescension, however admiring a tone Brod might take: "There remains to him only the capability for analysis and contemplation, and the capability to transform Pepička into a representative of an entire people."[77] From the point of view of Czech intellectuals, the pairing of a German man and a Slavic woman in Prague was always problematic, symbolic of the subjugation of Prague by Vienna. Einsteinová's isolation partook of this valence too.

Understandably, she was eager to get out. On 13 August 1912, an acquaintance named Lisbeth Hurwitz noted in her diary: "Today the Einstein family was over for dinner. He has been called to the ETH, and we got the impression that they would gladly leave Prague."[78] Once Einsteinová was back in her adoptive homeland of Zurich, you can almost hear the sigh of relief in her letters. "We are well and are all, big and small, very happy to have turned our backs on Prague. To me, in particular, because of the children, our stay there was so unpleasant," she wrote to her close friend Helene Savić. "Hygienic conditions there are such that sometimes it was really hard for me with my boy. You know already that there is no drinkable water there; milk is of just as doubtful quality; the air is always full of soot, there are no gardens, and there is very little free room, so that the children really had to live inside."[79] Zurich was certainly more pleasant for her, but the family situation did not suddenly turn into a paradise. Teddy's health did not improve, for one. There was also the matter of her husband's affair with Elsa Löwenthal, although it is not clear whether Einsteinová yet knew of it. Once Einstein got the call to Berlin in 1914, however, there was no escaping the truth.

At first the plan was for the whole family to move to the Prussian capital, but Einstein made it intolerable for Einsteinová to join him, a sordid story that is recounted in all his biographies. Already in December 1914 he was writing about her with disgust: "For me however it was a question of life or death; my nerves could not much longer withstand the years of heavy pressure this barbarian nature exerted on me."[80] "Barbarian" was strong language, and not especially frequent in Einstein's lexicon; the last time he seems to have used it was in his description of the Slavic inhabitants of Prague.

Consequently, while Einstein assumed his post in Berlin, Einsteinová stayed in Zurich with the children, and he came frequently to visit. Starting in 1916, under Löwenthal's increasing pressure, he began divorce proceedings. The debates between Albert and Mileva were protracted, acrimonious, and centered on two issues: access to the two boys and questions of money. Famously, a clause in the divorce agreement allocated to her the money from a Nobel Prize, should Einstein eventually win one. (He did, in 1921.) Perhaps appropriately for a marriage that had begun to disintegrate completely while in Prague, a persistent financial negotiation harkened back to the Bohemian capital. Again and again in the epistolary back-and-forth about finances, Einstein raised the issue of the "Prague money" (*Prager Geld*). This first cropped up in a letter from Albert in April 1914 as "an old tax bill [that] came from Prague for about 80 Kr. which I must pay."[81] This was linked to a bank account they had set up while he was at the German University, and he wanted the money in order to save it for the children: "I must thus insist that the money lying in Prague be transferred to my name. I will have it credited to the children and am convinced that the sum can *in this way* most securely benefit the children."[82] Two years on, the only invocations of Prague in the couple's letters concern this financial transaction, with Einstein proposing "to deposit my Prague money as well as 6000 Marks in the savings account created here for the benefit of our children in a place approved by both of us."[83] The deposit ended up coming to 7,583 crowns, and Einstein calmed down. The dissolution of the marriage was finalized on 12 June 1918.[84]

Einstein married Elsa Löwenthal, returning her last name to her maiden designation of "Einstein," on 2 June 1919, a few days after the eclipse expedition that would in time transform Einstein into a global celebrity. For her part, Einsteinová, now going by "Einstein-Marić," raised the two children in Zurich and battled with severe depression. Pleasant reminders of her former life would occasionally crop up. During the First World War, which began as a conflict between Serbia and Austria-Hungary, Marie Curie arrived in Switzerland from Paris on a vacation with her daughters. She and Einstein-Marić went hiking in the Alps together, two Slavic women raising families without their husbands. They got along very well.[85]

* * * * *

It would be a mistake to come away with the impression that Prague was all gloom, misery, and heartbreak for the Einstein family. The physicist was happy with his work on general relativity, even though he eventually abandoned the static theory. He enjoyed walks in the surrounding countryside. He came to appreciate the status of an ordinary professor, and he made some good friends. Fittingly, the most lasting of these friendships, the one that meant the most to him, was with someone who simply passed through the city. Paul Ehrenfest, the Viennese-born physicist who had married a Russian physicist and was now hoping to settle in Central Europe from St. Petersburg, visited the region seeking support during Einstein's time at the German University. Einstein invited him to Prague, and this proved the beginning of a vibrant and intellectually rich relationship.[86]

At first, making the arrangements for the visit was difficult, as in late January Einstein's guest room was occupied—the editors of his collected papers speculate that Einstein's mother, Pauline, or one of his friends from college was in residence—but by mid-February the long-anticipated meeting could take place.[87] "Tell me the date and hour of your arrival and which train station you will come to, and you will stay with me so that we can use our time well," Einstein wrote.[88] Ehrenfest arrived on 23 February 1912 from Brünn (today Brno), where he had been visiting his old university friend Heinrich Tietze. When Einstein had an appointment to play violin, he dropped his guest off with Anton Lampa. (Neither left a record of how that went.) Ehrenfest got along particularly well with Hans Albert.[89] The two physicists maintained a highly collaborative exchange, largely through correspondence, until Ehrenfest's death in 1933 at age 53.

Einstein must have been in a good mood during the entire visit. He wrote to his friends the Sterns on 2 February that "two days ago I was (hallelujah!) offered a position at the Polytechnic in Zurich and have already communicated my Imperial and Royal resignation here."[90] By the twelfth of the month the agreement with the ETH had been inked.[91] Einsteinová was overjoyed. Ludwig Hopf, Einstein's former assistant and fellow participant in the Fanta circle, expected it had been an easy decision: "Leaving Prague will surely not hit you too hard. Or have you in the meantime come to like it?"[92]

Not really, although now that he was going he did not want to leave unpleasant impressions behind him. When rumors surfaced attributing Einstein's 25 July 1912 departure to poor conditions at the German University and especially blaming the Czechs for making the physicist uncomfortable in Bohemia, he felt it important to "emphasize that I had in Prague no grounds for dissatisfaction." He wrote a lengthy letter extolling the city and its people that was published on the front page of both the *Prager Tagblatt* and the Viennese *Neue Freie Presse* on 5 August. In it he wrote:

> As I recall the Ministry was accommodating to me in the most extensive manner, and also during my activities in Prague I have had no difficulties with the education department; on the contrary, various small matters, including also those of a financial nature, were always resolved according to my wishes. My institute in Prague was completely sufficient for my purposes and in every connection satisfactorily equipped. Besides it is incorrect that here in Zurich an enormous and richly financed institute will be set up just for me. . . . My decision to leave Prague is simply to be attributed to the fact that I already before leaving Zurich had promised that under acceptable conditions I would happily return.

He concluded, somewhat at variance with what he had reported in his letters to Zurich from Prague during his first months there, that the reason he was leaving was "the more favorable living conditions that Zurich has over Prague. I do not allude to Prague's national relations—that never bothered or disturbed me—but rather I mean only the favorable location of the city on the lake and amid the mountains, that naturally greatly tempts a family man. These are the true reasons for my departure from

Prague. Of any confessional prejudice that one might suspect I have felt and noticed nothing."[93]

The ministry in Vienna and the German-identified locals in Prague were delighted with his valediction. In the final lines of the newspaper article, Einstein praised his successor, the Viennese Philipp Frank, as an excellent theoretical physicist. When a new rector of the German University assumed his post on 28 October 1912, he asked the prorector, Heinrich Rauchberg, to give a report on the preceding year. Rauchberg mentioned, matter-of-factly and in passing, that Einstein had left for the ETH and that Philipp Frank had arrived.[94] It was a very smooth transition. Einstein's legacy in Germanophone Prague—and, as it happened, in the broader intellectual community in both Europe and North America—would be closely linked to Frank. Einstein's Prague story now became his.

CHAPTER 4

Einstein Positive and
Einstein Negative

Überhaupt I must say that the hospitality of Prague towards wandering philosophers much surpasses that of Berlin and Leipzig. In great capitals it is more difficult to give one's time to strangers . . .

—*William James*[1]

On 6 January 1921, Philipp Frank stood waiting on a platform at the main train station in Prague, a few steps from the top of Wenceslaus Square, the heart of the New Town. In the past eight and a half years, Frank had watched his world become engulfed in the brutal destruction of the Great War and the Habsburg Empire, in whose capital he had been born and studied, disintegrate into a handful of separate states. He had moved countries and changed citizenships while staying put. In the fall of 1912, he had assumed the position of professor of theoretical physics at the German University in Prague, a large but somewhat provincial city in the northwest region of the Austrian half of Austria-Hungary. It had been quite an honor for the young physicist, who had been elevated as Albert Einstein's successor at the eminent physicist's own recommendation. Now, in 1921, he had the same job with the same title, but he lived as a citizen of the capital city of the new land of Czechoslovakia. There was an apartment crunch in the booming postwar metropole, so he and his newlywed wife, Hania Gerson—they had married on 16 November 1920—were bivouacking in Frank's office at the Physical Institute on what was now known as Viničná ulica, the same room Einstein had occupied and that had seen the early drafts of the static theory of general relativity. (The Franks would obtain their first apartment in the coming fall.) Here he was, standing by the tracks of what had once been Franz Josef Station but was now known as Wilson Station, after the American president who had supported the new state's independence.

The man he was searching for descended from the train "and still looked like an itinerant violin virtuoso," he recalled.[2] Einstein had come back to Prague. In order to avoid the paparazzi who now followed his predecessor—the eclipse expedition that had launched Einstein into celebrity was barely a year and a half in the past, and there was no sign of the enthusiasm abating—Frank brought him to the office where he had worked less than a decade earlier. Einstein was delighted. "After a pleasant trip I arrived happy and healthy in Prague, where I was expected by Frank and his young wife, Pick, and certain friends," he wrote the next day to his own newlywed, Elsa Einstein, back in Berlin. "Yesterday evening I was invited with local friends to Winternitz's, where it was very festive. It pleased me beyond description. I live with Frank at the Institute [and] his young wife, a former Russian student (very nice) in private digs—there's a shortage of housing in Prague. People live here somewhat better than in Berlin. It is very comfortable."[3]

Einstein spent only two nights in Prague before heading off to Vienna. It was a visit with a purpose: Frank and others had invited their hero to return to Bohemia and participate in an event at the Prague Urania, the local planetarium and astronomy lecture hall. After an evening out with old friends, he was prepared for his public talk and an open debate on the stage. This was big news even in the rapidly transforming capital city, and the *Prager Tagblatt* enthused about the visit. The lecture hall filled up well in advance of the event, and when it began, Einstein was formally greeted by the rector of the German University, professor of geology Franz Wähner. Next came the venerable city counselor Josef Eckstein, who offered effusions linking Einstein to German nationalist pride:

Relativity theory—as it has convinced the Czechs of the fruitful functioning of the German University in terms of its outstanding cultural and economic significance—attempts, in the space of the city of Prague and in the difficult time in which we live, to squeeze these down to a *quantité négligeable* whereby the law of inertia is never taken into consideration. Alone this proves the greatness of Einstein and the grandeur of German science, which commands the attention of even a national opponent, that I must not fear that even the mayor, if he should learn of this greeting, would disavow me.[4]

This was not entirely unanticipated: German-identified residents of the city had shifted from being a tiny local minority within a polity that they dominated to being a tiny minority in a country run by the newly minted "Czechoslovaks." Their expression of Teutonic pride was predictable, and Einstein made no comment about it. Instead, his letters to Elsa and friends focused on the debate he had with Oskar Kraus, a professor of philosophy at the German University. In recent years, Kraus had emerged as the most vocal Bohemian representative of a broader Germanophone attack on Einstein's relativity theory. Now Kraus had an opportunity to confront Einstein in person.

"My lectures here, and also the discussion evening, are already done," the physicist wrote to Elsa soon after the conclusion of the debate. "The latter was very amusing (also with Kraus)."[5] He told his good friend Paul Ehrenfest almost two weeks later that the Kraus affair had been "a generally droll circus performance; he was however serious about it."[6] Kraus had followed Einstein's lecture, which had focused on special relativity and the speed of light, with a witty and cutting broadside against the fundamental postulates of relative motion. Einstein had barely given it a response. Instead, claiming he did not want to exhaust the audience, he had picked up his violin and played a Mozart sonata. The *Prager Tagblatt* reporter noted that he hadn't been half bad. Einstein had been a hit with the audience. He now beamed upon Prague. "The present stay was a great joy for me," he told Elsa. "I have a strong desire to shorten Vienna and in the interim remain here."[7] But it was not to be. He went to the train station with the Winternitz family, the Franks, and the mathematician Georg Pick. He was off on his world travels, leaving Prague behind forever.

Prague did not let go of him so easily. The disciples who had been grouped together in the philosophy faculty of the German University in

Einstein's day had been divided a few months before his brief visit into a philosophy faculty, including the humanities and especially its titular discipline, and a natural sciences faculty.[8] At the time of the performance at the Urania, Frank and Kraus, until recently close colleagues, were separated. For the next two decades, they would be sharp (though personally cordial) rivals in the pitched debates over relativity theory from within the walls of the German University. This chapter focuses on the two of them and their quest to define the universal philosophical significance of Einstein's relativity within the local confines of Germanophone Prague. Their struggle was over nothing less than the status of Einstein's theory for history.

Given the narrow nature of his social circle, it was common among Einstein's former associates in Prague to think of Philipp Frank and Oskar Kraus together, though their personal affections for the former and slight distaste for the latter colored their comments. The mathematician Gerhard Kowaleski, for example, recalled the pair in his memoirs in starkly opposing terms. "Philipp Frank exercised a strong effect in Prague. His lectures stood out for calm and clarity. At the beginning of each hour he gave a brief overview of what was last presented. Frank was also active in the area of philosophy of science," he wrote. "Besides that he was a linguistic genius. He knew fluent Arabic and Syriac. Also Frau Frank, née Gerson, was of superior cleverness. Both also appeared often at the lecture evenings of Frau Fanta"—since Bertha Fanta died in 1918, such joint visits would have taken place before their wedding. "I still remember an impressive lecture about relative motion that Philipp Frank gave there."[9] Indeed, Frank was already before the war making an international name for himself as an explainer of relativity theory and its potentially radical philosophical implications, a reputation that would grow in ensuing decades.

Kowalewski held Kraus in less esteem. "In Prague Einstein found a fierce opponent in the philosopher Oskar Kraus. . . . Kraus remembered only sparse remnants of natural scientific and mathematical knowledge from his school days. Thus armed he charged into battle with relativity theory, which he only knew from Einstein's short popular book," he wrote. "As a jurist he clung to words and looked to dissect contradictions between a certain sentence on page x and one on page y. In Prague he had several public disputations with Einstein"—history registers but one of these—"from which naturally nothing emerged." Kowaleski was pleased about the Urania clash solely because of the violin concert at the end. "That was the only pleasure that this evening brought the public," he declared.[10] In his canonical biography of Einstein, Frank echoed Kowaleski's assessment of Kraus's scientific acumen. The latter was "an acute thinker in the philosophy of law, whose conception of scientific discussions, however, was more like that of a counsel at a trial," he concluded. "He made no attempt to explore the truth, but instead wanted only to refute his opponent by finding passages that were contradictory in the writings of Einstein's supporters. In this he was successful."[11] He was also, although Frank did not mention it, a significant figure in the Habsburg and post-Habsburg intellectual tradition, cultivating the legacy (reputational and archival) of his mentor Franz Brentano into a dynamic and powerful philosophical school.

Kraus, no less than Frank, was a representative in Prague of a vibrant area of intellectual ferment: the multiple braided Austrian traditions in the philosophy of science. To a remarkable degree the major trends of this important domain of philosophy were crafted by people either originating from or working in the Habsburg Empire and its successor states. A partial recounting of some leading figures should suffice as an illustration: Bernard Bolzano, Ernst Mach, Franz Brentano, Carl Menger, Ludwig Boltzmann, Alois Höfler, Edmund Husserl, Hans Hahn, Otto Neurath, Rudolf Carnap, Moritz Schlick, Karl Popper, Ludwik Fleck, Friedrich von Hayek, Michael Polanyi, Imre Lakatos, and Paul Feyerabend. Almost every major philosopher working in this space contributed notably to the philosophy of science, something that cannot be said of their counterparts in Germany.[12] To the extent that this fact has grabbed the attention of scholars, the emphasis has been on the undoubted significance of Vienna as a concentrating force. That was, to be sure, the main locus of the field's energies, but it was not the only one.

Philipp Frank and (more rarely) Oskar Kraus are sometimes mentioned in passing as contributors to this dialogue, but their association with Prague has had the unfortunate effect of relegating them to the background.[13] Even Frank, in an oral history interview from 1962 with the later famous (and non-Habsburg) philosopher and historian of science Thomas S. Kuhn, talked about the city that way at times. "The German University of Prague was really no different. It was just a branch of Vienna. It was different by the fact that Einstein was there, but that was casual circumstance," he told Kuhn.[14] But Prague was far from just that. Through the complex interaction of personalities, political transformations, and contingencies, Frank and Kraus both separately and in conflict with each other transformed Prague into one of the key nodes of European philosophy of science. The vehicle that enabled them to do this was the spirited debate over the epistemological significance, or lack thereof, of Albert Einstein's theory of relativity that occurred following the First World War. By tracking these two figures and their engagement with relativity, we observe the centrality of Prague in shaping both the apotheosis and the demonization of Einstein in the interwar period.

* * * * *

Einstein chaired the committee that sought his successor, and he placed Philipp Frank in the first position, even over his close friend Paul Ehrenfest, who was desperately searching for a job at that moment; they were trailed by Emil Kohl, the Vienna *Privatdozent* who had also been in third position in the search that had led to Einstein's own call to the German University. Anton Lampa, Pick, and Einstein had clearly learned from the frustrations of that previous round, when the Education Ministry had flipped the order of the first and second positions, and so they "came to the view that among the theoretical physicists within the country some are found of such ability that it is natural to propose only Austrians." This conclusion implicitly recognizes that the difficulty with Einstein's own candidacy had been his citizenship status, not his religious affiliation: all three candidates for his successor were Jews by confession, but they were also—as the report explicitly noted—"of German nationality," a nod to the increasingly tense relations that were developing between the Czech and German institutions.[15] Frank came out on top in no small part because he had already begun to publish important articles on relativity.[16]

The committee also thought it significant to note "that Frank has written several original essays with epistemological content (The Law of Causality and Experience. Mechanism or Vitalism?), that give evidence of the versatility of the author as well as of his ambition to engage with the general problems of knowledge."[17] The causality article, published in 1907, had been a landmark for Frank, marking the beginning of a career interrogating the implications of causality long before the paradoxes of quantum mechanics had raised this issue to primacy. Einstein's first contact with Frank, in fact, had occurred when the then Zurich-based physicist had written the other man to praise him for his analysis.[18] In his 1932 book *The Law of Causality and Its Limits*, published in Prague, Frank would thank Einstein for many years of exchanges on this question, as well as cite local friends like Hugo Bergmann in ripostes to Einstein nemeses Hugo Dingler and (from a very different point of view) Henri Bergson.[19] And so, in 1912, Frank arrived at the German University, sporting a full beard and, as one of his former students recalled, "by far the youngest professor at the University."[20]

Frank was widely praised for his astute lectures in theoretical physics, and he performed his pedagogical duties in the field with aplomb, but it soon became clear that his interest in the philosophy of science was no minor matter, but rather a core aspect of his research. Especially at this moment, in the early days of relativity and quantum theory, the boundary between physics and the philosophy of physics proved particularly blurry. While a student in Vienna, Frank had studied with Ludwig Boltzmann (until his suicide in 1906) and other philosophically inclined professors, but found his main interlocutors, the mathematician Hans Hahn and the philosopher Otto Neurath, in a tight discussion circle that in later years came to be called "the first Vienna Circle," a precursor to the famous movement in logical empiricism. The close-knit group met regularly from 1907 until 1912, when Frank departed for Prague. Together, they read cutting-edge works by philosophical scientists like Pierre Duhem, Henri Poincaré, and Abel Rey, but their main fascination was with the retired Viennese professor Ernst Mach.[21]

We have already encountered Mach as the rector of what was then the Charles-Ferdinand University, reluctantly negotiating the settlement that would lead to the split of that institution into the Czech and the German Universities in 1882. It was Frank's devotion to Mach's ideas and legacy that rendered his epistemological emphasis not a defect in the eyes of his colleagues but a great asset. For them Frank's investigations represented—much as Einstein's might have done had he stayed at the German University—a line of continuity that reached back to the great experimentalist and philosopher of science. It is impossible to understand Frank's impact on the philosophy of relativity theory, or the intellectual culture surrounding epistemology in Germanophone Prague, without placing Mach at their foundation.

Ernst Mach was born on 18 February 1838 in the town of Chirlitz (today Chrlice), just outside of the city of Brno in Moravia. His family stemmed from Bohemia and Moravia, which meant he had ancestors on either side of the German–Czech language divide and grew up speaking both languages at home and during his education at the *gymnasium* in Kremsier (Kroměříž).[22] After obtaining his doctorate in physics in 1860 from the University of Vienna, Mach took a position at the University of Graz before moving to Prague, where he would spend the longest stretch of his career. Given that Mach eventually left Prague in frustration—with both his assistant Gustav Jaumann and his counterparts at both universities—one might suspect that he experienced his time in Bohemia as a kind of exile from Vienna, to which he returned in 1895. In fact, he turned down other job offers (from Graz in 1876, from Jena in 1882, from Munich in 1885, and from Graz again in 1890) to stay and pursue his experimental program. The bulk of his publications across his career stemmed from his time in Prague, especially his authoritative experiments concerning the Doppler effect and his acoustic research that led to the dubbing of speeds above the sound barrier as "Mach 1," "Mach 2," and so on. (Christian Doppler, as it happened, had come up with his eponymous theory of the shift of frequencies—at first light, and later, more famously, sound—as a function of the velocity of its source while serving as a professor at the Polytechnic in Prague a generation earlier.)[23] Prague was also where Mach became interested in the history and philosophy of science (due to reading during a bout of illness that kept him from the laboratory), which he would incorporate into his teaching repertoire.[24]

Mach's seminal contributions to epistemology developed from a series of articles on perceptual psychology and empiricism and his landmark historical monographs, *The Principles of the Theory of Heat* (1896) and especially *The Science of Mechanics* (1883).[25] Even in his more abstract pieces about the analysis of sensations, which primarily aimed to break down the notion that there was a unitary perceiving subject, one can find traces of his local context, such as a vivid invocation of a walk with his three-year-old son "upon the walls about Prague."[26] The guiding principle of Mach's research, the insight that grounded the tremendous influence his work would have—especially upon Philipp Frank—was an implacable hostility to "metaphysical" thought in the sciences. For Mach, good science was about observations made by the senses, and *only* about that. Contrary to caricatures by later opponents, Mach did not reject all conceptual tools in scientific reasoning; rather, he considered such notions essential to give "economy" to scientific analysis. If, however, scientists were unable to point to concrete perceptions and observations that undergirded a concept, then that concept should be discarded from their toolkit. By reforming how they talked about ideas, Mach insisted, physicists could build a metaphysics-free science in which every theory was bound by ironclad links to the empiricism that validated all epistemology. The American philosopher and psychologist William James, upon visiting Mach in Prague in 1882, wrote to his wife in admiration: "I don't think anyone ever gave me so strong an impression of pure intellectual genius."[27]

Someone else who thought so, at least at first, was the young Albert Einstein. While living in Bern and working in the patent office, Einstein and a set of close friends formed a group they called the "Olympia Academy," which was dedicated to reading and discussing philosophy, physics, and also *Don Quixote*. Ernst Mach's *Mechanics* was a central text; its sharp epistemological analysis of concepts and their relation to experience clearly left an imprint on Einstein's 1905 declaration that since the luminiferous ether could not be detected by experiment, it was "superfluous" to electrodynamic theory.[28] As his career became more established, Einstein reached out to Mach in a flattering letter of August 1909:

"Besides I naturally know your major works very well; among them I especially admire the one about mechanics. You have had such an impact on the epistemological views of the younger generation of physicists, that even your present opponents, such as e.g. Mr. Planck, would without doubt be called 'Machian' by one of those physicists as they generally were a few decades ago."[29] However complimentary Einstein was of Mach's lucid criticisms of physical theory at this early stage, he did not hesitate to point out where he thought the venerable professor had slipped. One of the key points of this letter was to open a dialogue to persuade Mach that his well-known opposition to atomism as "metaphysical" was misguided.

It was a friendly exchange, and Einstein sent another letter 12 days later stating that he was "very delighted that you are pleased with the relativity theory."[30] This understanding of special relativity as essentially "Machian" was widespread among certain circles, especially in Prague. Georg Pick, Einstein's colleague at the German University who had studied with Mach decades earlier, assured his former teacher that there was nothing amiss about the strange ideas of relativity: "That the physicists who have formed the new theory were essentially influenced by your thoughts I learned, were it not objectively clear to me, from one of them himself: Einstein, through personal communication."[31] (Mach's pacifism and vocal opposition to anti-Semitism were also very congenial to the younger scientist.[32]) Einstein's actions at this period suited his words: not only did he continue to correspond with Mach about his current research on general relativity—a frustratingly undated letter mentions his work on a "new theory," but it is unclear whether this was the Prague static theory or the Zurich *Entwurf*—but he also signed onto the program of a recently founded "Society for Positivist Philosophy" dedicated to the antimetaphysical epistemological program. Ernst Mach was another signatory.[33] Upon Mach's death in February 1916, Einstein wrote a generous obituary, though he was more impressed by the encomium to the sage of positivism penned by Philipp Frank.[34]

Frank served as a pivotal mediator between Einstein and Mach in ways that the founder of relativity theory was unaware of at the time. Although Mach's critical philosophical and historical works were at the radical vanguard of physics when he penned them in Prague, his return to Vienna in 1895 brought misfortune and a degree of incomprehension. In 1898, at the age of 61, Mach suffered a severe stroke that forced his early retirement in 1901. Largely incapacitated, he retired to a Vienna suburb and engaged with the world of physics through correspondence and pilgrims who came to visit. Sometimes, these pilgrims were summoned. Around the time that Einstein first wrote him, Mach became aware of the buzz around relativity theory (already four years old) and how others thought it was linked to his own concepts. He himself could not follow it at first, and in June 1910 he invited a specialist from the University of Vienna to explain it to him: Philipp Frank. (Frank would four decades later describe Mach as "a man with a gray, somewhat wild beard, who looked like a Slovak physician or lawyer.") Frank attempted to articulate how the view of spacetime propounded by Einstein was a brilliant confirmation of Mach's approach to mechanics.[35]

Mach was not entirely convinced, especially as Hermann Minkowski in Göttingen began to interpret the theory in terms of four-dimensional "world lines" and Einstein himself proclaimed the equivalence principle and sought a theory with curved spacetime—both of which sounded suspiciously metaphysical. The crucial turning point seems to have been 1911, when Berlin physicist Max Planck initiated a full philosophical assault on Mach's antimetaphysical position; coincidentally, this was the same year that Einstein moved to Prague and began working in earnest on general relativity. Although the documentation of Mach's thinking in these final years is not as comprehensive as one would like, it seems reasonably clear that he began to turn against relativity theory in a rearguard effort to defend what he saw as proper physical reasoning. He did not include a discussion of relativity in updated editions of his *Mechanics*, and his posthumous *Principles of Physical Optics* (1922) included a foreword dated "July 1913" that attacked Einstein's theory as wrongheaded.[36]

After Mach's death, the contingent of his admirers in Prague attempted to rescue the legacy of their now-deceased icon from what seemed like a complete repudiation of their own conviction that relativity and "Machism" were one and the same. Anton Lampa, for example, wrote a brief but hagiographic biography of Mach that took on this issue in 1918. ("Hagiographic" here is meant literally; this is how Lampa describes first meeting Mach: "It was something entirely different than what I had until then met with in outstanding minds, something entirely distinct from the great researchers, the great poets; I was standing before a saint, who has overcome the final remnants of earthly troubles and from whose eyes beamed the infallible goodness of understanding everything.")[37] With grateful acknowledgment of Frank's assistance concerning the material related to relativity—although Lampa had brought Einstein to the German University in Prague and considered his work revolutionary, he was not especially familiar with the details—he went on to claim that "Mach is the forerunner of general relativity theory and had prepared the ground for it long before Einstein and Minkowski," even if Mach did not himself appreciate the theory.[38] This view was not dissimilar to Frank's own admiration for the clarity of Mach's conceptual analysis, which he continued to laud throughout his career, even as he began to diverge with the orthodoxy of Mach's views and develop his own intellectual position.[39]

Einstein himself was not so generous. Already during World War I he had begun to express exasperation in his letters to Michele Besso about the Machist gloss on general relativity being propounded by their mutual friend from their school days, Friedrich Adler, whom Einstein had come to regard as "a rather sterile Talmudist, pig-headed, without a sense for the real. Ultra-selfless with strong stitch of the self-torturing, even the suicidal. A true martyr's nature," someone who "rides Mach's nag to exhaustion."[40] Adler was a stubborn interlocutor, no doubt, but Einstein had also become disillusioned with Mach himself. Although he had failed to persuade Mach to relent on atomism, he thought he had made progress in other areas and wrote in 1921 that he was "amazed that Mach was not in favor of relativity theory. This course of reasoning lies precisely and entirely in his line of thought. I am curious how he comes to his standpoint of rejection."[41] (He still retained enough respect to endorse the construction of a monument to Mach in 1926.)[42]

With further reflection, Einstein deepened his own epistemological framework and came to see the problem as inherent to Mach's dogmatic insistence that everything had to be tied directly to sense experience, that concepts were only valid when they served as generalizations of sense data introduced to economize on mental exertion. By 1946, he was "convinced that even much more is to be asserted: the concepts which arise in our thought and in our linguistic expressions are all—when viewed logically—the free creations of thought which can not inductively be gained from sense-experiences."[43] For Einstein, scientists made their theories independently of experience, imaginatively, and without reference to sense data; only *after* one had the theory in hand would it be submitted to the acid tests of observation and experimentation. Mach wanted to reduce theory to an outgrowth of experiment, and the mature Einstein considered this too limiting. He began to criticize his own past allegiances to Mach, as, for example, in his 1947 autobiographical reflections:

> It was Ernst Mach who, in his *History of Mechanics*, shook this dogmatic faith; this book exercised a profound influence upon me in this regard while I was a student. I see Mach's greatness in his incorruptible skepticism and independence; in my younger years, however, Mach's epistemological position also influenced me very greatly, a position which today appears to me to be essentially untenable. For he did not place in the correct light the essentially constructive and speculative nature of thought and more especially of scientific thought; in consequence of which he condemned theory on precisely those points where its constructive-speculative character unconcealably comes to light, as for example in the kinetic atomic theory.[44]

In these postwar years, Einstein became increasingly vocal in arguing that Mach underestimated the power of human creativity, and he even began to rethink his own educational trajectory to elevate the influence of the eighteenth-century Scottish empiricist David Hume on his thought over and against the conceptual critique proposed by Mach.[45] Mach was good for tearing down old ideas but not for building up new ones. This was the lesson that the history of relativity theory taught Einstein, who in this way also cut one of the ties connecting him to the intellectual traditions rooted in the German University.

* * * * *

Despite his insistence on the lack of connection between relativity theory on the one hand and Machian empiricism and its descendant logical positivism (or logical empiricism) on the other, Einstein did not control the narrative that developed around his concepts. The dominant view among philosophers for the rest of Einstein's life was that relativity theory provided one of the greatest illustrations of the validity of logical positivism as an epistemological framework, no matter what Einstein might retrospectively offer in rejoinder. This solid connection between relativity and positivism owes its persistence to the unflagging intellectual and organizational work spearheaded by Philipp Frank.[46] It was a link forged in Prague.

Frank worked tirelessly on his defense of relativity theory, seeing its triumph as a vindication of the positivist position rooted in the framework first propounded by Mach. As he told Kuhn in 1962, "originally the relativity theory was based on positivism," and "it was Einstein who later more or less abandoned the positivistic conception of science. . . . But so to say in the birth, in the history of birth, the relativity [theory] has always been connected with positivism."[47] The best source for this connection, according to Frank, was Einstein himself. "It is indeed today generally known that Einstein's general relativity and gravitational theory grew immediately out of the positivist doctrine on space and motion," he wrote in a 1917 assessment of Mach's work that Einstein openly admired, "which Einstein himself described in detail in his eulogy for Mach."[48] The most influential Frank essay in this vein appeared in 1949, in the two-volume compilation *Albert Einstein: Philosopher-Scientist*, edited by Paul Arthur Schilpp, which featured other essays by positivist-affiliated thinkers that reinforced Frank's specific reading. "Einstein speaks here almost completely in the line of the logical empiricists," he stated, "which is not surprising, inasmuch as logical empiricism is, to a considerable extent, a formulation of the very way in which Einstein envisaged the logical structure of his later theories, e.g., the theory of gravitation."[49] The connection could not be severed: Einstein based his thinking on Mach, and the later positivists who followed Mach in turn based their thinking on Einstein. "Briefly," Frank concluded, "I do not see in the question of the origin of the fundamental concepts of science any essential divergence between Einstein and twentieth century logical empiricism."[50]

But even if Einstein had not arrived at special and general relativity by following a positivist playbook, Frank believed that the theories served as admirable illustrations of the conceptual coherence that came from positivism's epistemological stance. Like his fellow logical positivists, Frank had modified Mach's initial prescription—all sensible statements are based on sense perception—by adding another class of allowable utterances: analytic statements of logical relations. If one could purge from the fusty halls of science all statements that did not restrict themselves to the logical connections among sense data, then one would have a modern science cleansed of metaphysics and clotted thinking. An excellent example of this, perhaps even the best example, was general relativity:

> In this theory Einstein derived his laws of motion and laws of the gravitational field from very general and abstract principles, the principles of equivalence and relativity. His principles and laws were connections between abstract symbols: the general space time coordinates and the ten potentials of the gravitational field. This theory seemed to be an excellent example of the way in which a scientific theory is built up according to the ideas of the new positivism. The symbolic or structural system is neatly developed and is sharply separated from the observational facts that are to be embraced. Then the system must be interpreted, and the prediction of facts that are observable must be made and the predictions verified by observation.[51]

Studying relativity not only would be a way for philosophers as well as scientists to learn the true nature of spacetime but also could offer an exemplar of rigorous thought. This was all the more important, as Frank noted in 1938, because misinformed intellectuals (including scientists) felt a need to graft all sorts of metaphysical claptrap onto relativity:

> It is hardly possible to open a textbook on the theory of relativity—even if written by an otherwise competent physicist—without coming upon sentences of an entirely metaphysical character. Such sentences, wholly meaningless in physics, stand side by side with obviously physical sentences. It is therefore not amazing that great

confusion occurred among young physicists and also in wider circles of educated people, and that the opinion arose that the theory of relativity is of an entirely different character from all previous physical theories: that it is constructed less logically and contains many contradictions. One cannot blame the philosophers who sought enlightenment in the writings of physicists for believing that all statements in these papers were the outcome of physical research. But in spite of their sincere desire for scientific enlightenment, they assimilated from those papers on physics a host of purely metaphysical sentences.[52]

(As we shall see, this is a reasonably good description of Frank's views toward Oskar Kraus.) After Frank moved to the United States, he took this message to the American public in a work called *Relativity: A Richer Truth*. (Einstein wrote the foreword.) For Frank, relativity theory not only was unrelated to moral relativism, but was the antidote to precisely the kind of pernicious thinking that lay behind both the fascism of the past and the Stalinism of the present.[53]

Fascism, much more than Stalinism, was much on Frank's mind during the 1930s in Prague. Before then, he dedicated himself not just to publicly advocating for Einstein's relativity and the philosophical school of logical positivism he saw as connected with it, but also to working within the local German University and the transnational German Physical Society to make them friendlier to a progressive philosophy of science. Although many faculties at the German University greeted the advent of an independent Czechoslovakia with some apprehension—would the institution survive the creation of a Czech-dominated state, and would their rights as minorities be protected under the regime of President Tomáš Masaryk? (yes and yes)—for Frank the interwar period was in many ways the height of his career. It is true, as we will see later, that the status of the German University changed symbolically in 1920, declining such that the university become a somewhat subordinate institution in the new capital city, but the same reforms also created the new faculty of natural science, and this was a boon for Frank. Not least it meant that, while still being a physicist and teaching his usual classes, he was now free to act more forcefully as a philosopher. In the pre-independence faculty of philosophy, where he had been yoked to mainstream philosophers (like Oskar Kraus), Frank had had to worry about treading on toes. Now he was unleashed upon epistemology. He trained a number of students, including Josef Winternitz, the son of the Indologist who had been a good friend of Einstein's during his time in Prague (and who came to pick up the physicist on the train platform in 1921). Under Frank, the younger Winternitz wrote a sophisticated dissertation on epistemology and relativity with a neo-Kantian slant—referencing, among other topics, the ongoing disagreements between Frank and Kraus—that was reviewed favorably by Einstein himself, who encouraged Frank's explorations in this area even if he did not always agree with them.[54]

Frank's pedagogical and organizational talents—he served as rector for the German University in 1925–1926 and then as prorector the following year—also attracted the attention of other leaders of the logical positivist movement, who saw Prague as a potential solution for their own underemployment or unemployment. Rumors traveled fast among this group. Moritz Schlick, later the guiding light of logical positivism, wrote to Einstein in 1920 from his post in Rostock on the Baltic littoral, angling for some insight into what was happening in Prague: "Then supposedly at the German University in Prague they have the intention of dividing the philosophical faculty and hiring a special philosopher in the natural-sciences section. They supposedly have already thought of me for it. That would be truly great! Because in its geographical situation and its intellectual life Prague has great advantages over Rostock."[55] Einstein was still tied in many of his colleagues' minds to Prague, and they thought he would have inside tips. Schlick soon moved (via Kiel) to greener pastures in the Austrian capital, but others hoped for a haven in Prague. In 1925 philosopher Hans Reichenbach lamented his uncertain career prospects and aspired to emigrate to the United States—he would do so in 1938 to escape the Nazis—but was instead encouraged by Einstein to wait until Frank could put together something suitable in the natural sciences faculty in Prague.[56] (In the end, Einstein made a similar arrangement for Reichenbach in Berlin.) Frank did eventually fill his post with none other than Rudolf Carnap in 1931, drawing on the personal support of President Masaryk (himself a former professor of philosophy at the Czech University). Carnap was still there in 1933 when the young American philosopher W. V. O. Quine came for a visit that he later claimed transformed his entire intellectual trajectory—another important chapter in the philosophy of science midwifed by Frank.[57]

The most prominent organizational triumph for Frank, however, was the institutionalization of logical positivism as an international movement linked to the sciences, marked by the establishment of the Vienna Circle (which was inaugurated, ironically enough, in Prague). Once again Frank was able to take advantage of his dual identity as a physicist and a philosopher. In September 1929, the Fifth Congress of German Physicists and Mathematicians, sponsored by the German Physical Society, was set to take place in Prague; Frank served actively on the local coordinating committee. The strong presence in Czechoslovakia of this German society, founded in 1845 before there was such a country as Germany, was a response to the shock that the Central European academic community had suffered following the Triple Alliance's loss of the Great War. A boycott of German and Austrian scientists engineered by the Belgians, British, and French (and, somewhat desultorily, by the Americans) had cut off German physicists—with the important exception of the world-famous pacifist Albert Einstein—from participation in international conferences and travel. It was therefore especially important to strengthen networks among the outcast nations, even extending them in the mid-1920s to the pariah state of the Soviet Union.

Frank had become a member of the German Physical Society on 28 June 1918 through the nomination of Peter Debye, a Dutch (and later American) physical chemist and physicist who had himself been Einstein's successor at the University of Zurich when the latter decamped for Prague's German University. Within a few months of obtaining this membership, Frank's residence had shifted from his native Austria-Hungary to Czechoslovakia without him stirring from the camp bed in his office at the Physical Institute on Weinberggasse. Hoping to retain the German Physical Society's contacts in the newly independent states while also decentralizing the organization by reducing the dominance of Berlin, Society president Arnold Sommerfeld began in 1920 to create a series of so-called *Gauvereine* (regional societies). The first, not surprisingly, was established in Munich, Sommerfeld's residence, with the next appearing in Vienna. The fifth of the eventual 10 *Gauvereine* was founded in Prague on 2 February 1922. Frank served as its chairman until its dissolution on 7 March 1934, at which point all the members simply became out-of-town affiliates of the larger umbrella organization. At its inception, the Prague division seemed extraordinarily healthy—with 56 members in 1922, it was larger than its elder sibling in Munich—but its rolls quickly hemorrhaged physicists, and it soon became the smallest, half the size of the next largest one. By 1929, the year the Fifth Congress came to Prague, it had only 35 members.[58]

Frank was determined to make the most of the opportunity presented by the 1929 meeting. He successfully petitioned to attach a conference entitled "Epistemology and the Exact Sciences" to the Prague event, bringing together the groups that had been developing the philosophical framework known as logical positivism from both Berlin and especially Vienna, the latter being an offshoot of the small discussion circle he had formed with Neurath and Hahn in his days as *Privatdozent*. The cohort from Vienna, now under the leadership of Moritz Schlick, called itself the *Verein Ernst Mach* (the Ernst Mach Society), and Frank took advantage of the local connection to catapult the epistemological movement to wide recognition among the visiting physicists.

The *Prager Tagblatt* covered the proceedings in detail, dispatching none other than the novelist and art critic Max Brod—Frank's Fanta circle colleague and, as we shall see in the next chapter, an indispensable interlocutor for various circles of the Prague intelligentsia—to report on the event, as well as a caricaturist to capture Frank and Sommerfeld among the physicists and Neurath and Reichenbach among the philosophers. "Symbolically the lecture hall in which Ernst Mach taught was chosen as the gathering place for the scholarly discussion," Brod observed. "The institute on the Windberg was built according to Mach's plans, his instruments are displayed in the vestibule."[59] Brod had long been fascinated with the philosophy of causality, a topic he had at times discussed with Frank, and he recorded the discussions with informed sympathy and a *soupçon* of satire:

> If someone were to enter the hall in the middle of Reichenbach's lecture, he [or she] would have marveled how the curves and equations of chances of winning the adventure of roulette, *rouge* and *noir*, were discussed before a ring of serious scholars—until it became suddenly clear to this clueless listener through a turn by the lecturer that the discussion was not about Monte Carlo but of a much larger establishment and casino, as the once so venerable cosmos now seemingly presents itself to modern physicists.[60]

The debates were not without their fireworks: it was here that Sommerfeld openly confronted Frank with evidence that Einstein had in fact distanced himself from Mach's epistemology over the past decade, a surprising setback for the Prague physicist, who also used the occasion to stress the legacy of Mach in what many of the visitors had come to view as a Slavic backwater.[61]

The most significant consequence of this meeting, however, was that it erased the Bohemian setting from the history of positivism. At this meeting Hahn, Neurath, and Carnap, all of the Ernst Mach Society, presented a manifesto entitled "Scientific World Picture" to their leader, Moritz Schlick, laying out the fundamental tenets of logical positivism: the organization of sense perception through logical relations, the quest to eradicate metaphysics from philosophy and science, and the importance of modeling future thought on the sharp analysis manifested most clearly in the work of Albert Einstein on special relativity. This would serve as the foundational document of what came to be known as the Vienna Circle, and indeed the text took every opportunity to extol the Austrian capital. Buried in a parenthetical statement in the middle was a recognition of the lineage of its teachings in Prague: Mach to Einstein, Einstein to Frank.[62]

The 1929 Prague meeting became a model for a number of conferences, such as the "International Congress for the Unity of Science," which met in various cities over the coming years—including Prague in 1934 and

1937—and the social core of the enormously influential logical positiv-ists.[63] Prague proved crucial again in the 1930s as positivists fled from Nazi Germany and Vienna's rightwing dictatorship and took refuge in the democracy of the First Republic, where Philipp Frank exercised his superhuman capacity to find temporary perches for them before they de-parted to the United States or other safe havens. Looking back from the Anglo-American tradition, Prague was the friendliest of cities for a philosophical tradition that considered itself an extension of Albert Ein-stein's innovations in physics.[64]

* * * * *

Not everyone in Prague was delighted by this advent of logical positiv-ism. In 1930, a review of the Vienna Circle's manifesto appeared under the signature of a professor of philosophy at the German University. He did not care for it; in particular, he believed that to reject metaphysics was to annihilate the very essence of philosophical inquiry: "This meta-physics, which due to its value is the first and highest philosophical dis-cipline, is however at the same time the seal and the crown of the entire structure of knowledge, of the physical and psychical. It is scientific worldview and world-conception."[65] Nonetheless, he recommended that philosophers everywhere read the manifesto. It was important to know what your enemy was thinking. The review was penned by Oskar Kraus.

Despite the enthusiasm for logical positivism in some corners of the German University, if you had wandered its halls in the interwar years overhearing what the philosophers in the faculty of philosophy were say-ing, there would have been no way to avoid Kraus's hostility. He was convinced that Einstein's relativity theory was the harbinger of a loom-ing obscurantism in the world of thought, a cancer of which the Vien-nese manifesto was but a symptom. "Kraus" is not a name that features prominently (if at all) in histories of philosophy, or even of the German University, but in the 1920s he was a force to be reckoned with, as dem-onstrated by his inclusion on the stage in 1921 to debate with none other than celebrity scientist Albert Einstein. Kraus's path to vociferous anti-relativism reveals another way in which Einstein's legacy was mediated by the local environment in Prague. Kraus's views would in the long run not be nearly so influential as Frank's, which for a long time represented the central popular understanding of the conceptual significance of relativity, but for these two interwar decades Kraus's anti-Einstein activism accompanied the Germanophone intellectual world into its darkest hour.

Kraus was born on 24 July 1872 in Prague, and his whole life revolved around the city—at least the German-identified parts.[66] He came from a solidly middle-class Jewish family: his grandfather Ignaz Kraus had worked as a midwife, and his own father, an observant Jew, had gone into trade; his mother, Clara Reitler-Eidlitz, also hailed from a merchant family. Seven years older than Einstein, Kraus followed a path away from religiosity that was remarkably similar to that trodden by the physicist, even down to the same book—Ludwig Büchner's *Kraft und Stoff*—that served to foment an interest in science and rationalist atheism. (This interest did not last; Kraus later described his career as "the metamorphosis from atheistic pessimism to a theistic optimism.")[67] Sent to one of the elite *gymnasia* of the city, he distinguished himself well in studies but perhaps rather more in his lampooning of his classics teacher. The mock-Homeric epic he composed at age 16, the *Meyeriade*, was first printed in 50 copies and passed from hand to hand, but was soon reissued in multiple editions, becoming a staple of comedic reading among Prague's youth. (Franz Kafka loved it.)[68] Kraus was not fated to be a classicist.

In 1890 he enrolled at the German University to study law, producing some theoretical writings on the theory of obligations and contracts and earning his doctoral degree in 1895.[69] He fully expected to continue in this field, in accord with his parents' wishes, but became both intellectually and then professionally sidetracked by philosophy. While taking courses he met his future mentor, Anton Marty, the chief disciple of Franz Brentano. Engagement with Brentano's legacy—and to a slightly lesser extent Marty's—would be the single most important guiding thread of Kraus's career. Alfred Kastil, himself a "grandchild" of the Brentano school, called his fellow Brentanist Kraus "indisputably the finest mind of the second generation of students, and its most productive."[70] The two of them (though mostly Kraus) labored for years editing Brentano's mostly unpublished manuscripts, under the direct patronage of Tomáš

Masaryk, who himself had been a student of the master. (Regrettably, Kraus's editions are highly problematic and of limited scholarly use today.)[71] Kraus's attempts to follow Brentano's teachings to the letter colored everything he did, not least his opposition to Einsteinian physics, even though a charitable reading of both his mentors' works indicated some comfort with the notions of relativized space and time on metaphysical grounds.[72] This was not how Kraus read his Brentano. (Brentano was aware of the problem, lamenting in 1908 in a letter to Ernst Mach, of all people, that he despaired of "even the good Kraus, who—without saying anything false—makes me seem like a mystic.")[73]

Brentano is known today mostly by specialists in intellectual history, overshadowed by the overwhelming influence of his most famous student, Edmund Husserl, the founder of phenomenology. At the turn of the twentieth century, however, Brentano was a powerful force in philosophy in the Habsburg Empire. Born in the northern German states to a distinguished literary family, he studied theology and entered a Catholic seminary in Munich and then Würzburg, before being ordained as a priest in August 1864. His specialty in those early years was the scholastic philosophy derived from Aristotle, and he began lecturing at Würzburg to great success. There he met Marty, who was also an ordained Catholic priest. In the early 1870s, however, Brentano had a crisis of profession, though not of faith, resigning (as did Marty) his clerical office in protest of the declaration of papal infallibility by Pope Pius IX. He continued to lecture at Würzburg and Vienna, expanding his range to just about every area of philosophy. Controversy soon struck, however, when he became engaged to be married in 1880. Austrian law held that a priest, even one who had left the calling, could never be married, and so Brentano resigned his ordinary professorship at Vienna and was only allowed to return as a *Privatdozent* after a protracted legal case. In 1895, shortly after his wife's death (and also the year Mach moved to Vienna), he retired to Italy and later Zurich, dying in 1917.[74]

Kraus was captivated by Brentano's thought and personality, and his commitment to defending the reputation of his teacher's teacher forced a radical change in his career. In 1895, while still a jurist, Kraus applied to habilitate in the law faculty at the German University. Professor

Horace Krasnopolski, whom Kraus had criticized publicly in protest against the interpretation of canon law that had justified Brentano's demotion, blocked Kraus's advance. Kraus took up a position working for the city for several years, until Marty proposed to the aspiring academic that he habilitate in philosophy instead. In 1902 he was promoted under Marty's guidance with a dissertation on Jeremy Bentham's theory of value while still working as a lawyer, in the evenings attending Bertha Fanta's first philosophy discussion circle at the Louvre Café. His career was once again blocked, this time by his Jewish confession. The former priests Brentano and Marty both counseled that he convert to Catholicism for professional reasons, and in 1908 Kraus was baptized.[75] The following year he was appointed an extraordinary professor in the philosophy faculty of his alma mater. Marty died in 1914 and Kraus took over his teaching load, earning a promotion as Marty's successor in 1916 as ordinary professor of philosophy.[76] He became prorector in 1918–1919 and again in 1922–1923, and served as dean of the philosophy faculty in 1921–1922. This career trajectory indicates that for the entire time Albert Einstein was professor of physics at Prague, he and Oskar Kraus were colleagues.

That fact did not seem to register for either figure in 1911–1912, nor did Kraus pay much attention, at least in his publications, to Einstein during the remainder of the second decade of the twentieth century. Both before and during the Great War, his research focused on topics relatively distant from epistemology or the philosophy of physics. Kraus deepened his work on Bentham's philosophy—continuing Brentano's push to introduce British empiricists to Germanophone philosophy—applying it to questions of international law. In 1915 he published a commentary alongside Camill Klatscher's translation of Bentham's *Principles of International Law*, a controversial choice in the middle of the war both because it hailed from the English foe and also for its arguments in favor of pacifism and nonviolent confrontation.[77] His continued writings on the philosophy of punishment, a legacy of his time as a lawyer, arguably served as a resource for his former student Franz Kafka, who might have drawn on some of Kraus's theories in his posthumous novel *The Trial*.[78] Kraus's most sustained area of research, however, was the theory of value, in which he advanced the notion that "preference" rather than "utility" should be the basis of valuation, essentially claiming economics as a subdivision of applied psychology.[79] His arguments were in line with those of the emergent school of thought that would later be called "Austrian economics," granting Kraus a lasting, though largely unrecognized, legacy.

Kraus wrote fairly consistently about the theory of value over his entire career, but in the 1930s his publications began to include arguments against relativism (long a bugbear) that were simultaneously explicit attacks on relativity as a physical theory.[80] The general homogeneity of Kraus's thought over time—his unwillingness to revise his views in part a consequence of his unwavering fealty to what he understood as Brentano's philosophy—leads one to suspect that perhaps as far back as the early 1910s, he might have been already harboring a distaste for the new theory of spacetime. And in fact, this suspicion would prove true. In October 1913, Kraus wrote a letter to the Berlin experimental physicist Ernst Gehrcke with a strikingly heartfelt confession:

> People are suffering from extreme fatigue, and an irritability that is due not least to the absurd theories of the relativists. I have a burning desire to see the *source of error* revealed for all of the absurdities that you yourself, honored sir, have accurately characterized. I also see that you have already revealed internal contradictions and absurd consequences multiple times. But where is the *source of error*? Because despite my calculation errors, I am still able to recognize the fact that the theory of relativity is false.[81]

This was the first of a flurry of letters he sent to Gehrcke. So would begin a connection that would increasingly tarnish Kraus's good reputation over the decades to come.

* * * * *

Ernst Gehrcke was the head of the optical department at the Reich Physical-Technical Institute—the German Empire's bureau of standards—and also a professor at the University of Berlin. Gehrcke began publicly attacking relativity in 1911, and over the next several years he would blame mass suggestion and hypnosis for the so-called hysteria that he believed blinded people to the physical and philosophical incoherence behind relativity.[82] His publications and organizing efforts behind the scenes assembled the core of what would eventually blossom into an antirelativity movement in Germany. What might seem like an esoteric complaint fostered, through the strategic use of mass media and by playing on the antipacifist and especially anti-Semitic right-wing politics that flourished after Germany's defeat in World War I, the emergence of a surprisingly powerful cabal. The smoldering embers of resentment burst into hot flames upon Einstein's sudden ascent to stardom after the 1919 eclipse expedition—now the defense of Newtonian spacetime could be a shibboleth for incipient *völkisch* politics.[83] This band of vituperative polemicists and hypernationalist conspiracy theorists brought together by Gehrcke welcomed Oskar Kraus, the baptized Jew from Prague, into their ranks.

Kraus fired his opening salvo at a meeting of Lotos, a German-identified natural history society that had been founded in Prague in the wake of the 1848 rebellions and dedicated itself to popularizing knowledge in its journal of the same name. (Ernst Mach had been elected to Lotos in 1870, serving as vice president and as a member of the executive board, but resigned from the group in 1884 in the aftermath of the split of the university.)[84] At a meeting in late February 1920, as reported on the front page of the *Prager Tagblatt*, Kraus picked a fight with Philipp Frank, the leading local representative of the pro-Einstein, pro-positivist line, by mocking relativity with a parable. In the story, Kraus told of a virgin who wanted to always stay young and sought the help of a magician, who advised her to run quickly so that time would slow down. The report in the newspaper—which also documented Frank's objections from the floor—is quite amusing, but unfortunately did not include Kraus's original wording.[85] In his archive, however, there is an undated manuscript almost certainly hailing from this period, entitled "A Miracle of Science: A Shrove Tuesday Joke by Prof. Neinstein," which not only bears a similarity to the reported story but recalls the *Meyeriade* of Kraus's schooldays. The tale is set in Prague and involves the romance of Kasper and Kasperline, the Devil, and the latter's grandmother. Once again, the lady wants to be young forever, and the Devil explains that relativity could help fulfill her desire. "Why then for example are little boys and girls so young?" he asks. "Because they are constantly running, jumping around, and making a lot of movement."[86] Shenanigans follow, and the Devil concludes with Kraus's philosophical clincher: "Einstein says: Motion is relative. It is all the same whether Earth turns around the Sun, or the Sun around Earth; he couldn't care less whether you are calmly sitting and the others are running or whether you are running and the others are sitting. The effect turns out to be the same. Become skinny and remain young."[87]

In more sober terms, Kraus published his major objections in the society's journal, lamenting the corrosion of the philosophical foundations of physics as represented by claims about the relativity of simultaneity and other concepts that he insisted must be, on Brentanist grounds, absolute. "All *this* is in truth meaningless. All of modern physics suffers from this meaninglessness, indeed its counterintuitive nature," he maintained. "Even those opponents of Einstein who concede relativity for rectilinear uniform movement but deny it for the general fall into it. No! Relative motion alone is either quite simply nonsensical or it pertains just as much to rotational motion as to rectilinear uniform motion."[88] Introducing a frame that he would invoke repeatedly over the next two decades, he declared all of Einstein's reasoning to be a "fiction": "Einstein and his followers misinterpret the formulas and symbols of their theory when they consider the so-called relativization of time, the four-dimensionality of the spacetime continuum, and the bending of space to be anything other than symbols and fictions that are perhaps appropriate to provide certain services as calculational, descriptive, and heuristic auxiliaries for theoretical physics."[89]

Frank responded soberly in *Lotos*, dismantling the idea that motion cannot be understood as relative. The problem with Kraus's reading was that he did not understand the difference between a stated and an unstated comparison: "If the predicate of a judgment is a relative concept and the object of comparison or relation is not named, the judgment remains incomplete and all the consequences about incompatibility, indeterminacy, and absurdity are invalid."[90] Irate, Kraus repeated his earlier charges essentially unmodified, adding that the classical relativity principle and the principle of the constancy of light—the relationship between which Einstein himself had come to modify during his time in Prague, unbeknownst to Kraus—were incompatible.[91] Frank left the matter there, while Kraus picked a fight with Hans Reichenbach in the Berlin periodical *Umschau* in 1921, assuming that the latter had been implicitly ridiculing him in a previously published piece, and the philosopher Benno Urbach reignited the dispute in *Lotos* in 1922.[92] That last exchange ended with a calm but final note from the editors: "With the preceding article we close the discussion about this contentious matter in our magazine. The form in which it has been published here in its personalizing acerbity departs from our usual practice, the responsibility for which is the author's alone."[93]

Kraus continued to actively defend his view, earning support from the most militant and bigoted opponents of Einstein. Kraus provided them needed cover: they could not be anti-Semitic if Kraus joined them, and they could not be philosophical ignoramuses if someone with Kraus's reputation argued alongside them. Kraus's debut as a prominent antirelativity critic outside of Prague came with an invitation from Hans Vaihinger, the important neo-Kantian philosopher, for Kraus to come to the German city of Halle in late May 1920 and contest not so much relativity as the positivist Machian reading of it. "Professor Kraus from Prague will . . . give a lecture here in Halle against Einstein's doctrine of relativity. He explicitly wanted to express the wish that you," Vaihinger wrote to the ultra-Machian defender of relativity, Josef Petzoldt, "be invited to this lecture. . . . You would be welcomed by all, and we would be especially pleased by your participation in a debate about Kraus's lecture."[94] Petzoldt was willing to come and represent special relativity and positivism.

Einstein was unwilling to represent either. The previous year, he had refused to give Vaihinger an article on the interpretation of relativity for his prominent journal *Kantstudien*, but that did not stop the organizers of the Halle meeting from inviting Einstein and putting his expected attendance on the program.[95] The psychologist Max Wertheimer, himself born in Prague and a graduate of the German University, wrote to Einstein in outrage:

> In April, in Prague, I heard about the "imminent *important congress about Einstein, where Prof. Kraus* (!) *has been granted the chief role*, which will now (finally) in public reveal the elementary absurdities of the Einsteinian theories before a philosophical tribunal, so that it will become clear, how—."
>
> Here I find an invitation from the Kant Society: Mr. Einstein will also be in Halle! . . .
>
> Heavens alive, what kind of publicity racket have you gotten yourself into?! Would physicists of comparable caliber risk something like this?! For the most part feeble-minded, sluggishly cud-chewing, squabbling mediocrities and some like Kraus: insolent; and obsessed with publicity—yes, my Lord, if one could at least think that it made *any sense at all*: that something could be honestly advanced in the "conference" or even just treated seriously—but you, my good man, don't you know what these people are like and what they want—?![96]

Wertheimer was not the only one who pressured Einstein to stay away, especially given the outburst of right-wing protests against relativity at other similar meetings. "I believe that I'll turn Halle down," Einstein wrote to his wife Elsa a few days later, "because the jabbering will sit poorly with me."[97] A good thing, too. Petzoldt would later report that the whole affair had been a shambles, and that the "philosophers there were not at all even clear on the experimental foundations of the theory."[98]

Kraus was in full flower at the event, elaborating again on the fictive status of relativity: "Fiction: a representational image which corresponds to reality in nothing, proving its worth as an auxiliary for calculation and research, as a trick for thinking. This is also, in my view, the case with Einstein's so-called relativity theory," he wrote in the article expanding on his speech, "but these fictions are not always recognized as such—including by their author—they are treated not as though they were auxiliary assumptions in the sense of conceptual constructions without a value in reality or merely analogical symbols, but rather as though these imagined things existed. So fiction becomes reality and vice versa reality a fiction."[99] The relativity of simultaneity was a case in point, because it was a patent absurdity that something could be at once both simultaneous and not simultaneous. Rather, Kraus contended that one should take these ideas *as if* they were true, adapting his host Vaihinger's conception of an "As-If" (*Als-ob*) philosophy, a notion that he claimed Einstein's former colleague, the mathematician Gerhard Kowalewski, had introduced him to.[100]

As it happens, Einstein had already refuted this argument in a charming 1918 dialogue he composed between a relativist and an imagined opponent, eerily foreshadowing these controversies: "Thus still none of the magnitudes dependent on choice of coordinates correspond, for example, to the components of the gravitational field at a spacetime point; so nothing 'physically real' corresponds to the gravitational field *at one place*, let alone to that gravitational field in combination with other data. One can thus neither say that the gravitational field at a point is something 'real,' nor that it is something 'merely fictive.'"[101] Kraus either never became aware of this piece or chose to ignore it.

Soldiering on, Kraus became a highly sought-after member at antirelativity gatherings, though he was careful to avoid association with the more disreputable segments of the scene. When writing a diatribe against relativity at the invitation of a Viennese newspaper, he insisted that he wanted no part of the xenophobic and racist demagoguery that had characterized the attacks on Einstein and general relativity at Bad Nauheim in 1920.[102] For example, he withdrew from a proposed lecture series at the Berlin Philharmonic scheduled for 2 September of that same year when he learned that it was going to be a largely anti-Semitic event, despite already appearing on the program. The organizers, hoping to cover up Kraus's real reasons and at the same time stoke the resentment of the conspiracy-minded, claimed he had been denied a visa by the German government, which was seeking to protect the tender feelings of their prized physicist.[103]

Kraus was never one of the most prominent voices in the chorus of Einstein's attackers—those roles were played by Gehrcke and Nobel laureates Philipp Lenard and Johannes Stark—but he continued to promote his views in a variety of venues that treated him as a more respectable face of what was becoming an increasingly unhinged movement. Although he continued to publish articles developing the same limited set of arguments in local newspapers and occasionally academic journals—when they would entertain the increasingly polemical tone of his writing—his main intervention that decade was the release in 1925 of a short book entitled *Open Letters to Albert Einstein and Max von Laue.*[104] (Kraus included von Laue because the latter had once rejected one of his pieces for a physics journal, an act of professional gatekeeping that Kraus saw as aggressive censorship.) The book makes for unusual reading: the text is studded with boldfaced terms, italics, and exclamation points, and the tone veers occasionally well beyond the threshold of politeness. "The edifice of special relativity theory is nothing other than [the] structure of all mathematical deductions that can be drawn out from the—in itself absurd—postulate of the invariance of the speed of light," he proclaimed to his principal adversary. "From your invariance postulate, whose correctness is assumed from the outset, is deduced an unforeseeable abundance of consequences and even if the entire system based on this statement were subsequently completely 'free of contradiction,' then it would not achieve the slightest thing for our knowledge of the physical world; it is and remains a deduction from impossible premises, a mathematical concept-fiction."[105] To forestall rebukes (such as Frank had made, quite accurately) that his physics knowledge was too meager and his understanding of relativity theory based on popularizations, Kraus announced that Lenard and Gehrcke had proofed his text to make sure the physics was correct. Very reassuring.

Needless to say, neither Einstein nor von Laue registered a response to these charges. There was little need. The consensus on the validity and utility of special relativity had been almost unanimous among the physics community by the time Einstein had arrived in Prague in 1911, and by the late 1920s even the last reputable naysayers within the mainstream physics community had given general relativity their stamp of approval in the face of astronomical findings reported from London and California. The anti-Einstein group continued to write, but both the changed intellectual environment and the intensifying ugliness of the movement's right-wing affiliations all but discredited it. In a last-ditch salvo, a group of authors produced a polemical booklet, *One Hundred Authors Against Einstein*, consisting mostly of quotations from Einstein's more and less legitimate critics over the years, typically out of context. Oskar Kraus's name was featured prominently on the title page, and he penned some new paragraphs to reiterate his old points. The Prague philosopher continued to provide cover, as did the lead editor Hans Israel (and two other Jews among the 28 contributors), for the thinly veiled anti-Semitism of the chief actors.[106]

This focused examination of Kraus's antirelativity polemics could give one the impression that he was single-mindedly focused on this one issue, and that he had come to be broadly recognized as a crank. Yet Kraus did not live by relativity alone, and he continued a broad range of activities across the Prague intellectual spectrum. At the same time he was writing his *Open Letters*, for example, he produced a sensitive assessment of Albert Schweitzer's theology that bordered on the hagiographic.[107] (In December 1928, he arranged for the famous physician and humanitarian to receive an honorary degree from the German University.) He also concentrated on more philosophical work. He was active in the famed Prague Linguistic Circle, the polyglot collection of scholars who were transforming the science of language in midcentury Europe. Although he objected to some presentations delivered to the Circle, such as that of Rudolf Carnap on 20 May 1935, he gave his own lecture to the group a few months later and entertained an active exchange of ideas on a wide range of conceptual issues (all from a Brentanist perspective, of course).[108] It is possible that Kraus was mellowing as he neared retirement. Einstein's long-time correspondent and Bertha Fanta's son-in-law, the philosopher and Zionist Hugo Bergmann, ran into Kraus on a visit to his hometown from Palestine in late September 1935. "On the final evening I also visited Oskar Kraus and the Brentano Archive," he noted in his diary. "Kraus was actually very touching when he paused in the middle of the conversation and with closed eyes demonstrated Brentano's reduction to the act on a certain occasion, and when he conceded that it was Husserl's achievement that he articulated that an *a priori* psychology existed."[109]

Next to Einstein, Husserl was one of Kraus's nemeses, so this praise stood out.

Even Einstein, as it turned out, benefited from Kraus's late style. In 1938, Kraus published a piece, "On the Misinterpretations of Relativity Theory," in an edited volume commemorating the centenary of his master Franz Brentano. A reader who had been following Kraus's decades of publication attacking the relativity of simultaneity would have noticed a sharp change in both tone and content. "This concept of psychological simultaneity is the inevitable assumption of the concept of any physical measurement," Kraus wrote. "Against Einstein one can say that his 'heaven-storming titanic work' is not pernicious only because the immediate views about space and time *remain untouched in the unattainable heights of* a priori *necessities of thought*."[110] By introducing the notion of psychological simultaneity, Kraus was able to come to terms with Einstein: relativity of simultaneity cannot hold *psychologically* for any observer, Kraus believed, but it might actually be a physical truth. For several years, Kraus had been having conversations with Reinhold Fürth, one of Philipp Frank's students who had been appointed to the natural sciences faculty of the German University. Fürth had patiently succeeded where Frank had failed in explaining to Kraus that his understanding of relativity theory was inaccurate.[111]

Kraus felt he needed to make amends to those he had maligned. As his widow recalled in an undated note preserved in his archives in the Austrian city of Graz: "Many years after" the 1921 public debate at the Prague Urania, "Kraus realized that he had been wrong on a certain point. He summoned together all the students of Charles University and all the professors of the various faculties. He demonstrated before the assembled auditorium with a colleague from the natural sciences faculty the aforementioned problem. Then he stepped before the platform and said very plainly: 'I made a mistake.'"[112] He even wrote a private letter to a party he had only previously addressed publicly. "In recent years it became clear to me that it is possible to interpret the sp. theory of relativity in such a way that it does not contradict any self-evident truth," he wrote (in English) to Albert Einstein in July 1940. "I should be glad if you, after perusal of my Essay, could agree that my interpretation is possible."[113] He included a copy of the 1938 article. Einstein never responded.

* * * * *

Kraus's apologia came rather late, for him and for Prague. He had retired from teaching at the German University in December 1938, a citizen of the amputated state of Czecho-Slovakia. For reasons known only to him, he remained in Prague until March 1939, when Nazi forces rolled into the city. He was arrested and held by the Gestapo for six weeks. Granted an entry visa to Great Britain, he fled first to Scotland and then to Oxford, where he died of cancer on 26 September 1942. Kraus's dogged philosophical defense of strict adherence to the views of Brentano had already contributed to his eclipse among mainstream philosophers, who had gravitated toward the phenomenology of his rival Edmund Husserl and the latter's student Martin Heidegger. Philosophers of science ignored him as well due to his association with the reviled anti-Einstein cohorts. His works on the topic of relativity, including his recantation, were no longer read, though a cryptic aside from Philipp Frank in 1962 attacking readings of relativistic theories as "fictions" indicates that even decades later the sting of his barbs had not entirely faded.[114]

Frank had left Prague with his wife in fall 1938, at first intending only to engage in a limited lecture tour in the United States but eventually permanently settling in Cambridge, Massachusetts.[115] It was a difficult time for the couple. The Great Depression meant that jobs were scarce, and the earlier wave of refugees from Central Europe—admitted reluctantly and placed with difficulty—had taken the few academic positions available. Frank's world had crumbled around him. Two years earlier, the dictatorship in Vienna had shuttered the Vienna Circle in the wake of the assassination of Moritz Schlick by a former student. Otto Neurath, one of the original members of the discussion circle with Frank and Hans Hahn, had been in exile since 1934 in the United Kingdom. (He died in 1945 in Oxford, three years after Oskar Kraus.) Rudolf Carnap had also fled, first to Chicago and then to the University of California at Los

Angeles, assuming the mantle of leadership of the logical positivist movement there along with Hans Reichenbach. The center of gravity had moved to the West Coast and the Midwest of the United States. Frank now stood on the sidelines and had to fend for himself.[116]

He had some help from friends who specifically valued him for his philosophy of science more than for his achievements in physics. At the tail end of his lecture tour of 20 colleges from October to December, he ended up at Harvard, where he gave a talk on "Philosophical Interpretations and Misinterpretations of Quantum Theory." None of his supporters wanted to send a Jewish scholar back to Prague, where it was only a matter of time before Hitler's vise clamped shut. Astronomer Harlow Shapley raised $2,000 and managed to parlay that into a one-year position lecturing on philosophy and physics at Harvard. Frank grew very active in philosophical communities in the United States, becoming the first elected president of the Philosophy of Science Association in 1946 and establishing in 1949 the Institute for the Unity of Science—an American cousin to Neurath's intellectual program. After his temporary position at Harvard ended in 1953, the Institute became his only affiliation until his death on 21 July 1966. He continued to publish widely on the philosophy of science, especially relativity, and kept body and soul together through soft money scrounged up here and there.

Einstein, now similarly a Germanophone one-time resident of Prague marooned on the eastern seaboard of the United States (though with an infinitely more secure perch at the Institute for Advanced Study at Princeton), proved vital in rescuing the fortunes of his former successor as professor of theoretical physics at the German University. Frank needed to earn money, and he turned to Einstein with a plan. "I conceived the idea of taking advantage of this physical proximity to prepare an account of his life and work," he wrote in the preface to his 1947 biography of his predecessor. "When I told Einstein about this plan he said: 'How strange that you are following in my footsteps a second time!'"[117] The advance on the biography from the Knopf publishing house largely financed Frank's post at Harvard.[118] The book was composed in German but was first published in a slightly abridged form in English. The German text explains a bit more about the composition: Frank began writing it in

1939 in New York, continued it in Chicago the following year, and then concluded the first draft in Boston in 1941. The only part that was composed originally in English and had to be rendered into German concerned Einstein's relationship to the atomic bomb, a section that Frank added to his manuscript after the public revelation of that device in 1945.[119]

Frank's *Einstein: His Life and Times* is far and away the most influential book ever written about its titular subject. Every subsequent biography has derived significant material and much of its structure from the authorial choices Frank made in the early 1940s.[120] Einstein's school friend Michele Besso read it with pleasure and wrote to Einstein discussing its finer points.[121] No small part of its status draws, understandably, from the fact that Einstein supported Frank's efforts and even intended to write a preface for it (which regrettably never materialized).[122] However, for all its wonderful first-person texture, that feeling of the reader being actually beside Einstein, we have no idea what sources Frank drew upon in assembling it. He did not use footnotes and never attributed quotations to verifiable sources. One scholar who knew both Frank and Einstein reported that the book "is known to be based largely on epistolary correspondence," but he in turn did not attribute this claim to an actual source.[123] There is no way to uncover these letters, assuming they existed. None remain in Einstein's rather carefully curated archive, and Frank's own papers were inadvertently destroyed by a relative after his death. As Frank told Thomas Kuhn in 1962: "Some of the Einstein letters I brought from Prague, but most letters I lost, really, on the way."[124] All we have is the text he left, composed in the wake of a tumultuous 20 years battling for the philosophy of science in Prague.

Frank's biography of Einstein remains an engagingly readable book and finds a new audience in every generation. Some sections had greater impact on Einstein's legacy than others. Frank's extended discussion of relativity physics from a logical positivist perspective was for decades the dominant way this theory was understood in the Anglophone world. But the part of the book that has been carried over, almost unchanged, into the lion's share of later biographies is Chapter 4, "Einstein in Prague." Frank's position as Einstein's immediate successor in that city granted him absolute credibility in narrating this period, almost as though he had been an eyewitness to what had happened. An archival manuscript detailing the history of Einstein's time in Prague by local philosopher Emil Utitz, for example, consists almost entirely of recountings of Frank's stories.[125] Uncovering Einstein's life in Prague requires digging behind Frank's accounts—some real, some apocryphal, and some seemingly projected from Frank's own decades of experience at the German University to Einstein's three semesters there—to find different narratives and follow alternative pathways through history.[126]

CHAPTER 5

The Hidden Kepler

In the year 1608 there was a heated quarrel between the Emperor Rudolph and his brother, the Archduke Matthias. Their actions universally recalled precedents found in Bohemian history. Stimulated by the widespread public interest, I turned my attention to reading about Bohemia, and came upon the story of the heroine Libussa, renowned for her skill in magic. It happened one night that after watching the stars and the moon, I went to bed and fell into a very deep sleep. In my sleep I seemed to be reading a book brought from the fair. Its contents were as follows.

—*Johannes Kepler*[1]

These words open the *Somnium* (*The Dream*), often considered one of the first works of science fiction, by Johannes Kepler (1571–1630), considered even more often as one of the crucial figures in the development of Copernican astronomy. Kepler worked on the manuscript—published posthumously by his son Ludwig in 1634—throughout his life, but the most thorough revisions took place while he was the imperial astronomer at the Habsburg capital of the Holy Roman Empire: Prague.[2] It is a brief, fanciful tale. In his dream, the narrator meets an interlocutor, an Icelandic boy named Duracotus, who has been educated by the Danish nobleman astronomer Tycho Brahe (1546–1601). Duracotus's witch mother, Fiolxhilde, puts Duracotus in touch with demons who present him with a description of what the universe looks like from the point of view of Levania, known to us as the Moon. The narrator wakes up just as Duracotus gets to the creatures that inhabit it. While the *Somnium* begins in Prague with a book, a book that interacts with local mythic lore to spawn a story that grows ever larger and more impressive, it ends with a disappointing jolt. There is no more fitting place to begin this chapter, which is about a book from Prague, a dream of Kepler, and an effort to uncover a world hidden in plain view. All three became entangled with Einstein in unexpected ways.

The book in question is by Max Brod, whom we encountered earlier with Einstein at Bertha Fanta's salon, and was published in 1916. Entitled *Tycho Brahe's Path to God* (*Tycho Brahes Weg zu Gott*), the historical novel concentrates on the last year in the life of the imperial astronomer in Prague, Tycho himself—finally settled after scrambling for patronage following the ignominious collapse of his fortunes in Scandinavia—and in particular his chaotic family environment in the suburban palace of Benátky nad Jizerou (Benatek in German, the language Brod used) and especially his newly arrived, mercurial assistant, Johannes Kepler. Strange as it might seem, this novel has been treated by generations of historians and physicists as one of the most important sources about Albert Einstein's life in Prague. This is, as we shall see, a total misunderstanding, but no less influential for being erroneous. Tracing the origins of this error and uncovering the historical story of *Tycho Brahes Weg zu Gott* will reveal more about early-twentieth-century Prague—and, as it happens, about Einstein and his world—than the mythology that has encrusted it.

It is frequently impossible to trace the genesis of historical fantasies, but in this instance we can locate it precisely: Philipp Frank's 1947 biography of Albert Einstein. It stands to reason that given Frank's extensive personal contact with his subject and those who had known him at the German University, his rendering of Einstein's Bohemian period would be taken as especially authoritative. Surprisingly, in addition to some personal anecdotes and rumors that he relates—here, as always in Frank's *Einstein*, without any attribution of sources—the biographer fills a sizable portion of his account of this period with lengthy extracts from Brod's novel. As he justified this choice:

It was often asserted in Prague that in his portrayal of Kepler, Brod was greatly influenced by the impression that Einstein's personality had

made on him. Whether Brod did this consciously or [half] unconsciously, it is certain that the figure of Kepler is so vividly portrayed
that readers of the book who knew Einstein well recognized him as
Kepler. When the famous German chemist W. Nernst read this
novel, he said to Einstein: "You are this man Kepler." ["*Dieser Kepler,
das sind Sie.*"][3]

While it is true that Einstein's relationship with Walther Nernst was at
times fraught—Einstein's Hungarian collaborator Cornelius Lanczos
characterized it as a "curious love-hate relationship"[4]—it seems that both
eminent scientists remained cordial, and no one besides Frank seems to
have recorded any mention of the novel by Nernst. Shortly after the physical chemist's death in late 1941, Einstein wrote a heartfelt but equivocal
valediction for his former colleague in the *Scientific Monthly*: "Although
sometimes good-naturedly smiling at his childlike vanity and self-
complacency, we all had for him not only a sincere admiration, but also
a personal affection. . . . He was an original personality; I have never met
any one who resembled him in any essential way."[5]

   Einstein did not feel the same way about the author of *Tycho*. Not
much of a reader of fiction, he did—upon the encouragement of his close
friends the physicist Max Born and his wife Hedwig—read the novel. His
judgment of the work was mostly positive: "I have read the book with
great interest. It is without doubt interestingly written by a man who
knows the shoals of the human soul. As it happens I believe that I met
this man in Prague." His judgment of the author was a bit harsher: "He
apparently belongs to a small circle there that was contaminated with philosophy and Zionism, loosely grouped around the University philosophers, a medieval-seeming small flock of impractical people that you have
met through reading the book."[6] He did not mention the depiction of
Kepler and certainly did not suggest that he saw a representation of himself, however altered, on its pages.

   Nevertheless, with only two exceptions to my knowledge, every major
subsequent biography of Einstein has equated the physicist with the depiction of Kepler in *Tycho Brahe's Path to God* and then quoted Brod's
descriptions of Kepler as though they were straightforward reportage of

Einstein's character.[7] Interestingly, not a single pre-Frank biography makes any mention of the novel, although they occasionally invoke Kepler for other reasons.[8] Not only is this unreflecting quotation of Frank's quotation of Brod spurious historical reasoning, but Brod's Kepler is not an especially accurate representation of the Einstein we have seen in these pages.

As should by now be clear, there is no straightforward way to trace Einstein's experience in Prague, especially when it comes to the later ramifications and memories of that relatively brief period. Most often, the Prague period is overlooked as too short, too scientifically unimpressive, or too marginal to be discussed in detail. *Tycho Brahe's Path to God* is the exception that exemplifies this rule. When people today, following what they find in the biographies, want to learn about Einstein's character as a young man, one of the places they will likely turn to is a novel by Max Brod set in early-seventeenth-century Bohemia. The book, like Einstein's Prague period, has succumbed to the inevitable warpings and permutations that accompany the flow of the past into recorded history. When we slow things down, examine the origins of the book's association with Einstein, and then trace out the documentable relations of Einstein with its author, as well as the author's composition of this particular novel in the first place, we instead enter a completely different domain, one yoked to Einstein's life story by coincidence, but no less part of the narrative for that.

Brod's novel depicts the relationship between Tycho and Kepler, itself a salient episode in the long history of science in Prague. The major figure the book informs us about is its creator, Max Brod, who was during Einstein's time in Prague the center of a vibrant coterie of Germanophone Jewish writers but today remains largely unknown, or known only as a shadow. Brod was the friend, supporter, and later literary executor of Franz Kafka—he published the latter's works posthumously (despite the author's demand that the manuscripts of *The Trial*, *The Castle*, and other texts be burned)—and Kafka's reputation has largely left Brod in the shade. The conventional interpretation of *Tycho Brahe's Path to God* further eclipses this figure by foregrounding Einstein (and Kepler) rather than the man who wrote it. What, however, might we learn about this novel and its relationship with Einstein if we start not from Frank's quite possibly apocryphal invocation of Nernst, but instead with Brod?

Brod enjoyed the blessing and the curse of outliving most of his generation—he died in December 1968, a few months after the Soviet-engineered invasion of his beloved Prague. He thus had the opportunity to read Frank's biography of Einstein, including its interpretation of the astronomical novel he had published in the midst of the Great War. Brod had known Frank personally, having sat in on some of his physics lectures at the German University, indulging an interest in the science that was blossoming even during Einstein's time in Prague (he published on the philosophy of physics in 1913).[9] He continually, as we shall see, denied that there was any significant mimetic relationship between his fictional Kepler and the real Einstein. In fact, in 1956 he maintained that it was a mistake in historical fiction in general (and his in particular) to bypass the main figures and instead put too much weight on "the supporting cast, with whom (as for example with Kepler in my *Tycho Brahe*) at times much takes place that contrasts in the first order with the chiseled sculpture of the hero and does not so much serve as a representation of his opponent."[10] For Brod, the main character was Tycho; for us, it will be Brod. First, I will turn to Brod's place in the literary world of Prague, and then I will move to the real Tycho he researched and the fictional Tycho he depicted. These routes will, in the end, bring us back to Einstein, but not in the manner Frank indicated. By following the course of the novel's production and reception, we will perceive a Prague that coexisted alongside the Prague of Albert Einstein and that did much in later years to shape his legacy.

* * * * *

Max Brod was born on 27 May 1884 in the heart of Prague, and this connection with the city would form the single most important aspect of his life. His childhood was distinguished by both an excellent education and a severe affliction with kyphosis—excessive outward curvature of the spine, addressed by his mother through extraordinary treatments that prevented severe permanent deformation. Both distinctions accentuated his tendency toward intellectual pursuits.[11] From his youth he was drawn to a literary career and to languages: he developed a strong command of Czech alongside several other tongues, which he used frequently. (His Hebrew—an important marker of self-identification for Zionists in and around the Fanta group—apparently remained weak, with Hugo Bergmann complaining: "When even one who stands so close to us as Brod writes about Jewishness without reading Hebrew!")[12] His father wanted young Max to study law, and so that is what he did, enrolling at the German University in 1902 and earning a doctorate in the subject in 1907. He wrote literature the entire time, and he selected his career as a civil servant in the post office precisely so he would have more time for composition. (He and Kafka discussed this challenge of selecting appropriate day jobs at length.)

Though his commitment to literature would dominate his long life, he simultaneously lived rich personal and political lives (the two were often intertwined). In 1913, when he began composing *Tycho*, he married Elsa Taussig, the daughter of the major Prague merchant Eduard Taussig. They never divorced despite the fact that his turn toward the erotic in fiction was accompanied by personal infidelities; his treatment of Elsa was a source of tension in his circle. His public life was divided between Jewish politics and German newspapers. Guided by his friendship with both Hugo Bergmann and Martin Buber, he became increasingly active in Prague's Zionist movement, leading him to co-found the Jewish National Assembly (*Jüdischer Nationalrat*) on 22 October 1918—in parallel with the declaration of independent Czechoslovakia—and serve as an emissary to the new president, Tomáš Masaryk, to lobby, successfully, for the recognition of Jews as a nationality in the Czechoslovak constitution. (He did not visit Palestine until 1928, though Tel Aviv would in time become his second city after Prague.) In 1924 Brod quit the civil service to work as a music and theater critic for the *Prager Tagblatt*, the most important German-language paper in Prague, which after its 1875 creation came to be widely distributed throughout Austria-Hungary and after 1918 strongly identified with support for Masaryk's regime at a time when identity politics pushed against Germanophone cooperation with the state.[13]

The *Tagblatt* would form the major outlet for Brod's later critical writing, partially replacing literary journals, but it did not change the extraordinary level of his output. Bibliographies of his work list dozens of novels and biographies, hundreds of reviews and essays of criticism, poetry, drama, musical scores, translations, and more.[14] He was fêted by his colleagues and admirers throughout his life, resulting in not one but two memorial collections on separate significant birthdays extolling his impact on numerous quarters of cultural life.[15] His pivotal role among the so-called "Prague Circle" of writers—a role partially self-constructed and touted in his own historical accounts—enabled him to see that friends' works found the right publishers; this position of authority raised his stature while he was in Prague and then diminished it once he left.[16] His literary reputation started out high, but it was never unequivocal. In 1917 an assessment of the entire circle of Germanophone writers in Prague proclaimed that "Max Brod is perhaps not the most gifted of his colleagues, but he is certainly the most multifaceted."[17]

In the early years of the century, much under the influence of the philosophy of Arthur Schopenhauer, his works focused around heroes who embodied what he described as "indifferentism": a renunciation of ambition and a removed, ironic approach to everyday life (as exemplified by William Schurhaft of *A Czech Serving Girl*, discussed in chapter 3).[18] Already by the time of Einstein's arrival in Prague, Brod had transitioned out of the expressionist phase that had made him the toast of Berlin critics to pen several novels on Jewish themes and then make a decided turn to historical fiction, of which *Tycho* was the first full expression—though the book's saturation with Jewish motifs makes discriminating between these two literary periods difficult.[19] From *Tycho* onward, issues of mediation and distance—between "Czechs" and "Germans" or between the Jews and everyone else—began to dominate Brod's novels, culminating in his theorization of the idea of "*Distanzliebe*," or the kind of love he felt for German culture even though, as an increasingly Zionist writer, he believed himself to be outside of it.[20] Even recognizing the personal transformation of his aesthetic and philosophical ideas over the course of his productive career, it is hard not to concede that most of his novels have not commanded a persistent readership. The exception to this is *Tycho*

Brahe's Path to God, which some critics consider "perhaps his best work," though its popularity seems to have something to do with the presumption of its connection to Einstein.[21] This is surely the reason why it, alone among his many translated novels, remains in print in English.

The writing of *Tycho,* which began in 1913 while Brod was still at the post office, consumed him, as he devoted his afternoons and evenings to researching and producing his first historical novel and his longest work to date.[22] When the Great War broke out, the hunched back left by his childhood kyphosis kept him out of the conflict and allowed him to finish the book. *Tycho* was a new departure for Brod not only in style and scope but also in ambition, a decision he signaled by shifting from his previous publisher, Axel Juncker, to Kurt Wolff's house. "I don't want to give him"—Juncker—"*Tycho Brahe,*" he wrote to Wolff in June 1914. "I don't have another book, since I do not want to publish anything small or imperfect now. *Tycho Brahe* ought to be decisive and grant me victory."[23] Wolff gladly took the book and promoted it heavily, though with some trepidation that there would be no market for a work set among seventeenth-century astronomers in the midst of the European carnage. It was serialized in the prestigious monthly *Die weißen Blätter,* which published Kafka's novella *Metamorphosis* at the same time, and in complete form appeared in 1916, notwithstanding a copyright date of 1915 on the title page.

* * * * *

"With ever more urgent letters the great Tycho Brahe, as soon as he felt himself in a secure position at the Prague court of Rudolf II, invited the young astronomer Johann Kepler to visit him. The correspondence had been carrying on already for several years."[24] So begins the novel, with Kepler's journey in 1600 to visit the imperial astronomer Tycho Brahe planting the reader firmly in the context of early modern Prague. Though we meet Kepler first, it is evident from these lines—if the title has not already alerted us—that Tycho will be the agent who drives the narrative, even though he is a compromised protagonist, reduced to persistent supplication to bring a young apprentice to his doorstep. We soon learn that Tycho is looking for more than an assistant: he wants "a friend! A truly worthy comrade and brother!"[25] The character of Kepler is clear enough from the start, as he immediately dismisses astrology—an important duty of the imperial astronomer in Rudolf's occult-friendly court—to the two astronomers who pick him up in Prague to bring him to the palace at Benatek: Hagecius (the Latinized name of Tadeáš Hájek z Hájku) and Frans Gansneb Tengnagel.[26] Already Kepler is in conflict with the normal order of things.

That normal order is chaos, some of which Kepler exacerbates. He walks into the heart of an extended family—Tycho Brahe's wife and several children, the household dwarf Jeppo, and a retinue of astronomical assistants (principally Tengnagel and Longomontanus)—and disrupts it through the very contrast of his personality with that of his new master. Where Tycho, once we meet him, is voluble, impulsive, generous, defensive, apprehensive, and in general constantly reacting from moment to moment, Kepler is stasis personified. He is interested in understanding the mechanics of the heavens and persuaded that Copernicus's postulation of the sun as center of the universe is correct. This is bad news for Tycho, who wants his new apprentice to help defend his own cosmology, which features a stationary central Earth orbited by the Moon and the Sun, around which all the other planets circle in turn. Brod's Kepler is not moved by the hubbub around him or the wishes of his patron. He is, always, just himself:

In this gaunt man with the small, as though unripe, undeveloped little face lived an idiosyncratic insistence, an entirely simple direction of all his braced mental forces, that sealed him off entirely from everything external, made him inviolable, but also unable to absorb anything that did not concern his science. His entire gift and, corresponding with that, his entire passion was directed to only one goal, to the scientific comprehension of the world, as the next step of which he held before his eyes the elucidation of the laws of the heavens so exclusively that a friend once could remark: If there were from a certain moment onward no stars, there would also no longer be any Johannes Kepler.[27]

Kepler is oblivious: he "truly noticed nothing; he possessed a happy blindness to everything that distracted him from his scientific goals."[28] At times, this astronomer seems utterly inhuman: "That was Kepler. He had no heart. And thus indeed he had nothing to fear from the world. He had no feelings, no love. And thus he naturally was also safe from the aberrances of feeling."[29] These are precisely the characteristics, as Tycho explains to Tengnagel, that make Kepler so essential for the future of his science: "Kepler is indeed no longer a man, but a ghost. Kepler is nothing *outside* of us, as I now understand it, no, each of us has his own Kepler inside him and has to withstand against him, against his inner Kepler, the hardest test of his soul. . . . Kepler is both our Devil and our Savior, both in one, my Tengnagel."[30] These are precisely the kinds of passages that biographers recite as firsthand descriptions of Einstein in Prague, even though they are a rather poor characterization of the physicist.

There are many loops and intricacies to the plot—Tengnagel's affair with Tycho's daughter Elisabeth, the sad fate of the dwarf, a military-style siege that occurs when Tycho's sons' attempt to rout Tengnagel[31]— but the central arc of the novel involves the tension between Tycho and Kepler and the personal realization (the "path to God") that emerges for Tycho as a result. Roughly a third of the way through the book, a simmering dispute breaks out between Tycho and Nicolaus Reimers Baer (1551–1600), also known as "Ursus" via the Latinization of his last name, who served as the imperial astronomer before Rudolf enticed Tycho to Prague.[32] This controversy, which has been well treated by historians and was well chosen by Brod for the heart of his novel, began in 1588, when Ursus published a version of the geoheliocentric system that Tycho claimed (with some justification) had been plagiarized from his own notes while Ursus had visited Tycho's majestic observatory of Uraniborg on the island of Hven. Ursus's resulting fame led, in part, to his appointment as imperial astronomer in 1591. When Tycho learned of Ursus's claims, he denounced his rival as a thief and discredited him both in print and through his extended network of patrons, with the end result that Tycho displaced Ursus in Rudolf's court. Ursus responded by publishing *De astronomicis hypothesibus* (*On Astronomical Hypotheses*) in 1597, an astonishingly slanderous and vulgar work—including adulterous innuendos about Tycho's wife—so toxic that most copies were pulped through the efforts of Tycho and members of the court.[33] (Brod compressed the chronology and pushed it back so that Ursus's book appeared over two years later, tightening the story's drama.)

Considering that this was essentially an academic squabble, Brod handled the suspense fairly deftly. When reading Ursus's screed, it occurs to Tycho, through a suggestion from his daughter Elisabeth (who is trying to intercede for Tengnagel, who is at that moment in the doghouse), that "Kepler would be the most appropriate person to put Ursus in his place once and for all. Much more suited than the decent Hagecius, who engaged with science still only as an amateur. Kepler's basic style by contrast would rip out the weak counterarguments of his opponent by the roots."[34] The idea brings him solace and then almost immediately rage—for as Tycho reads Ursus's work (at almost the exact midpoint of the novel), he discovers that Kepler is also in it, and that Ursus has praised the young astronomer almost as much as he has degraded Tycho. The Dane is puzzled:

> Then he found Kepler's name again behind the frontispiece. This time especially boldly printed . . . "To the most famous mathematician Raymarus Ursus!" . . . This was a letter from Kepler. A forgery! No, such a clumsy audacity was not to be ascribed to Ursus. . . . It was an authentic Kepler letter. With his signature. And what a letter! Veneration, admiration, mastery . . . in each line a mess of praise. . . . Tycho wanted to scream; his tongue became very thick, it would not stir. His heart beat so fast and hurriedly that it had already hammered out an empty space in his chest . . . Kepler thus belonged to Tycho's worst enemy, that was the explanation.
>
> "Betrayal" . . . mumbled Tycho softly.[35]

Tycho is devastated, and he turns to this treacherous assistant on whom he had pinned his hopes and demands atonement: Kepler *must* write a decisive refutation of Ursus's libel, now more than ever in order to expiate Ursus's additional crime of including the letter without authorization. Even though Ursus dies a few chapters later, Tycho never relaxes his demand that Kepler not work on the orbit of Mars, or on any cosmological system, until he accomplishes this vitally important task. In the novel, Kepler's *Apologia Tychonis contra Ursum* is presented to Tycho on the latter's deathbed, although it was in fact completed in April 1601, a few months before Tycho's death. (It was not, however, published until 1858; it remains a masterly discussion of the history and philosophy of astronomical hypotheses.)[36] The novel's plot, in short, hinges on two divergent personalities negotiating the boundaries of loyalty and betrayal.

God does come into it. This struggle over earthly reputation and terrestrial friendship among scholars of the heavens serves as the psychological impetus for Tycho's reconciliation with his daughter, with Tengnagel, with Kepler, and with the Almighty. After many conflicts with Kepler, Tycho decides to forgive him and petition the emperor for his appointment as the next imperial astronomer. While in the royal anteroom, Tycho encounters the famous rabbi Judah Löw ben Bezalel—also known as the Maharal, and widely identified since the nineteenth century via an apocryphal attribution as the creator of the famous Golem of Prague—and the two begin conversing.[37] Upon seeing Löw, Tycho recalls his own fall from grace in Scandinavia and his wanderings that led him to finally settle in Prague, and Brod makes a direct connection with the fate of the Jews:

> And now the Jewish people truly appeared to him—as homeless and transitory as he was, always demonized as he was, in their teachings as misunderstood as he was and yet nonetheless holding firmly to them, as robbed and wounded as he was, this people of misfortune— formally as a symbol of his own way of life. He recalled that he already earlier once had compared himself with Ahasverus, the Wandering Jew.[38]

Tycho, a Danish Lutheran, in Brod's rendition stands next to Löw as the most "Jewish" character in the book. It is therefore fitting that he takes spiritual counsel from the religious sage. "God does not exist for the sake of the righteous person, to serve him and support him," Löw tells Tycho, advising him to reconcile with his family and his God, "but the righteous one exists to serve God and to support him."[39] At this point the reader might recall that the epigraph of *Tycho Brahe's Path to God* comes from

Genesis 32, which describes the battle of Jacob and the angel at Pniel, bringing the theme of exile and the connection of Löw's theology with the contemporary ideas of Martin Buber to the fore. Rather than being a departure from Brod's recent novels *Arnold Beer* and *Jüdinnen*, *Tycho* continued his exploration of Jewish identification.[40] The novel ends with Tycho's death on 24 October 1601. He has finally found his path to God: not through astronomy, but through reconciliation.

This was the novel that Philipp Frank presented as based on the historical Albert Einstein of 1911–1912 as transcribed by the pen of Max Brod from 1913 to 1915. Obviously, a lot more happens in the storyline than reflections on Kepler's ethereality. Most striking for a reader who comes to *Tycho Brahe's Path to God* looking for a *roman à clef* about Einstein in Brod's Kepler is that the novel is not really about Kepler at all: as the title indicates, Tycho Brahe dominates the narrative. Today, when telling the grand drama of the development of heliocentric astronomy, it seems obvious that Kepler should be the central figure, but this narrative did not come into its own until the publication of Arthur Koestler's immensely popular *The Sleepwalkers* in 1959, itself based on Max Caspar's detailed biography of Kepler from 1948.[41] None of Caspar's work was available to Brod when he crafted his novel—but he did have access to a host of sources on the historical Tycho and his final two years in Prague, sources that (despite the nitpicking of a few later historians at Brod's poetic license) he used quite thoroughly.

Chief among these was John Lewis Emil Dreyer's 1890 biography *Tycho Brahe: A Picture of Scientific Life and Work in the Sixteenth Century*, which came out in a corrected and expanded German translation in 1894.[42] By contrast, the only biography of Kepler available to Brod was an 1871 Latin account. Brod could have consulted it—he knew Latin well enough to translate some Catullus into German—but the novel hews so closely to Dreyer's outlines that it seems he did not.[43] We even know how Brod became aware of Dreyer's work: Gerhard Kowalewski, the mathematician who frequented the same Fanta gatherings as Einstein, was an enthusiast of the history of mathematics and mentioned Dreyer's biography to the novelist.[44] Numerous other late-nineteenth-century sources on Tycho (and more rarely on Kepler) would have been accessible to Brod in Prague.[45]

But even beyond the available sources, it was overdetermined that for someone in Brod's place and time, the hero of this story must be Tycho. An 1872 source on the Kepler–Tycho relationship explained this most clearly: "Tycho's name is still popular today in Prague, although he barely lived 2 years there; everyone knows his grave in the Týn Church, and Ferdinand's pleasure castle is shown to every newcomer as Tycho's observatory. But as for Kepler, who lived for 12 uninterrupted years in Prague and also later was in constant communication with it, the vast majority here scarcely know his name, and many fewer his relations with Bohemia."[46] The fact that Kepler could be identified as "German" while Tycho was neutrally "Danish," as well as the convenient symbol of Tycho's grave in the famous Utraquist church on the Old Town Square, no doubt cemented his importance for Czech nationalists.[47] (Even among Germans, however, the name "Johannes Kepler" was so obscure that one reviewer of Brod's novel consistently referred to him as "Keßler.")[48] Brod wrote within a relatively recent but robust tradition of historical novels about Tycho in Prague. In 1908, Julius Kraus's novel *Prag* featured Tycho as a character, and in 1916—the same year that *Tycho Brahe's Path to God* appeared—Auguste Hauschner published *Der Tod des Löwen* (*The Death of the Lion*), in which Tycho played a prominent role.[49]

The obsession with tying Brod's book's secondary character Kepler to Einstein has torn the work not only out of the rich history of astronomy in Prague, but from its immediate context of Germanophone writers. The problem stems from thinking of it as a book about *science*. The amount of astronomy in the novel is quite meager, and the dominant themes stand out as religious and interpersonal. Contemporary readers did not understand the novel to focus on science at all. Instead, it became the center of heated discussion as a work of literature and as a commentary on literary politics.

* * * * *

Obscure today except among literature specialists or Einstein biographers, *Tycho Brahe's Path to God* was one of the sensations of the season when it first appeared, sparking intensive commentary across the

Germanophone public sphere far beyond Bohemia, as well as the Czechophone sphere within the central city of Prague. From the start, the book sold exceptionally well, even given wartime conditions in Austria-Hungary and Germany, the two major markets. The first print run was 4,000 in 1915, which sold out quickly. The following year an additional 5,000 to 8,000 copies were sold, and by 1917 between 26,000 and 32,000 copies were in circulation. In 1920 the publisher, Kurt Wolff, counted more than 52,000 texts in print. Translations followed apace: Czech in 1917, English in 1928 (translated by Felix Warren Crosse under the misleading title *The Redemption of Tycho Brahe*), Italian in 1933, Hebrew in 1935, and Danish in 1950.[50]

Brod was ecstatic. He wrote breathlessly to Kurt Wolff in February 1916:

> *About my work*:—I believe that you are mistaken if you see no possibility for success for Tycho Brahe during the war. *I need no longer argue about it, because the facts have already proven me correct.* Unfortunately I have no direct news from Mr. Meyer; but I learn from his advertisement in the last issue of the "Jüdische Rundschau" that Tycho Brahe (the first and second thousand copies) have already sold out and are no longer in stores, that one must now already be referred to a reprinting which is announced as "appearing soon." The three largest booksellers in Prague tell me that the demand here is entirely extraordinary. There is already a dearth of copies, orders from the field can no longer be fulfilled, etc. . . . . [T]o date none of my books has found such a unanimously enthusiastic reception as this one. . . . Also all of my earlier works are unimportant to me set against Tycho Brahe. Only this book can achieve success for me and, what is more, an effect on mankind.[51]

(Ironically but characteristically, his metric for his international success appears to have relied on chatting with Prague locals.) Brod told Wolff that he was even getting unsolicited praise from people he did not know, including a young Berlin critic named Rudolf Kayser, who asked permission to publish an analysis of the book.[52] (Kayser, coincidentally, would marry Ilse Löwenthal within the decade, making him the son-in-law of

Albert Einstein.) As late as 1968, Brod remained proud of this novel: "I stand firm (I am so vain) to my *Tycho Brahe* and do not hold it against anyone if he praises this book. Even if he knows nothing else by me besides *Tycho*, written more than half a century ago."[53]

The response of the contemporary German-language literary press justified Brod's excitement. For the most part, the reviews were glowing, although it must be stated for the record that several of the early reviews were written by his close friends, who were deeply embedded in the Prague milieu that birthed the novel. Brod's Zionist colleague and intimate Hugo Bergmann, for example, reviewed the book in a Jewish journal immediately upon its appearance, stating that "Tycho's path to God is like the staggering of a drunkard, he hurtles from one eruption to another," and emphasizing the symbolic significance of the epigraph about Jacob wrestling with the angel.[54] Lifelong comrade Felix Weltsch in 1917 recognized the important theological awakening that strikes Tycho but continued to analyze the secondary characters as well. "Kepler will indeed go further on his path to God," he wrote, "he will meanwhile still find his great 'laws'; but he will never attain this nearness to God in which Tycho spent the happy unraveling of his entire moral journey."[55] Likewise, Otto Pick, part of Brod's close "Prague Circle," offered a detailed and intelligent reading of the work in the influential *Die neue Rundschau*, combining his analysis with a literary biography of Brod to date, a portrait of the artist as a thirtysomething growing out of indifferentism.[56] Many reviews from those further removed from Brod's immediate milieu likewise praised the work.[57]

There were, to be sure, also negative evaluations. Jesa d'Ouckh panned the novel in *Das Reich* in 1916–1917, charging that Brod was simply incapable of persuasively presenting psychological states. This had the ironic effect of making Kepler a more compelling character than Tycho, precisely because Brod did not try to supply him with any rich interiority. "These are not the artistic qualities that make this book worthy of notice," d'Ouckh continued. "On the contrary: through an expansiveness of psychological foundation and development of character, bordering on artistic tactlessness, through a widely distributed exhibiting of side characters and their fates distant from the guiding idea, the reader must work his way through painfully mediocre language to the essential."[58] Wolfgang Schumann the year before had likewise found little to praise in the prose: "Peculiarities and disharmonies of language, inner turmoil of the colorful story are the consequences; I read the book with excited involvement, but it soon faded in memory, since this noble metal is only half forged and hurriedly so."[59]

These criticisms were overwhelmed by the positive assessments in the Germanophone press and dwarfed in their acidity by the largely negative evaluation that Czech-language authors gave to both the German original and its Czech translation of 1917 by Adolf Wenig, available at a price of 4 crowns and 29 heller. The speed with which Wenig's translation came out, preceding as it did any other rendition by a decade, is not especially surprising given Brod's close connection with and patronage of Czech literary circles. The Czech edition featured a lengthy foreword by Wenig that articulated the significance of both the author and the book for Czech readers—including a specific connection of *Tycho* with Brod's novella *Ein tschechisches Dienstmädchen*, which indicates this was the most likely touchstone for readers in Czech—and included a substantial statement by Brod himself about his motivation for writing the novel. Brod carefully framed the work in a Bohemian context, stating that he had long been obsessed with two Prague stories: the Tycho narrative, and the Goetheborg episode in the life of Bedřich Smetana, which had led to his creation of Czech nationalists' talismanic opera, *The Bartered Bride*.[60] Despite the amount of research we know that Brod put into *Tycho*, in his letter to Wenig he downplayed his erudition:

> Of all the symbols of spiritual oppression which had impressed me in the various epochs of life, the symbol of Tycho Brahe finally gained the greatest power over me and called for representation.—Here I can now actually reveal a surprising fact: I built the entire core of the Tycho novel only from my superficial schoolboy knowledge, I created in my own mind the character and life course of Tycho Brahe before I went into the literature. And now my great amazement came! When I was so distant and saw before me in crude outlines the course of the history I wanted to create—when I finally resorted to the scholarly books

which afterwards proclaimed the events, environment, and biography of its hero—I gathered that actually everything agreed down to the least remarkable details with the image that I had created without any kind of real foundation other than intuitive feeling.[61]

Brod clearly wanted to emphasize his intuitive genius, the way that he (and the local legends he had been raised on as a child) contained an artistic truth that was deeper than historical research and could only be buttressed, not supplanted, by the latter. "Even today, three years after finishing the novel," he continued, "it happens that I got hold of a collection of Tycho Brahe's letters which I did not know about, and in them I found expressions and views on certain subjects in exactly the manner in which I had composed them."[62]

If Brod believed that these confessions would endear his novel to Czech critics, he was mistaken. Most of their reviews were negative. Perhaps the most salient critique—predictably, given the intensification of the separatist movements within Bohemia during the war that would soon culminate in independence—was a nakedly nationalist one that appeared in the journal *Zvon* in 1918. The main flaw of the work, the author K. M. argued, did not have to do with its major character Tycho Brahe or even his foil Johannes Kepler but rather concerned a minor character who appeared in the first scene and doddered onstage periodically: referred to always as Hagecius in the novel, the reviewer called him Hájek, and we have already met him. Observing that Dreyer's biography of Tycho was the main source that Brod had followed closely and that this work included much laudatory material on the close intellectual friendship between Hájek and Tycho, K. M. fumed that Brod had manipulated Dreyer's account to the disfavor of a historical personage solely with this character—"the *only Czech* who plays a large role in the novel! If one were to compare it with the truth—so be it!" He went on:

> There were at the Rudolfine court enough vile schemers that a novelist could utilize. But he did not even settle on an inaccurate misrepresentation at large; when the matter came to a Czech he settled in the end on the figure of a completely extraordinary character, on the Czech scientist Tadeáš Hájek z Hájku! The personality of this great

scholar and glorious Czech is little known today and is consistently undervalued; still, however, from the sources it is clear that he was not only the greatest Czech scientist of his era but also the most ingenious of all Czech astronomers of past centuries. He was known as a first-rate scholar to all contemporary scientists, he was forward-looking, he crushed superstition and mistaken prejudice when it was a matter of prognostication from celestial phenomena, and also was thus accused of wizardry; and he was persecuted because he worked on the scientific improvement of the calendar, so that he could remove senseless mistakes from it; he lived with Brahe in equitable friendship for twenty-five years but surpassed him in progressive views, adopting the Copernican system that Brahe rejected.[63]

Instead of the real scientific hero, Brod generated "a monster of the court, a two-faced schemer, a double-dealer toward Brahe, a smooth courtier, a scientist of dismissive impertinence, a vapid pseudohumanistic aesthete, a harmful scoffer and hypocrite, a scientific amateur (!!—p. 110), a Catholic-clerical spy, a garrulous 'manikin'—in brief a person comical and without character in every situation. So here Max Brod, whose sympathy to Czechs is emphasized, simple-mindedly, dishonestly, plainly crookedly made from a glorious figure of Czech cultural history" a travesty.[64] This was not the only review to focus on Hájek in particular and the depiction of Bohemian astronomy in general.[65]

Other reviews were less acerbic in tone and engaged more thoughtfully with the implications of the whole work than had the German-language reviews. The most important literary critic writing in Czech, František Xaver Šalda, perceptively homed in on the Jewish themes in the novel, linking it to Gustav Meyrink's *Golem*, as did many other Czech reviewers. While he considered it "first a good, artistic work," he also noted that it was a "Prague novel," like Meyrink's, only in a very limited sense: "In Brod's work Prague is not a subject of interest, but background like theatrical scenery: it is the *exiled soul* which lost its direction, that unconscious confession in God's world, and contributes substantially to its suffering."[66] (Another review noted that most of the novel took place at the Benatek/Benátky palace; when Prague did appear, it was mostly as a force for the good, as in the scene with Rabbi Löw.[67]) Instead of advancing the nationalist vision put forward by the *Zvon* review, Šalda exonerated Brod of the main vice that reviewer had accused him of: "Here is the key to Tycho Brahe. It is not a *German* novel, it is a *Jewish* novel in its symbolic fateful power. . . . And this Jewish central essence is so beautiful in Brod's book, and at the end of the day it is rather close and comprehensible to serious Czechness."[68]

F. Marek's review in *Cesta* in 1919 concentrated even more closely on the plot, as opposed to symbolism, and oriented its discussion around the Tycho–Kepler conflict over what to do about the treacherous Ursus. Tycho's need for a defense against Ursus's slanderous work, Marek noted, made him even more dependent on Kepler than he had been when he summoned him to make use of his mathematical virtuosity. Likewise, his dependency on imperial patronage required Tycho to engage in astrological forecasting, in which he did not believe. He was a person compromised by his vanity and desire for high position, and this produced intellectual shortcomings: "The clearest thing that emerges is his faint-heartedness in the argument with Kepler: when his views touch on the composition of the heavens, and on neither one side nor the other is there anything certain—Brahe keeps to his system because Copernicus allegedly contradicts the Bible and does not want to offend the majesty of his emperor, where Kepler stands only for the truth."[69] At every stage, Tycho is blocked from achieving full realization as a character, unlike the secondary figures in the novel. Only a turn to relations with God can save such a character. If the events of the novel were to happen in our present, they would stimulate our sympathy.

* * * * *

Maybe they *were* happening in the readers' present, the Prague of the second decade of the twentieth century. As the later enthusiasm for Philipp Frank's invocation of the novel in his biography of Einstein indicates, and as was hinted in several of the reviews and commentaries about it, there was an abiding sense that this historical novel about astronomy in the age of Rudolf II was a *roman à clef*, a (perhaps thinly) veiled exposé of various actual people that could be understood if one decoded it correctly. The magnetic quality of Kepler's character—his seeming modernity, his abstraction from the world around him—as well as his increasing salience in the narratives of the "scientific revolution" that began to proliferate after the Second World War understandably trained attention on him. Who was Brod's Kepler? But what if, on the other hand, we postponed that question for a moment and instead tried to uncover the real personage who might be lying behind the villain who never actually appears in his own right: Ursus.

Answering this question unfolds one of the central interpersonal dramas in Austrian literature at the dawn of the twentieth century and along the way will also unscramble the Kepler riddle. To many of the Germanophone reviewers who knew Brod personally, the true identity of the scurrilous Ursus was easy enough to parse. Paul Adler stated it baldly: "This spirit of the *Fackel* is in Tycho de Brahe's world the pamphleteer 'Ursus.'"[70] When searching for someone in contemporary Austrian letters who printed slanders and provoked geniuses into a frenzy, there was only one place to look: the Viennese writer and critic Karl Kraus (no relation to the Prague philosopher Oskar Kraus), whose journal *Die Fackel* (*The Torch*) was a mainstay of fiery (and hilarious) mockery of the mores of the day and their main defenders.

Max Brod loathed Kraus.[71] The feud had begun in 1911—incidentally, during the first months of Einstein's professorship at the German University in Prague—in Berlin, where the two writers were engaged in a proxy war between feuding journals: Brod was involved with Frank Pfemfert's *Die Aktion* and Kraus patronized Herwath Walden's *Der Sturm*. Though fully aware of Kraus's barbed pen and superhuman ability to hold a grudge, Brod picked a fight with the critic 10 years his senior. Most writers would have avoided a tangle with Kraus or would have not responded to snipes from his direction, but Brod issued an ad hominem comment that compelled Kraus to engage in hostilities. Kraus retaliated in a punningly devastating salvo in the 8 July issue of *Die Fackel*:

The esteemed Max Brod in Prague is something different. He writes a polemical essay in order to establish why he considers the polemical essay to be a lower form of art. He remains correct. He complains that I call him "Mr." He is right. He cites bad jokes I have made; truly, they are bad when they are quoted by Brod. Because a word arrives through the air in which it is breathed, and in bad air even those of Shakespeare are snuffed out. Spirit spread on Brod's bread turns into lard [*Geist auf Brod geschmiert ist Schmalz*]. "The simplest similarity of words becomes for him a great experience," says this Brodcrumb and doesn't know how right he is.[72]

Things escalated from there. Two years later, Brod wearily recounted "that I once badly criticized a book by Karl Kraus in the *Literarisches Echo* and that since then he has poked fun at me at every opportunity."[73] But he would not stand down. Kraus stood for everything Brod did not (and vice versa). Where Brod represented a pluricentric Germanophone literature from provincial Prague, Kraus represented the haughtiness of Habsburg Vienna. Where Brod was an increasingly active Zionist, Kraus, though of Jewish birth, was an anti-Dreyfusard and (unbeknownst to the public at the time, including Brod) a recently baptized Catholic. "Thus I feel put off by such types of my race as Karl Kraus as the most repellent," Brod wrote to his friend Richard Dehmel in the same 1913 letter, "because I see embodied in them what has degraded my people for millennia. There exists thus an implacable enmity between me and Karl Kraus, of whose oh-so-clever and entirely unrepentant satires I also battled in a personal essay in the *Aktion*."[74] He refused Dehmel's subsequent offer to broker a peace, "because I see very well that between him and me hostilities must reign *because of principle*."[75] Kraus harbored identical feelings, threatening to pull his work from Kurt Wolff's publishing house on the grounds that Wolff was now printing work by Brod—"one of the most unfortunate hysterics that has ever tarnished me in love and hate."[76] Kraus died in 1936, but that did not stop Brod's sallies. Even in the late 1940s, he was still lobbing Janus-faced assessments at the Viennese critic: "Karl Kraus admittedly attacked and made fun of much evil and many bad authors, but without discrimination also did so of great poets like Heine, Werfel, Hofmannsthal, George, or Kafka."[77]

The duration of this antagonism indicates that its primary audience was not the deceased writer but others who were still alive, rather closer to (Brod's) home. Though Kraus was inextricably identified with Vienna in the popular mind, he actually traveled widely in search of talent. A large number of Prague authors published in the *Fackel*, and Kraus himself gave many readings in the city. Of the roughly 700 lectures Kraus gave in his lifetime, 95 (or about 13 percent) were in Bohemia, of which 57 were in Prague itself.[78] Kraus also patronized many Prague writers whom Brod considered his own clients and discoveries. The most famous of these triggered Brod's starkest statement of his animus toward Kraus and explains the 1911 onset of their conflict and its Berlin setting. As Brod described it in 1937, a year after Kraus's death: "Thus Karl Kraus shortly after the publication of *Weltfreund* came out effusively for Werfel (yes, that's how it was before the verdict flipped) in an article, using the moment to speak out against me."[79] Franz Werfel, one of the greatest German-language poets of his day, had famously been unveiled as a major talent when Brod read some of his protégé's poems from his forthcoming debut collection *Der Weltfreund* during an appearance in the German capital.[80] Werfel reciprocated Kraus's overtures of friendship, hopping from Brod's small pond to Kraus's larger one.[81] Betrayal by poaching one's star pupil? Yes, Kraus was Ursus.

This could mean only that Kepler was none other than Franz Werfel. (As with the abortive Kepler–Ursus relationship, Werfel and Kraus later fell out.)[82] Brod was obsessed with Werfel from the moment he learned of the younger poet's gifts. Brod's memoirs, penned 20 years after Werfel's 1945 death in exile in Beverly Hills, begin not with the story of his own childhood, but with Werfel. After recounting the course of their friendship, Brod added: "A few years later I took this remarkable relationship as the basic motivation in my novel *Tycho Brahe's Path to God*."[83] This was something Brod had been saying for years. In the same 1937 article where he described the conflict with Kraus over Werfel's soul, he confessed that "for me from the experiences around the discovery of the young Werfel, about a few of which I have only hinted at here, and from the conflicts that I then and later survived, there arose my book, that I consider my most painful and that many consider my best: *Tycho Brahe's Path to God*."[84] Those in the know in Prague, like Paul Adler, saw the novel as an explicit rebuke to Kraus not only personally, but aesthetically, in its abandonment of trivial barbs in favor of high-minded philosophy.[85]

Brod claimed in his memoirs that he came to realize how transparent the code was only later. He now recognized that "Werfel" and "Kepler" contain the same number of letters, the same vowels, and the same intonation. More than that, "Brod" begins with the same consonant cluster as the name of Kepler's patron "Brahe," and, he added, the unusual letter "x" in his first name was paralleled by the "y" in "Tycho."[86] (He might have added to this list the assonance between "Ursus" and "Kraus.")

Once one reads *Tycho* this way, the identifications seem overdetermined. The clues are there in the very dedication of the book: "To my friend *Franz Kafka*"—which happens to be one of the earliest mentions of the reluctant writer's name in print.[87] The novel is, at root, a book about loyalty and friendship; this is why Tycho insists that Kepler defend his honor and why he later intercedes for him with Rudolf II. Kafka understood such messages clearly: he had dedicated his first collection of stories, *Betrachtung* (*Contemplation*, 1913), to "M.B.," and Brod was delighted that he "was able to return thanks by dedicating my novel *Tycho Brahe's Redemption* to him."[88] Kafka understood the gesture well enough, but he was not necessarily happy about it. For one, it put him under an even stronger obligation to Brod, who had nurtured his literary ambitions for many years. "That you want to dedicate *Tycho* to me is the first happiness immediately touching me for a long time. Do you know what such a dedication means?" Kafka wrote Brod in February 1914. "That I (and even if it is only an illusion, some side rays of this illusion do in reality indeed warm me) will be lifted up and included in *Tycho*, so much more alive than I. How little will I encompass this story! But how I will favor it as my ostensible property! You are doing me an undeserved good turn, Max, as always."[89]

To his fiancée, Felice Bauer, he gave a different interpretation. "We (for the sake of certainty once again: Max and I) are no longer so close due to my fault. He does not feel it so much in his innocence," he wrote two months after his grateful acknowledgment to Brod, "and for example has also dedicated to me his new novel, *Tycho Brahes Weg zu Gott*, his most personal book, in fact a torturous self-tormenting story."[90] The aesthetic criticism is not too surprising. Kafka had already begun to diverge strongly from Brod a few years earlier; in his diary he was quite critical of the latter's 1911 novel *Jüdinnen*, a sentimental *Bildungsroman* that marked Brod's shift to Jewish themes.[91] Nonetheless, he assured Bauer a year later that "the new novel from Max is very dear to me."[92] Franz Werfel had been castigated for spurning Brod's affection; the other Franz would be the beneficiary. (All this has not prevented some from insisting that Kepler was actually Kafka.[93] In that case, the dedication would have been rather gauche.)

Given the documentary evidence supporting the identification of Kepler with Werfel, not to mention the literary resonances it gives to the historical novel, the persistence of the Einstein–Kepler association is somewhat remarkable. The most it has going for it, aside from what Frank records Nernst reputedly saying, is that the book concerns astronomers, and Einstein is one scientist whom we know Brod met. But Brod was familiar with many scientists—he was seemingly acquainted with most of the compact Germanophone intellectual community in Prague, in addition to the Czech-identified scientists in his circles—and he could have modeled a scientist character on any one of them as easily as (or more easily than) the violin-playing physicist he had met on half a dozen occasions.

For his part, Brod was horrified by Frank's claim that the Kepler character in *Tycho* was a thinly concealed portrait of Einstein. That allegation was made in 1947, when Einstein was still alive, and Brod would work before and after the physicist's death in 1955 to dispel the association. He devoted several pages to the question in *Streitbares Leben*, his memoir. First, he wanted to stress how much he admired Einstein specifically as a *well-rounded* individual, who enjoyed Kant and music as well as physics: "Einstein was open-minded to all stimulations, his paths of thought sometimes took completely surprising, in certain circumstances pro-Kantian, directions. In general one had the impression of standing opposite a great man without prejudices."[94] But to admit a connection between his Kepler and the real Einstein, he argued, would do violence to the characters in his novel. He began by observing how Einstein "would experimentally change his point of view in discussion, knowing how to position himself to test also the the opposing viewpoint, and would then consider the entire situation in a wholly new way from a changed angle. . . ."

He never locked himself in, he did not run off in a bunch of different directions and therefore, jokingly and with virtuosity, remained always securely and creatively on task." According to Brod, it was this "characteristic of his of scientific courage and always beginning anew I modeled in my novel about Tycho Brahe in the figure of my Kepler, while in Tycho Brahe himself I wanted to depict a more rigid scholar, insistent upon his system." (Of course, in the actual book Kepler is quite inflexible and Brahe more willing to compromise on hypotheses, but no matter.) Brod insisted that the "decent Professor Frank misunderstood" his intention "and later published somewhere that I had wanted to present in my Kepler a certain egocentric drive that was (supposedly) made evident in Einstein's person. I have never noted such an egocentric drive in Einstein; on the contrary, I always found him kind, helpful, and astonishingly open-minded. I also wrote about this to Einstein in America when Frank's interpretation appeared and excited a certain stir. Einstein accepted my explanation with a cheerful joke—and thus the matter was resolved between us."[95]

In the end, Brod reiterated that "my friend Werfel, as already hinted, had contributed many more essential and more painful points to the figure of Kepler than had Einstein, who was after all only a fleeting guest on the path of my fate."[96] Brod and Werfel had drifted far apart after their initial quarrels over Kraus. The former ended up in Tel Aviv, escaping World War II and the Holocaust to become a mainstay of the Germanophone émigré community in Israel. Werfel had headed west, settling in the United States with his glamorous wife Alma, who had previously been married to composer Gustav Mahler and architect Walter Gropius. He settled in California, but in a trip to New York in 1945, shortly before his death, he happened to meet the most famous Germanophone émigré of all: Albert Einstein.[97]

* * * * *

While Brod's engagement with Einstein is not recorded—at least not in any straightforward manner—in *Tycho Brahe's Path to God*, the book does capture the aftermath of an important conflict in Austrian letters. Yet because so many have wished to see Brod's relationship to Einstein encoded in the 1915 novel, they have missed the more obvious point: the two men who had once played music together in Bertha Fanta's salon on the Old Town Square maintained an epistolary intellectual exchange for decades after Einstein left the city. Brod's views about Einstein (and Einstein's about Brod) are to be found not in novelistic metaphors and allusions, but in a forthright philosophical exchange about the place of ideas and intellectuals in the modern world.

This conversation would play out in a different world, a world that had emerged over two decades after the publication of *Tycho*, after the foundation of the First Republic of Czechoslovakia in 1918, and after the rise of the National Socialist Party to power in Germany in 1933. In this new world, we find the two men confronting a Europe that had abandoned democracy and plunged into fascism. Albert Einstein had settled at the Institute for Advanced Study in Princeton, a quiet oasis in a quiet town from which he observed the distant European cataclysm. He was also exerting himself strenuously on behalf of displaced refugees, writing affidavits himself or connecting the unfortunate who appealed to him with those who could help. His reputation as a lifeline for asylum-seekers in the United States had spread broadly, and it was with this in mind that Max Brod wrote to the former Prague professor on 30 November 1938.

The date is telling: this was not a pleasant time to be in Prague as a German-speaking Jewish intellectual identified as a fervent supporter of the liberal regime of Tomáš Masaryk. (Masaryk had stepped down in 1935, succeeded by his deputy Edvard Beneš, and died two years later.) A month before Brod's letter to Einstein, the leaders of France and the United Kingdom had negotiated a pact in Munich with Adolf Hitler, ceding substantial territory around the rim of Czechoslovakia to the German Reich. The truncated remnants of what was now known as Czecho-Slovakia (or the Second Republic) shifted under the new leadership of President Emil Hácha toward closer alignment with its National Socialist neighbor, which had incorporated Austria into its bulk earlier that year. Czech-identified residents of Prague spurned German-speakers, German-identified residents anticipated future dominance, and both sides lashed out at Jews. For the first time, Max Brod felt vulnerable in the only home he had ever known.

"My position here becomes more intolerable from day to day. I can no longer write what I think. . . . I also feel myself immediately threatened," he told Einstein. "So, e.g., yesterday the *Völkischer Beobachter*"—the newspaper of the Nazi Party—"carried a large open attack on me, with a photograph. The occasion was offered by certain erotic passages from my youthful works written decades ago. But the *Völkischer Beobachter* is now in Prague one of the most widely circulated newspapers and the reflexes will not fail to materialize." Perhaps surprisingly for someone who had been considered one of the most visible Jewish nationalists in Prague and who had repeatedly advocated on behalf of Zionist causes, Brod was unwilling to depart for Palestine, where many of his close friends from the Fanta circle had already emigrated. "I have decided to emigrate to America while there is still time," he informed Einstein.[98] He had enough money to establish himself; what he needed was a visitor visa. He asked Einstein's help to secure an invitation from an American university so he could immigrate above the U.S. government's meager quota. Brod insisted that he was a good catch: he could lecture on a large number of topics, including Czech politics, Czech music (especially Janáček), Jewish matters, and his own work. Besides, he declared, "I would bring all of the still unpublished manuscripts of Franz Kafka with me, edit them there, and establish a Kafka archive."[99] (One can only imagine the alternative future for American-based Kafka scholars.) These would be the only Kafka manuscripts to survive, with Camill Hoffmann unable to save those in Berlin from Gestapo destruction.[100]

I found no record of a response from Einstein, or from the other influential Americans Brod must have contacted. On the night of 14–15 March 1939, with the Kafka manuscripts tucked in a briefcase, Max Brod fled Prague as German forces invaded the Second Republic. He was on the last train to leave the city before the Nazi occupation, and he was able to see German troop movements from the window of his carriage as he hurtled toward the Black Sea. There he boarded a Romanian ship to travel through the Dardenelles and arrived in Tel Aviv.[101] He was finally in

Palestine, a member of the Jewish settlement of the Yishuv, together with other colleagues from Prague like Hugo Bergmann. For all his professed Zionism, however, he was not content. In June 1940 he again wrote to Einstein—on the stationery of Habima, the Yishuv's theater company, where he worked as a dramaturg—to reactivate an invitation that the famous author (and now Princeton resident) Thomas Mann had arranged for him from Hebrew Union College in Cincinnati.[102] Again, there is no record of a response from Einstein.

The nine years following the Munich pact constituted a period of intense suffering for Brod, and the prodigiously prolific writer went silent. His wife of 30 years died in 1942 in Tel Aviv. He was shocked by the suicide of his friend, the writer Stefan Zweig—who had penned a foreword for a reissue of *Tycho Brahe's Path to God*—that same year. In 1944 he learned that his brother and collaborator Otto, who had stayed behind in Europe, had been murdered in Auschwitz along with his family. His close Prague friends Oskar Baum, Ludwig Winder, and Franz Werfel— estranged though they might be—were sick or in exile, and soon many were dead. The only writing he committed to paper in this period was an unpublished Hebrew drama titled *Shaul*. (Brod never published a major work in Hebrew.)[103] His first significant publication after the war was the two-volume philosophical work *Diesseits und Jenseits* (*This Life and What Lies Beyond*), in which he engaged heavily with Philipp Frank's theories of causality and repeatedly discussed Einstein's controversial position on quantum theory: "There is thus, taken strictly, no unified modern physics today at all. But rather there are two physicses, in which one (Einstein) holds to the causal explanation of the world in the midst of all revolutionariness, while the other, the theory of discontinuity and of essential indeterminacy, pretends to no longer think causally."[104]

After the foundation of the state of Israel in 1948, however, Brod apparently found his fictional muse again. That year he published a new novel, *Galilei in Gefangenschaft* (*Galilei in Captivity*). It was the first novel he ever wrote that was not set principally in Prague.[105] "For me Prague is my home," he had written to a Czech newspaper in 1930. "And I do not have any other real home. My family, as far as I can determine from its past history on my father's side, lived in Prague. And Prague is not only in some, but in *all* of my novels the main scene of the story."[106] Like Brod himself, his fiction had become homeless.

In an afterword to the novel, Brod proclaimed *Galilei* to be the conclusion of a trilogy of historical novels that he called "The Battle for Truth." *Galilei*'s immediate predecessor was *Rëubeni, Prince of the Jews* (1925), a chronicle of a historically documentable ostensible Jewish messiah in the Renaissance who hailed, in Brod's invented account, from the Prague ghetto. The work won the Czechoslovak State Prize in 1930—a rare honor (although two years earlier the laurels had gone to none other than Franz Werfel, for *Barbara*, the first German-language work to win).[107] A decade earlier than *Rëubeni* the first book in the trilogy had appeared: *Tycho Brahe's Path to God*.

Brod wanted the opinion of the world's most famous scientist on his new creation. He sent a copy to Einstein in June 1949. "I would be very much delighted if you were to like my *Galilei* as much as you did my *Tycho Brahe* in its day," he wrote. (Alas, I could not locate any trace of what Einstein said about that work to Brod.) "It seems to me personally an artistic and conceptual advance over my youthful work, it also happens that—unfortunately—its theme, freedom of thought, is today because of threatening totalitarian systems more timely than ever. If you found the time to read the book, your judgment would be very important to me."[108]

Galilei in Captivity is a sprawling work, longer than either *Tycho* and *Rëubeni* by a substantial margin, though it focuses on a very narrow period of time: the early 1630s, when the astronomer attempted to secure papal sanction for the heliocentric theories of Copernicus, despite the fact that he was proscribed from defending those theories by the terms of an Inquisitorial decree of 1616.[109] In many ways, the book intensified the approach to historical fiction first seen in *Tycho*. Galileo Galilei has not one foil, like Tycho's Kepler, but several: the sculptor Gian Lorenzo Bernini, the Jewish astronomer Simon Delmedigo (Galileo always refers to him with the non-Jewish moniker "Giuseppe"), and Galileo's daughter, the cloistered nun Maria Celeste. All of the significant characters in the book, including the pope, deliver verbose monologues about philosophical topics, often touching on theological concerns about this world and the next. Stylistically, though, the novel is a departure: chapters alternate between an omniscient third-person narrator and Galileo's increasingly frantic first-person voice. The climax of the book comes when the pope—angered that the strawman Simplicio in Galileo's *Dialogue Concerning the Two Chief World Systems* utters some of his own arguments against Copernicus—arranges to have Galileo censured and placed under house arrest. In the midst of these machinations, in the last quarter of the novel, Delmedigo interjects a fantastical plan, led by Jews and inspired by the legacy of Rëubeni, to help Galileo escape.[110] The whole bizarre episode seems a reflection of Brod's romanticization of Irgun and Haganah para-military operations in the British Mandate of Palestine. Thankfully for the credibility of the novel (and historical accuracy), Galileo declines to participate. He ends a broken man, but one who draws inspiration from his daughter's tranquility, while Delmedigo carries his master's manu-scripts to Amsterdam to continue the pursuit of truth. Most notable in Brod's account of Galileo's resignation is his adherence to Catholic faith and his acceptance of his punishment, a contrast to the arrogance and cravenness he oscillates between earlier in the text.

This time, Einstein's detailed and fairly critical response to Brod has been preserved in the archives. The letter of 4 July 1949 reads, in full:

> Finally the much announced books have arrived. I have already read about a third of *Galilei*. To me it is entirely incomprehensible how one can achieve such a deep look into the bustle of people that composes what one usually calls history. When it is a case of the distant past, the effect is indeed all the more improbable and more senseless.
>
> In what pertains to Galilei himself, I had in fact imagined him rather differently. One can have no doubt that he was eager to know the truth in a manner that was rare. But it seems to me difficult to think that he as a mature person found it worth the effort to attempt, against so much opposition, to assimilate the truth he found to the conscious-ness of the superficial and petty interests of the surrounding masses. Was this effort really so important to him that he wasted his last years for it? His coerced recantation was indeed actually unimportant, because Galilei's arguments were after all accessible to anyone

seeking knowledge, and every somewhat initiated person must have known that the official recantation could only have been coerced.

Though it does not suit my view of the contrarian inner independence of the old Galilei if he really went without urgent necessity to the lion's den of Rome in order to scuffle there with clerics and other politicians. Certainly I cannot imagine that I would have undertaken such a thing in defense of relativity theory. I would think: the truth is incomparably stronger than I, and it seems to me laughable and quixotic to want to defend it with a sword and Rocinante. It will also be difficult for me to believe that Galilei internally heeded Catholicism and the Church. But you will of course know why you presented him that way.

The description of the milieu made a big impression on me. It must take extraordinary energy to form out of sparse information the bustle of people in so lively and convincing a manner.[111]

Most telling about this reaction is how little it was actually about the novel on its own terms, and how much it represented Einstein interpreting his own experiences with hostility to relativity theory (addressed in the last chapter) and his relationship to religion (coming in the next). When Einstein claimed he imagined Galileo differently, what he really meant was that he imagined himself differently; he referenced nothing specific about Galileo that Brod had misrepresented and in fact praised the vividness of the historical reconstruction.

In the meantime, *Galilei in Captivity* sparked another controversy in Israel. In 1948, the year of independence, Brod was awarded the highly prestigious Bialik Prize for Literature from the city of Tel Aviv for this novel, written in German and published in Winterthur, in the canton of Zurich. The prize citation read: "The book is saturated with genuine Jewish spirit and the eternal ideals of the people of Israel. From that perspective, the book possesses the highest educational and public value for our generation."[112] On both aesthetic and especially political grounds—the recency of the Holocaust, the tone-deafness of the timing, Brod's discomfort with the Hebrew language for literature (though he used it orally)—a legal case was brought against the city arguing that this novel was not "Hebrew Literature." (The Hebrew translation by Dov Sadan was also attacked.) Brod was scarred by the ferocity of the reaction, and he omitted any mention of the Bialik Prize in his memoirs.

In the wake of that debacle, getting a response from Einstein was a welcome balm. Brod both expressed a desire that Einstein read on to the end—as far as we know, he never did—and defended his historical research: "But he was really the way I have presented him. His letters, which we have in a beautiful edition in the university library here, testify to this."[113] He continued to try to elicit Einstein's approval. Later that year he sent the physicist a musical composition he had written, referencing in his cover letter the evenings at Bertha Fanta's.[114] He also implored Einstein in the strongest terms to read his 1952 novel *Der Meister* (*The Master*), a fictionalization of the life of Jesus. He especially wished for a public endorsement: "All the more valuable for me would be a supportive word now about the Jewish issues also from your side concerning this, as I believe, important, indeed earth-shaking, situation."[115] It appears that Einstein's engagement with the fiction of Brod, who would outlive him by more than a decade, was over. The writer died on 20 December 1968 in Tel Aviv. According to legend, the Prague telephone book from 1938 was found on his desk.[116]

* * * * *

Although *Tycho Brahe's Path to God* is not really about Prague, being largely set in the suburban palace of Benatek, or about Einstein, who does not really lurk—despite repeated assurances to the contrary—lightly veiled behind the character of Johannes Kepler, a close examination of the book and its context reveals multiple points of contact that radiate from Einstein's time as a professor at the German University of Prague.

The Einstein–Kepler association has proven durable not only because of the persistent repetition of this misidentification, but also because it is patently *believable*. Kepler was a fascinatingly innovative thinker of his own era, one who moved from place to place looking for support for his radical ideas. Einstein, for one, returned to Kepler time and again in his reflections on the philosophy and history of science, taking several opportunities to express his thoughts on the astronomer in writing. He admired Kepler as one of "those few who cannot do otherwise than openly acknowledge their convictions on every subject," as he stated in the introduction to a volume of Kepler's translated letters—though here, too, he made the same assertion as he had with Brod's Galileo, maintaining that Kepler's "life work was possible only when he succeeded in freeing himself to a large extent from the spiritual tradition in which he was born."[117] (Kepler, even more than Galileo, was profoundly religious.)[118] In a 1951 letter to his school friend Maurice Solovine, Einstein even defended Kepler's belief in "a reasonable astrology," a view he considered "not at all that strange, because the suspicion of an animistic causal connection, as is characteristic almost everywhere among primitives, is not in itself unreasonable, and would be given up by the natural sciences only gradually under the pressure of systematically obtained results. Kepler's researches naturally contributed to this process, which was played out in his own spirit as a hard inner struggle."[119] Einstein understood himself as a beneficiary of a scientific worldview that he attributed to Kepler's internal conflict.

More distant observers were content to draw analogies between Einstein and Kepler based on little more than their shared commitment to mathematizing the universe, flattening the centuries and worldviews that separated these two Germanophone physicists who happened to share the experience of living for a time in Prague. This was enough for Einstein's former collaborator, Cornelius Lanczos, to see in his friend "a true disciple of Kepler."[120] This version resonates with Philipp Frank's identification of Brod's abstracted astronomical assistant with the creator of relativity theory, even if the novel's author had intended to describe the ethereal distractions of the flamboyant poet Franz Werfel.

Perhaps the strangest Einstein–Kepler comparison was made by a prisoner in Munich speaking to a journalist named Dietrich Eckart in 1924. Loquacious in his rage, and apropos of nothing at all, the prisoner began to pontificate about how "the physicist *Einstein*, whom the Jewish press lets themselves marvel at as a second Kepler, clarifies that he has nothing to do with Germanness." Shifting from the putative linkage between the two scientists to the question of identification, the prisoner scoffed at

Einstein's being lauded by Jewish writers while he distanced himself from them. "The custom of the Central Union of German Citizens of the Jewish Faith—only the religious community of the Jews, not however to also parade their nationality—[Einstein] finds 'insincere.' A white raven? No. Only a person who thinks his people are already over the hill and thus doesn't consider it necessary any more to pretend."[121] The prisoner in question was named Adolf Hitler. He and the anti-Semitic movement he fostered would in time come to shape, more than anything that preceded it, the question of how Einstein would, or would not, identify himself as Jewish. That, too, is a story that begins in Prague, at a time before either Einstein or Hitler had given much thought to Johannes Kepler.

CHAPTER 6

Out of Josefov

On these three constant elements of one's experience—
homeland, language, and tradition—one builds the feeling
of the individual's belonging to a community, one that is
broader than the originally given community of the family
and the elective community of friends. One feels oneself to
belong to those who have the same constant elements of
experience, and feels their totality on this level as one's
people.

—*Martin Buber*[1]

The myths surrounding Albert Einstein have acquired such solidity that
aspects of his character can seem ahistorical. The most prominent of these
is his genius: the brilliance (and quantity and diversity) of his contributions to physics are such that the popular image can make it seem like
he was always that way, that as a toddler he had the capacity for tensor
analysis, or that his scientific intuition was not enabled by years of
comprehensive, rigorous training. Other characteristics that are just as
inseparable from the Einsteins that grace posters and coffee mugs are
revealed upon inspection to have been matters of conscious choice and
self-fashioning. For example, the mismatched (or absent) socks or the
frazzled hair. When younger, Einstein—who had sported a moustache
since his teenage years—was more polished and, sartorially speaking,
rather conventional. As he grew older, he chose a look that resembled
what people would enjoy calling "bohemian," lower-case. Although it
evolved over time, this trademark seems timeless in retrospect, as though
it were always lurking behind a façade, waiting to burst through.

What, then, of Einstein's "Jewishness"? As with all the ascriptions of identity that we have seen in these pages, it is more complicated than it appears. For many who have thoughts about the matter—ranging from devoted followers of the laws of the Torah to rabid anti-Semites—being a Jew is a core matter of identity; it is not an outfit that one can don or discard at will, nor is it a sliding scale whereby one can opt for greater or lesser identification with Jewishness in different contexts or at different times. Einstein remains the modern world's most famous Jew. But should we think of him as "being a Jew"—something that inhered in him by virtue of descent, an ineradicable part of his identity—or should we see his Jewishness as an identification that was more variable, a label he defined in a panoply of ways and that he persistently refused to consider stable? This question is one that he himself confronted repeatedly, and that confrontation became sharply visible and charged during his unexpectedly brief tenure as a professor of physics in Prague. Prague and Jewishness would remain intertwined strands in almost all of his mutable positions about his and others' Jewishness, though the connection was so submerged that it was perhaps not even apparent to him.

The link between Prague and Jews should not be surprising; indeed, the city's association with Judaism is one of the most widely known aspects of its history outside the Czech Republic, due in no small part to the worldwide stature of Franz Kafka, the Bohemian capital's most famous writer, German-speaker, and Jew. Kafka himself had a complex relationship to the religion of Judaism and the cultural communities of Jews, although one that was significantly closer to everyday understandings of what it means to be "Jewish" than Einstein's.[2] In what follows, I again will leave off scare quotes in referring to Jewishness, as I have mostly done for "German." The central aim of this chapter requires that they be kept always in mind, however, as the label was alternately adopted by individuals or ascribed to them, and it rarely meant the same thing in any two instances. Following the links, both personal and public, between Einstein's Prague and Prague's Einstein will bring to the fore the way the interaction of man and city generated a conflation between what it meant to be a Jew *and* what it meant to be a German, a conflation that often subliminally structured many of Einstein's (and others') reactions to such crucial matters as Zionism and the Holocaust.

The topic of Prague's Jewishness, its Germanness, and the way these shaped Einstein's mature attitudes toward personal belief and nationalist identification comes so late in this book not because it is marginal but because it is central. My claim is not that Einstein definitively "became Jewish"—whatever one understands that to mean—in Prague, but rather that an examination of the repeated juxtaposition of Praguers, pacifism, and Palestine throughout Einstein's life offers a more substantial way to parse his complex views on Jewishness than has been generally recognized. In 1911–1912, the various factors operated largely independently in Einstein's life, but they did not shed their Bohemian roots until the scientist's death and became silently incorporated into the myth afterward.

* * * * *

The most salient Bohemian story, featuring prominently in almost every biography of Einstein, has to do with an oath. It begins in Zurich with a piece of paper and ends at the Presidial Chancellery of the Viceroyalty in Prague. We can do worse, as with so many of the important legends about Einstein, than to paraphrase the version that appears in Philipp Frank's 1947 biography of his predecessor at the German University.[3] This one happens to be verifiable through contemporary documents. When Einstein received his offer for the ordinary professorship, he was made aware that he would have to swear loyalty to the Habsburg emperor Franz Joseph, a condition for employment as a civil servant anywhere in Austria-Hungary. He also had to inform the Viennese bureaucrats about his religion.

The dilemma lay at the juncture of these two matters. Einstein did not then see himself as having religious beliefs, and he did not want to declare himself—certainly not on an official form—as having an allegiance to a religion. Einstein's reluctance to affiliate with organized religion dated back to before his teenage years. His parents identified as Jews and, though not devout, adhered to some of the outward customs of the tradition. As a boy, he became curious about going deeper into the rites, at least for a time. This is how the old Albert in 1949 briefly described the young Albert's rapidly evolving attitudes in his "Autobiographical Notes":

> As the first way out there was religion, which is implanted into every child by way of the traditional education-machine. Thus I came—despite the fact that I was the son of entirely irreligious (Jewish) parents—to a deep religiosity, which, however, found an abrupt ending at the age of 12. Through the reading of popular scientific books I soon reached the conviction that much in the stories of the Bible could not be true. The consequence was a positively fanatic [orgy of] freethinking coupled with the impression that youth is intentionally being deceived by the state through lies; it was a crushing impression.[4]

He chose to forego a bar mitzvah and—in the manner of teenagers—believed he had decisively turned away from the religion.[5] When he came of age and could fill out his own bureaucratic forms as a university student in Zurich, and later as a patent clerk in Bern and a professor back in Zurich, he declared himself to be "without religion," or, in the German term, *konfessionslos*.[6] Yet many of his friends in Zurich identified as Jews, and it has been reasonably argued that despite his civil status, Einstein did not disavow the same casual identification for himself.[7]

When it came to the Viennese paper-pushers, Einstein therefore had a ready answer concerning religion: he would remain *konfessionslos*. The emperor, however, did not believe that one could honestly swear an oath unless it was vouchsafed by a belief in a deity—measured, by proxy, through adherence to a religious confession. Einstein wanted the job, so a simple substitution was made: *konfessionslos*, his original attestation, was replaced by "*mosaisch*," the faith of Moses. And thus, so the tale runs, through the unsparing precision of Austrian bureaucracy Einstein became Jewish again. On 23 August 1911, the physicist donned a uniform to make it official, something he laughingly depicted to his good friend Heinrich Zangger back in Zurich: "Yesterday I swore my solemn oath of office at the Bohemian governor's in a most picturesque uniform, whereby I helped myself to my Jewish 'faith,' which I had assumed once more only for this end. It was a droll scene."[8]

If you are of a spiritual bent, the half-comedy of the whole thing might belie a deeper seriousness. Who cares if it took filling in a blank on a form to bring Einstein back to his Jewish identity? The Lord works in mysterious ways, and for some observers this moment would become of pivotal importance as Einstein engaged more publicly with his Jewish identification.[9] Perhaps, although contemporary evidence indicates that Einstein did not take this official oath-taking especially seriously.

Consider his deep friendship with Paul Ehrenfest, whom Einstein met in Prague in early 1912 when the underemployed Austrian physicist made a grand tour of Central European universities from his temporary roost in St. Petersburg. (His wife, Tatiana Afanas'eva, also a physicist, was Russian; they had met at Göttingen and married in 1904.) Einstein found Ehrenfest an ideal thinking partner and exerted considerable effort to arrange for a position for his friend. Ehrenfest did not make it easy: he and Tatiana were committed atheists, and Ehrenfest would not fib on an official form and swear an oath to a god he did not believe existed. Einstein was perplexed. "It *irritates* me right off that you have this quirk of being *konfessionslos*; let it go for the sake of your children," he wrote him in April 1912, as the possibility of an Austrian position was foreclosed because Ehrenfest would not make the same choice Einstein had. "After becoming a professor here, you could by the by again return to this curious hobby horse—you only need to do this for a short while."[10] When he recommended Ehrenfest to Max von Laue for a position, he warned that "he is rigid about remaining *konfessionslos*, and thus cannot be selected to a position in Austria, and probably also not in Germany."[11] The only place for the "fanatically *konfessionslos* (curious)" was Switzerland, which did not demand any such oath.[12] (Ehrenfest ended up in Leiden, in the Netherlands, in a post Einstein declined, and which also did not require declarations of religious belief.)[13]

The casualness with which Einstein suggested donning and doffing Jewishness when convenient suggests that his newfound faith at the Prague Chancellery was not especially sincere. And so it would prove when he returned to Zurich in 1912; again he registered as *konfessionslos*. It is also worth noting the reports that his two sons were baptized Orthodox Christian in Serbia in 1913, and that after he moved to Berlin the following year he considered sending the children to a Lutheran school.[14] (They stayed in Zurich, so the question was moot.) More relevant in terms of conviction was Einstein's variable but generally peevish behavior concerning official forms required by the Prussian bureaucracy. In the deposition for his divorce proceedings in 1918, for example, he listed himself as a "dissident," that is, a nonbeliever, yet the actual divorce decree again registers him as *mosaisch*. (Then again, the same form declares the former Mileva Einsteinová, now Marić again, as also Jewish, which was never true.)[15]

Even when remarried to Elsa Einstein—a marriage in which both parties were of Jewish heritage—Einstein made difficulties. As in most European states at the time, in Germany a portion of one's taxes was distributed to state-sponsored religious institutions according to the proportion of the population who registered as belonging to each. Registering as a member of a religion therefore had fiscal implications for churches and synagogues, and Einstein refused to comply. The first exchange over the issue was farcical: Albert and Elsa's marriage was apparently registered as Protestant. Elsa's daughter Ilse wrote to complain about the error—presumably at her mother and stepfather's insistence: "Prof. Einstein and his wife have never belonged to a confessional community, although they are children of Jewish parents. . . . Since Prof. Einstein and his wife have never belonged to the Protestant church, they are naturally not in a position to present you with a certification of their leaving the congregation."[16]

It was a different matter when the Jewish Community (*Gemeinde*) came calling in 1920, presumably a result of the marriage being now registered as Jewish. By this point, as we shall see later, Einstein was quite vocal about identifying as a Jew and especially with aspects of Zionism. As much as that may have signaled to others his identification with the Jewish religion (and thus the Community), he refused to be officially considered as anything but *konfessionslos*. "On careful consideration I cannot resolve to enter the Jewish religious community. As much as I feel myself a Jew," he wrote, "just as much do I confront traditional religious forms as a stranger. Nonetheless, in order to show that Jewish matters lie close to my heart, I am very ready to give annually a certain contribution to Jewish charity."[17] The Community responded that "every Jew is by force of law a member, liable for taxation, of the Jewish Community of the region in which he lives. . . . The Community is therefore not authorized to ignore your assessment according to this rule."[18] Einstein balked:

> I explain to you <once again> definitively that I do not intend to enter into the religious community, <and that I do not consider it necessary . . .>, but rather remain *konfessionslos* as I have been up to now.
>
> To your letter I remark that the word "Jew" is ambiguous, in that it refers 1.) to nationality and origins, 2.) to religious confession. I am a Jew in the first sense, not in the second.[19]

Despite repeated pleas from letter writers that his membership in the Community would assist with Zionist goals, or would improve the status of Jews, Einstein demurred: "The Community is an organization for the practice of ritual forms, which lies far from my views. I must take it as what it now is and not as what one might wish to see it perhaps transformed into."[20] Nonetheless, a biographer writes that he joined the Jewish Community of Berlin in 1924.[21]

If he was indeed so mercurial about the issue of being *konfessionslos*, and so cagey about being officially labeled a Jew—though he did not mind the label outside of official circles—what accounts for the staying power of the Prague story we began with? The answer hinges on the politics of two newspaper articles in Austria-Hungary and brings us back to the importance of Einstein's stay in Prague for cementing this aspect of his public image.

The first article appeared in the *Prager Tagblatt*—one of the most widely circulating non-Viennese papers in the empire—on 26 May 1912, entitled "Einstein in Prague." Inasmuch as this piece, signed with the initials R. K., has a hero, it is our physicist, though it also discusses the parallel case of Robert Raudnitz, a distinguished professor of pediatrics. It is more noticeable for its prominent villain: Count Karl von Stürgkh, minister of religion and education, whom we earlier saw fiddling with the order of the three-person list submitted by the German University's hiring committee, thus offering the open post to Gustav Jaumann of Brno rather than Albert Einstein of Zurich. R. K. began by mentioning an earlier article that had announced that Einstein was going to be leaving his position in Prague. "Einstein? This name seemed familiar to me. For years all the physics journals have been filled with works about Einstein's famous relativity principle, about his velocity principle, and about his new system of four-dimensional mathematics."[22] Why was he leaving? According to sources contacted by R. K., the major issue was that "he had to, as they relate in University circles, first undertake a change of religious identity before he was considered capable of teaching mathematical physics in Prague. Einstein had apparently been *konfessionslos* in Zurich and the Austrian educational administration" insisted that he declare himself Jewish.[23] This demand, R. K. alleged, was not a foible of the emperor but the result of the machinations of von Stürgkh, a creature of the Viennese Christian Socialists, which in this case produced the ironic outcome of defending religion by creating another Jew. It was a shame that Einstein was leaving the city, ostensibly because of the Habsburg establishment's hostility toward Jews, not because R. K. was a personal acquaintance of the scientist's—"I have never seen or spoken to Einstein, he never paraded out in public and obviously did not belong among those scholars who appear with advertisements in the papers"[24]—but because "with Albert Einstein Prague is losing one of the most interesting phenomena of contemporary natural science."[25]

Two months later, on 29 July 1912, a very similar article, anonymous and headlined "A. Einstein," appeared at the top of the front page of the *Montags-Revue* in Vienna. In terms of biographical and scientific accuracy, the piece paled in comparison to its predecessor in the Prague daily, but its point was very similar: "Austria is losing him; foreigners do not come gladly to Austria or flee again at the first opportunity."[26] The article continued:

But now to Einstein. Who is he? A small bloke from Württemberg. What has he accomplished? The relativity principle and quantum theory—which in fact Planck formulated, but Einstein co-founded it— are his monumental achievements up to now, whereby it only should be stressed that modern physics does not any longer understand

by "theory" hypotheses but on the contrary the accomplished firm
basis upon which it can journey further by reliable steps.[27]

The illness affecting Austrian higher education was again diagnosed as
von Stürgkh, whose hostility to the Jews—in contrast to his Prussian
equivalent Friedrich Althoff's efforts to expand official quotas in order
to attract stronger faculty—deprived Austria-Hungary of a leading cultural position.

Both articles are tendentious and polemical. They are also significant.
It is important to remember that this was not 1919, when the eclipse expedition to confirm general relativity catapulted Einstein to global su-
perstardom. This was seven years earlier, when general relativity did not
even exist and Einstein was a physicist who was well-regarded by his peers
for matters that the general public did not interest itself in. Had these
stories appeared after the eclipse, the deployment of Einstein as a rhetorical cudgel against a politician would have been humdrum, even banal.
This early, however, the tactic should direct our focus to von Stürgkh and
Habsburg politics with respect to the Jews. Einstein was the most recent
Jewish professor to leave the German University, and so his story was
weaponized. The incident had the effect, however, of elevating the *konfessionslos* anecdote into one of the defining features of Einstein's time in
Prague: that was where he was *made into a Jew*. As we have seen, this association was not necessarily Einstein's own understanding of events,
but the mythology was available for hagiographers and journalists to turn
to when they needed to find information about the suddenly celebrated
creator of relativity. Looking through their archives, they found this story,
and a narrative was created.

* * * * *

As Einstein wrote to Emil Starkenstein, an extraordinary professor of
pharmacology at the German University in Prague, in July 1921: "One can
however very well be *without* religion without being disloyal to one's
people (at least in my opinion). I am also *konfessionslos* and consider myself a loyal Jew."[28] Such statements show the physicist straddling his

resistance to one kind of identification as a Jew and his embrace of another. That second form of identification would manifest most spectacularly in Einstein's relationship to the Zionist movement, a complicated micro- and macropolitical dance that was characterized by reactions ranging from initial hostility to fellow-traveling affiliation to public distancing.[29] All three of those phases would be intimately related to people Einstein met initially in Prague. Understanding the historical position of Prague's Jewish residents—especially the German-speaking and -identifying ones—is thus important in appreciating the nuances of the antinationalist Einstein's association with Jewish nationalism.

Our earliest documentation of a Jewish community in Prague reaches back to at least the year 906. Almost two centuries later there was a massacre of Jews during the First Crusade. We also know there was a synagogue in the Lesser Town before the twelfth century, because it burned down in 1142, at which point it moved north of Old Town Square, to the region that would become the Prague ghetto.[30] Such are the records of the presence of Jews in the early centuries of the capital of Bohemia: they show up in chronicles only when disaster befalls them.

As conflicts and eventually wars among Christians washed across the territory, significantly depopulating it, the Jewish community grew as a total percentage of the city, such that by the early eighteenth century Jews represented a stunning proportion of the population, making up nearly half of the residents of Old Town and about 28 percent across all districts.[31] As Bohemia recovered from the wars of religion, Prague grew substantially, diluting the Jewish population, which continued to expand mostly in the countryside.[32] The various communities still labored under heavy constraints and regulations, including an order of expulsion from Empress Maria Theresa on 18 December 1744 (later rescinded), some of which were alleviated by her son Joseph II in his *Toleranzpatent* of 1782. A significant consequence of Joseph's decree was the requirement that Jews cease to use Hebrew or Yiddish except in religious ceremonies and instead convert their everyday transactions into German, which decades later would lead to the extinction of the Jewish dialect of *Mauscheldeutsch* and also cement the strongly misleading image in the eyes of Czech-identified Bohemians that all Jews were "German."[33]

Emancipation came eventually: freedom of movement in 1849, and legal equality in 1867 after the *Ausgleich* compromise that created the dual monarchy of Austria-Hungary out of the Habsburg Empire. The most illuminating context in which we can situate Einstein's own position as a German-speaker of Jewish heritage in Prague is that of the university (and, after 1882, universities), which saw a dramatic expansion of students identified as Jews. In the eighteenth century, Jews seeking medical degrees had migrated out of Bohemia for their education to Halle and Frankfurt an der Oder before moving back to practice. Yet by 1863, they comprised 10 percent of the student body of the University of Prague, rising to 11.7 percent on the eve of the 1882 split (and 17.9 percent of Prague's German Polytechnic). The split produced a significant asymmetry, a consequence of the historic pressure for German to be identified as the language of Jews and a perception of its greater potential to facilitate upward mobility. Although a significant proportion of Bohemia's Jews were bilingual, for educational purposes they were concentrated in the German University: in 1885 the 404 enrolled there comprised 26 percent of the institution's students; by contrast, the 50 students at the Czech University made up 2.5 percent of its population. By 1910, on the eve of Einstein's arrival, the proportions had declined in both schools: Jews were 20 percent of the students at the German University and 2 percent at the Czech, with similar proportions at the polytechnics.[34]

Given the significance of the universities for the creation of elites within the city, one can see how Jews came to be identified as "German" by those who saw themselves as "Czechs," while some self-identified Germans could at the same time refuse to recognize Jews as fully equivalent except insofar as they helped to shore up the German groups against Slavic inroads.[35] The focus on language politics in Bohemia as *the* central issue after the weakening of liberal compromises in the first years of the twentieth century coincided with the rise of political anti-Semitism among pan-Germanists in Austria-Hungary and the German Empire. The Jews of Prague, although rather more of them identified themselves on the census as Czech-speakers rather than German-speakers by 1910, comprised over half of the Germanophones, which exacerbated nativist currents and prompted explosions of anti-Semitic actions. The most visible of these began in 1899—parallel in many ways to the notorious Dreyfus Affair then raging in France—with the arrest and trial of Leopold Hilsner, a 23-year-old Jewish vagrant, for the alleged murder of young seamstress Anežka Hrůzová in the town of Polná. The initial trial was riddled with bias and perjury, which led to its dismissal and a second trial, resulting in Hilsner being sentenced to death in 1900. The emperor commuted the sentence to life imprisonment, while intellectuals such as Tomáš Masaryk protested the railroading. Hilsner was pardoned by the emperor in 1918 in the last months of the Great War, which also happened to be the last months of the Habsburg Empire itself.[36] This, then, was one manifestation of the difficult position of Jews between Czech nationalist activists and the Habsburg state.

In juxtaposition was the simultaneous romanticization of Prague's Jews at the very moment that the old ghetto, the Judenstadt—renamed Josephstadt in German and Josefov in Czech in 1852, in honor of the eighteenth-century emperor's Edict of Tolerance, which permitted Jews to live outside its confines—was being modernized out of existence. By 1890, only 20 percent of the population of Josefov remained Jewish, but the emigrants were more than replaced by the destitute, who squeezed into the ramshackle and crowded apartments. Population density in the Old Town was 644 persons per hectare, but in the neighborhoods of Josefov it was just under three times that (1,822 per hectare). Citing public health concerns, an urban renewal plan known as *Assanierung* in German and *asanace* in Czech was proposed in 1885. Six hundred individual buildings were torn down in its wake, removing essentially all structures except for the Jewish town hall, the main synagogues, and the cemetery, and replacing them with wide boulevards in the Parisian style.[37] (The main boulevard, still a thoroughfare today, is called Pařížská.) By the time Einstein arrived in the city, Josefov was gone, memorialized in Gustav Meyrink's gothic novel *The Golem* (1913–1915), set during the *asanace*. Einstein's engagement with Prague's Jews would happen outside of Josefov.

A contemporaneous event associated with Prague Jewry would have longer-term (though subterranean) implications for the physicist's relationship with the religion. From 1909 to early 1911, just before Einstein's arrival in the city, the Austrian philosopher Martin Buber gave three public lectures to Bar Kochba, the Jewish students' association of the German University, and those lectures—later published and widely circulated—further intensified the enthusiasm of local Jewish intellectuals for Jewish nationalism and a cultural Zionism focused on building a community in Palestine.[38] Max Brod, for one, noted the electrifying effect of the speeches in his diary, and a week after one of them, on 4 May 1910, he and Kafka went to see a Yiddish theater troupe that proved transformative for both of their identifications with aspects of their Jewish heritage.[39] Buber's ideas would not have triggered a new activism among the incipient Zionists in Prague if the ground had not already been prepared for it by Bar Kochba and its journal, *Selbstwehr* (*Self-Defense*). The members of Bar Kochba would have an outsized influence on Central European Zionism despite their small numbers—there were never more than 200 members in all the Zionist student organizations in the capital, both German and Czech, despite there being over twice that many Jewish students.[40] The group's significance stems partly from its relationship with Buber, but perhaps most importantly from the activism of one of its most visible leaders: Hugo Bergmann.

We encountered Bergmann a few chapters ago, as the husband of Bertha Fanta's daughter Else and the student in Einstein's physics class who introduced the scientist to the cultural circle that gathered in Fanta's apartment. It was fitting that Bergmann served as the cultural vector connecting the new professor who had just arrived from Zurich to the largely Jewish social group, as Bergmann's gift in both Prague and, after his emigration in 1920, in Palestine and later Israel was in bringing Jewish intellectuals together.

Bergmann was born in 1883 in Prague and attended school together with Kafka, who (according to Brod) used to copy his friend's homework.[41] Bergmann and Kafka entered the German University together thinking they would study chemistry, a field where there was more potential for advancement for Jews than could be found in other disciplines they were more interested in.[42] The science was not to their tastes, and Kafka departed for law while Bergmann opted for philosophy, studying both science and Franz Brentano's ideas with Anton Marty, Christian von

Ehrenfels, and occasionally the master himself when he would visit. After doing very well in his doctoral work—which focused on atomic theory and, at the time of its completion in 1905, was quite current with the very developments that interested Einstein during that same year—he looked for a university where he could write his habilitation. He hit barriers because of his religion.[43] Brentano urged Bergmann to convert to Christianity, not least for the sake of professional advancement, but Bergmann had been a convinced Zionist since 1898, joining Bar Kochba in 1901, and refused.[44] Instead, he abandoned his habilitation plans, took a job in the university library, and married Else Fanta in 1908; the two took their first trip to Palestine in 1910.

Bergmann continued writing philosophy, and writing it in German, though it is clear from the footnotes of his work that he spoke Czech and generously engaged with scholarship written in that language. This is quite noticeable in his 1909 study of the early-nineteenth-century Bohemian philosopher and mathematician Bernard Bolzano, in which Bergmann also delved into Bolzano's views on relative and absolute space and time and made references to non-Euclidean geometry.[45] (There was a reason he chose to audit Einstein's courses in 1911.) The book even garnered a respectful review from none other than Oskar Kraus on the front page of the German-language newspaper *Bohemia* in 1910.[46] Despite his multilingualism—he also studied Hebrew in Prague and considered Zionists who did not learn the language a contradiction in terms—he strongly identified with the cause of maintaining German cultural institutions in his native city. As he wrote in a letter to the philosopher and former Brentano student Carl Stumpf in 1914:

> My mother tongue is German, I attended only German schools, speak and think in German. These are in general the signs by which in this country one judges membership to the German people or the Czech people. By these criteria I am thus German as much as anybody else. I am this to an even greater degree than others because I studied at a German university and that I *only* studied with Germans (if I leave learning from books out of the accounting), which I could do in my discipline. The interest that my German teachers had in maintaining

and promoting German culture is thoroughly my own. What pertains especially to the relation to Czech culture, I consider the German to be generally greater by contrast, and to be the one which—under equivalent circumstances—deserves to be recommended and promoted. For this promotion of German culture particularly in Prague I contributed according to my powers and want to do this all the more everywhere there, where I do not harm Jewish interests by doing so.[47]

Especially interesting here is the final comment about the potential conflict between being German and being Jewish—but also the potential synergy between the two. This conflation of Germanness and Jewishness was characteristic of Einstein too, as we shall see.

Bergmann remained active in Bar Kochba long after his student days— he had invited Buber to give his landmark speeches—and wrote extensively about the settlement of Jews in Ottoman Palestine. Although committed to Jewish emigration to the region, he was not at all sanguine about the impact it would have on the local Arab Muslim and Christian populations, putting him outside the norm of public statements on this topic. In 1911, the same year he met Einstein, he wrote an important article demanding that Zionists address their relationship with their neighbors: "Of the 700,000 residents of Palestine there are approximately 600,000 Arabs. Should we not honestly ask which problems arise for our settlement movement from the fact that Palestine has had for generations over half a million people in it who occupy land, consider it their homeland, and—at least now—are in charge there, that they determine the character of the land?"[48] His solution, a characteristically German one, was to cast the Jews as bearers of civilization and cultivation (*Bildung*), a move that comes across with hindsight as condescending: "We come to Palestine as bearers of culture [*Kulturträger*]. The Arabs should learn from us."[49] The worry animated his political engagement and prepared the ground for him to assume a noteworthy, albeit soon marginalized, political position within the Yishuv.

He held on to the library post, except during his military service in the Great War, until 1919, when he and Else made the decision to emigrate.

(They later divorced.) In 1936, now known as Schmuel (Samuel) Bergman, with one less *n*, he became head of the library and a professor of philosophy at the Hebrew University in Jerusalem, and continued to serve as a social node for Max Brod, Felix Weltsch, and other refugees of the Prague Jewish community until his death in 1975. Along the way, he would become a periodic touchstone of Albert Einstein's engagement with Zionism.

* * * * *

As far as one can determine from correspondence and the recollections of his friends in Switzerland, Einstein had no serious encounters with Zionism until he arrived in Prague. The initial acquaintance did not leave an especially deep or positive impression. The Fanta circle exposed him to some of the leading Zionists in the city, but it seems the major reaction he had to that group—as we saw in the previous chapter in the case of Max Brod—was a vague distaste toward their nationalist politics. He was friendly with people like the archaeologist Wilhelm Klein, a baptized Jew who was shunned by most of the other Jews in Einstein's circle, including the physicist's close colleague and friend Georg Pick.[50] Yet we know that within a little over half a decade, Einstein would publicly come out in support of several Zionist goals, and in 1921 he would travel to the United States at the invitation of Chaim Weizmann (later the first president of the State of Israel) on a trip to raise money for various Zionist causes—principal among them a university in Palestine. How did this change come about?

The series of excellent studies of Einstein's relationship with Zionism frame this as a Berlin story, deeply grounded in Einstein's experiences as a pacifist during the Great War.[51] In June 1921, Einstein published a noteworthy article called "How I Became a Zionist" in which he described his transition from ignorance about the situation of the Jews to his present activism: "Until seven years ago I lived in Switzerland and as long as I was there I was not conscious of my Jewishness, and there was nothing in my life that could have stimulated my Jewish feelings and enlivened them. That changed as soon as I moved my residence to Berlin."[52] Noteworthy in this narrative is the complete absence of Bohemia.

The reminiscences of those who were crucial to enrolling him in Zionist causes echo this account. Kurt Blumenfeld, the Zionist activist who would play a central role in this effort, noted in his memoirs that when they met, he "only knew that [Einstein] was considered a renowned physicist. I also did not know that Max Brod and Hugo Bergmann in Prague, where he had worked as a professor of physics in the years before the war, had already tried in vain with him." He continued: "Einstein told me later that he had been entirely immersed in cosmic problems during his time in Prague. Questions of nationality and of the relations of Jews to their surroundings seemed to him then as laughable minutiae."[53] Hugo Bergmann did not even remember exerting himself: "I do not recall having ever discussed *Judaism* with Einstein during those years of 1910–1912, although I took an active part in Zionism. I do not think Einstein was interested in Judaism at that time. . . . It was only ten years later that Einstein, under the influence of the Zionist leader Kurt Blumenfeld, came closer to Judaism, and in particular to Zionism whose faithful servant he was to become later."[54] Bergmann would not have predicted that he would run into this mustachioed violinist-physicist once again in 1923 on the latter's visit to Palestine.

There is no disputing that the Berlin wartime context was absolutely central to Einstein's changing attitude toward Zionism. Among the traumatic consequences of the Great War was the displacement of many Jews, largely rural, from the Eastern European borderlands between the German and Austro-Hungarian empires on the one hand and the Russian on the other. A sizable group of these *Ostjuden*, religious, traditional, and impoverished, made it to Berlin, where they looked rather different from the elite *Westjuden* of the Prussian capital. Einstein was evidently shocked by the discrimination that he witnessed against those whom he regarded as victims of war.[55] As he wrote in April 1920 to the Central Association of Germans of the Jewish Faith—in language rather reminiscent of his simultaneous dispute with the Jewish Community

over its plan to list him on its taxation rolls—he was by no means a
"German citizen of the Jewish faith":

> What indeed is Jewish *faith*? Is there some kind of lack of faith by force
> of which one ceases to be a Jew? No. In that designation hide, how-
> ever, two confessions of a "beautiful soul," namely:
>   1) I want to have nothing to do with my poor *Ostjude* brothers
>   2) I do not want to be seen as a child of my people, but only as a
> member of a religious community.
>   . . . But I am a Jew and I am happy to belong to the Jewish people,
> if I also do not consider them as some kind of chosen people.[56]

Einstein's own consciousness of his awakening interest in Jewish nationalism was thus derived from his pacifism and his horror at the tragedy
of modern warfare.

Yet even this juncture of pacifism and Zionism, centered in the Berlin
where Einstein was living at the time, demonstrates Einstein's extensive
personal connections with Prague. His opposition to military conscrip-
tion, for example, put him in contact in 1925 with leaders of War Resist-
ers International such as Hans Kohn, a Prague Zionist and pacifist who
was a close associate of Max Brod and others affiliated with Bar Kochba.[57]
In mid-October 1919, German Zionist Julius Berger informed the Zionist Organization in London that Albert Einstein—now *the* world-
famous scientist in the wake of the successful British-led eclipse expedition earlier that year—was despite his vocal antinationalism increasingly
amenable to certain projects and ought to be contacted by them directly.
The office delegated the task to Bergmann, who had left newly born
Czechoslovakia for London on his eventual path for Palestine and worked
for the Zionist Organization. Bergmann extended an invitation to the
physicist to become involved in the creation of a Hebrew University, and
explicitly linked it to their mutual interactions in Bohemia:

> I hope that you still remember the times when we had you in Prague
> and when I could attend your seminar. I left Prague in the spring to
> collaborate as executive secretary of the newly-established Education

Department of the Zionist Organization, organizing of public instruction in Palestine, and above all to participate in the preparations for the Hebrew University.

That is the question, this university, which induces me to write to you once again after such a long time.[58]

Einstein responded within two weeks quite positively: "I take a warm interest in the affairs of the new colony in Palestine and especially in the university to be founded there. I will gladly do everything that is in my power for that."[59] He did not mention Prague but did in this first letter note some unemployed Jewish physicists who might teach at such a Hebrew University. A few months later he prompted the *konfessionsloser* Ehrenfest, now in Leiden, to send recommendations for teachers at a university for *Ostjuden* in Palestine to "Dr. Bergmann from Prague" in London.[60] The tone and phrasing in the letter give the impression that Ehrenfest had encountered Bergmann during his visit to see Einstein in Prague.

Einstein's attraction to Zionism was in part fueled by his opposition to war, which was itself based on his aversion to nationalism. Commentators at the time and since have found the physicist's mixture of beliefs perplexing, given that Zionism was nothing other than a form of Jewish nationalism. How could one be for and against nationalism simultaneously? Einstein recognized the tension, as Blumenfeld recalled him saying one night as they were walking home together: "I am *against* nationalism, but *for* the Zionist cause. The reason has become today clear for me. If a person has two arms and constantly says: 'I have a right arm,' then he is a chauvinist. If a person however lacks a right arm, then he must do everything to substitute for that missing limb. Thus I am in my general attitude to humanity an opponent of nationalism. As a Jew, however, I exert myself from today forward on behalf of Jewish-national Zionism."[61] The clearest manifestation of this positive variant of nationalism came in the form of higher education, which Einstein understood as an unalloyed good that official anti-Semitism—in the form of the dreaded *numerus clausus* that imposed quotas for the numbers of Jews admissible to universities in many European states—frustrated. As his sister Maja wrote in unpublished biographical notes from the early 1920s: "His later advocacy and activities on behalf of Zionism came from this impulse, less in accordance with and on the basis of Jewish dogmas as from the inner obligation with respect to those ethnic fellows [*Rassengenossen*], for whose scholarly activity in the sciences an independent working place should be arranged."[62] Einstein wrote to his colleague Max Born that he "would consider it reasonable if Jews themselves gathered the money in order to offer Jewish researchers outside the universities support and opportunities for teaching."[63]

The Zionist Organization understood the draw they had in Einstein. His endorsement was a tremendous magnet for donations, yet at the same time his idiosyncratic views on war, nationalism, sovereignty, and the internal politics of the fledgling Hebrew University in Jerusalem (the campus opened in 1925) made him a problematic figure for the movement's leaders.[64] In this area, as in most matters, Einstein refused to stay on message when the message was dictated by other people. As the Yishuv (the Jewish settlement in Palestine) expanded and began developing more of the armature of a state-in-gestation, with increasingly violent confrontations between Arabs and Jews and hostilities against the British officials administering the Mandate, Einstein allotted his wavering support not to the mainstream trends in Zionism but to the Brit Shalom (Covenant of Peace) movement. Brit Shalom was never a mass affair, with about 60 members in the Mandate, 13 overseas, and 80 sympathizers, but one of the last group was Einstein. It was founded in 1925 by Arthur Ruppin, Georg Landauer, Hans Kohn, and Hugo Bergmann, the latter two Prague Jews. (Martin Buber was also involved.) The group was interested in building a peaceful, binational entity in Palestine and laid a particular stress upon Arabic–Hebrew bilingualism, a feature that drew from several of its founders' experiences living as Germanophone minorities in Bohemia and Poland.[65] For its pacifist and culturalist orientation, and also its personal connections that reached back to his time in Prague, Einstein found the group appealing and issued statements in support of it in various contexts.

The connection between politics in Palestine and the experience of being a German-speaking and German-identified Jew was relevant not only to the founders of Brit Shalom, but also to Einstein and his personal acquaintances. Repeatedly in Einstein's correspondence one comes across statements that merge identification as a Jew—whether or not a Zionist one—with identification as a German. "Indeed it lifts the heart and strengthens one's faith in the future of humanity," wrote Viktor Ehrenberg, a professor of German law at Leipzig and physicist Max Born's father-in-law, after the eclipse expedition, "when one learns that the researchers of all countries bow before a man of Jewish blood who thinks and writes in the German language, in full recognition of his greatness."[66] For some, support for Zionism was impossible precisely because their own national identification tilted toward Germanness, as in the case of Georg Schlesinger, who wrote to Einstein in 1921:

> I have as a German—my Jewish religion is my personal affair—for now a completely exclusive interest in the reconstruction of my cast-down German fatherland, to which I seek to contribute the sacrifice of all my meager powers. For new foreign enterprises—and I consider the Zionist-Palestinian effort as one in the most pronounced manner—I have neither time nor means. As a German national whose heart bleeds over the unconscionable mutilation of his fatherland, I wish from all my heart to everyone who avows a Jewish national state success in the erection of their long desired national-Jewish homestead in Palestine.[67]

The point I want to underscore is that for many in Einstein's milieu, and, as we shall see, for the scientist himself, the framework for thinking about identification as a Jew—in "ethnic" terms as opposed to religious ones—was often explicitly modeled on prior attitudes and affinities related to a certain conception of "Germanness" and also the German language.[68]

In the years to come, Einstein's relationship with the various Zionist leaders was largely mediated by that language. Bergmann is a good example. In 1923, on his only visit to Palestine—while returning from a lengthy visit to the Far East—Einstein gave an emotional speech on Mount Scopus in Jerusalem at the site of the emergent university. His travel journal recalls meeting with the university librarian, who was none other than Bergmann, "the sincere Prague saint, who sets up the library with too little space and money."[69] The two continued to correspond after this meeting, often about philosophical topics related to their mutual interests, in an extended conversation that had begun in Prague. In 1929 Bergmann published a book, in German, giving a Brentanist interpretation of the debates over causality in contemporary physics, a dispute in which Einstein stood firmly in favor of a rigid understanding of causality in the quantum realm against an emergent indeterminist interpretation crystallizing around the ideas of Niels Bohr and Werner Heisenberg. The book was dedicated to Anton Marty, who by now was long deceased, and, bore a foreword from Einstein, who approved of its tenor.[70]

More frequently, however, the topic of their exchanges shifted to contemporary politics in the Yishuv. The same year the book on causality appeared, Palestine erupted in violent protests by Arab residents against increasing colonization by Jews, uprisings that were brutally suppressed by the British. Einstein was appalled. "The events in Palestine seem to me again to have proven how necessary it is to establish a form of true symbiosis between Jews and Arabs in Palestine," he wrote to Bergmann in distress. "I understand by that the existence of permanently-functioning mixed administrative, economic, and social organizations. The divided state of being next to each other must lead from time to time to dangerous tensions. Besides this all Jewish children must learn Arabic."[71]

Bergmann wrote back an impassioned response in agreement about both the collapse of civil relations and the need to take conciliatory measures to restore confidence. Einstein was so pleased with this letter that he requested permission to have it translated into English and published, which Bergmann agreed to in hopes that a major English-language Jewish-affiliated venue would run it and help build external pressure on Zionist leaders in Palestine.[72] Before 1929, Einstein had seen Zionism as a way of fighting pernicious nationalism with a benevolent form, even as the Zionist leadership was baffled by his support of Brit Shalom. The rebellion brought the conflict between Einstein and Weizmann's group to a head. The physicist even appealed, without success, to the British high commissioner for Palestine to request that Arabs involved in the uprising not receive the death penalty.[73] Relations both between Jews and Arabs and between Zionists and Einstein continued to deteriorate. It was in a letter to Bergmann in 1930 that Einstein most forcefully articulated his grim prognosis for the future: "Only direct cooperation with the Arabs can achieve a worthy and secure existence. If the Jews do not realize this, the entire Jewish position in the complex of Arab lands will eventually become fully untenable. *It makes me less sad that the Jews are not clever enough to understand this, as that they are not righteous enough to want it.*"[74]

By 1936, Einstein had publicly distanced himself from Zionism, but not from the personal relationship with Bergmann that had drawn him into interactions with it. In 1947, Bergmann commissioned an article from Einstein on relativity theory for the *Hebrew Encyclopedia*; the scientist, increasingly loath to undertake such tasks, nevertheless sent in his contribution on time and without fuss.[75] In 1950 Bergmann invited him to come to the newly created State of Israel, but Einstein pled ill health and demurred. When the Israeli leadership offered him the presidency in 1952 upon Chaim Weizmann's death, he turned it down for the same reason, much to the relief of Prime Minister David Ben Gurion. Relations with the Israelis did not improve, but the connection to Bergmann remained warm. The philosopher visited the physicist in Princeton in 1953 and was crestfallen at Einstein's death in 1955.[76]

* * * * *

Between Einstein's enthusiasm for the Hebrew University and equivocation before the Jewish state lay the catastrophe of the Holocaust, and this event too had its Prague resonances for him. Initially, the fate of the Jews in Bohemia seemed optimistic as the new country of Czechoslovakia was carved out of the carcass of the Habsburg Empire in 1918. After extensive consultation and lobbying of its new president, Tomáš Garrigue Masaryk, by representatives of Jewish groups in the country, including Max Brod, the constitution of 1920 made Czechoslovakia the only state in Central Europe to recognize Jews as a separate nationality, granting them the rights and privileges of any other national minority. There was no *numerus clausus* in Czechoslovakia or legal discrimination against Jews, and Masaryk made efforts to combat anti-Semitism—dating back to his opposition to the Hilsner case—which gained him broad recognition among Zionist advocates across the continent. He even visited Palestine in 1927 and was guided by Hugo Bergmann.[77] (Both had been students of Franz Brentano, albeit in different generations.)

Einstein was one of those strongly impressed by the new Czechoslovak president. In 1921, before he himself had been so honored in physics, Einstein wrote to Norway nominating Masaryk for a Nobel Peace Prize: "Having been made aware by some friends about the petition of the Czech parliament to confer the 1921 Peace Prize to President Masaryk, I take the liberty of endorsing this petition most warmly. . . . Masaryk has earned the greatest merit as a protector of oppressed nationalities, especially the Czechs and the Jews."[78] Masaryk did not win, but Einstein's admiration continued. In 1930, he sent public congratulations on the statesman's eightieth birthday—Masaryk remained in office until 1935, when he resigned due to ill health, dying two years later—which were printed in the local newspapers: "Professor Masaryk is the living example of how one's love for one's own people can indeed be in perfect harmony with the outlook of a world citizen."[79]

Although Einstein was drawn to Masaryk's public persona because of his stance on questions related to Jewish identification, the Czech president also appealed to him because of his approach to topics related to pacifism, which should not be surprising given how intrinsically intertwined both issues were in Einstein's thought. On 13 April 1931, after receiving information from Prague pacifists Pavel Moudrý and Heinrich Tutsch, Einstein pled the case of an imprisoned conscientious objector directly to Masaryk: "The superior court of Brünn in your country has sentenced to a long prison term Mr. Přemysl Pitter, a man of high moral character who shares the antiwar sentiments just described. . . . I feel impelled to suggest that you exercise your powers of executive clemency in this case."[80] He received only a standard, boilerplate response on this instance, but something about the letter clearly nagged at Masaryk, and over a year later, on 22 July 1932, he penned a four-page, handwritten missive to Einstein revisiting the philosophical issues behind the case. The letter is worth quoting at length:

During the war in England I had the opportunity to observe the conduct of these *Conscientious objectors*: in many cases since then I have analyzed such *Objectors*. As president and commander in chief of the armed forces (formally of course) I must from time to time also here at home examine similar cases and rule on them. To put it briefly, in many cases they decide to openly refuse service not for any of the religious conclusions about moral motives, but rather because of anarchism, etc. Military courts are thus suspicious in advance and each individual case is painstakingly reviewed. So was Pitter's case. According to the investigation this man is not without faults in his motives to refuse military service!

I consider it fair that conscientious objectors perform some sort of other public service instead of military service, as long as we still have some kind of army; and I also deem it fair that the commander of the army does not tolerate any <u>anarchist</u> propaganda against military service.

Guided by this point of view I judge all the circumstances in which I must decide. Esteemed professor, there are two kinds of pacifism; be assured that we try hard, especially me, to see true pacifists not only in principle, but also in tactics.[81]

Einstein's admiration only deepened at this response, though his own views on conscientious objection were not swayed:

It is indicative of your deep feeling for humanity that you were good enough to write me on the subject at such length in your own hand. If all countries enjoyed the leadership of men such as yourself, the movement to abolish war would not appear as hopeless as, alas! it does today.

I find it hard to understand why so few people seem to regard it as shameful and unworthy of governments to coerce people into performing the very acts which the religions, taught and professed by those same governments, consider most evil—acts, moreover, that seriously imperil the very survival of world civilization.

Your letter was a rewarding experience and I thank you for it.[82]

As it happened, Einstein again participated in the mid-1930s in a multiyear campaign for Masaryk to be awarded the Nobel Peace Prize in recognition of his maintenance of Czechoslovakia as the only remaining democratic island in the fascist sea of Central Europe. The major competition was Carl von Ossietzky, an imprisoned German pacifist. Despite Einstein's support for Masaryk, he worried that if he won, it might distract attention from Ossietzky's case, which he believed was more crucial to highlight in the current political climate as a way to signal the barbarity of Nazi Germany. After many machinations in which Einstein tried (and failed) to stay in the background, Ossietzky won the 1935 prize. Masaryk died before his candidacy could be considered again.[83]

The collapse of democracies into fascism across Europe was obviously a matter that touched Einstein intimately, as he chose to leave the continent in 1932 for a visiting professorship at Caltech and then assumed a post at the Institute for Advanced Study in Princeton, New Jersey, instead of going back to a newly National Socialist Germany that displayed a particular obsession with demonizing him personally. The flood of Jewish refugees tormented Einstein, and he advocated as much as he could—he worried about expending his political capital and thereby hurting all the cases—to bring those fired, displaced, and threatened to the United States. He was especially successful with younger physicists. As he wrote to his lifelong friend Michele Besso from Princeton in 1937: "The beautiful thing here is that I can work together with young colleagues. It is noteworthy that in this long life I have exclusively collaborated with Jews."[84] The beauty of the moment would not last long, as Hitler's terrorizing of Jews would unfurl into slaughter on an unprecedented scale. Among the millions of victims, many touched Einstein personally. One of those recalled an awkward relationship from the German University in Prague and picks out from the vast catalogue of suffering a single story.

The man's name was Emil Nohel. He was born on 3 January 1886 to Heinrich and Julie (née Kallen) in the Bohemian village of Mcely, a moderate distance northeast of Prague, and his birth certificate bears the star of David stamp of the Jewish Community.[85] His farmer parents harbored ambitions for the boy and sent him to a *gymnasium* in the New Town of Prague on the Graben (Na Příkopě in Czech). He did well there and subsequently enrolled for seven semesters of physics and mathematics study at the philosophical faculty in the German University in 1904, supplemented by a single semester in the law faculty. He took courses from

Ferdinand Lippich and Georg Pick, as one might expect, but also with Anton Marty in Brentanist philosophy. He began working on a mathematical dissertation under Pick's supervision entitled "On the Natural Geometry of Even Transformation Groups," which he would defend in 1913, and in the meantime was hired in 1911 as a research assistant at the Institute of Theoretical Physics.[86] He was immediately assigned to work with the new professor of physics just arrived from Zurich, Albert Einstein.

It is not entirely clear from surviving documents what the relationship between the two men was like, but we know that it ended quickly. Understandably, given what came afterward, Nohel's family and Einstein's friends and biographers have painted a picture of affectionate interchange.[87] Nohel's son Yeshayahu would later recall that "the many hours Einstein and my father spent together in Einstein's study, his world view and character left a lasting impression on my father. . . . He was fond of Einstein's first wife and regretted their separation."[88] (His son also believed that Nohel had remained Einstein's assistant for all three semesters the latter was in Prague, which was definitely not the case.)[89] Nohel had large shoes to fill. He had been hired to replace Einstein's highly productive assistant and sometime collaborator Ludwig Hopf, who had accompanied Einstein from Zurich but soon left Prague for a *Privatdozent* position in Aachen. Nohel and Einstein never managed to forge a harmonious working relationship, and after a few months Einstein replaced him with Otto Stern, fresh from a doctorate in physical chemistry in Breslau, starting on Easter 1912. (Stern would follow Einstein back to Zurich and embark on his own glorious career, marked along the way by the 1943 Nobel Prize in Physics.)[90]

"After their ways parted the connection between them did not last. My father was too modest (or too proud?) to bother the famous man. It may be that from a professional point Einstein's influence on father was even negative," noted Yeshayahu Nohel. "I have heard it said that other students and physicists, too, were so overwhelmed by his sheer brilliance and his originality of thought, that they did not dare to measure themselves against such a model and gave up scientific research. I assume that the break in that connection saddened my father. His affection for

Einstein the man, however, stayed with him till the end."[91] Nohel continued to teach and research in Prague for a few years but soon moved to Vienna as a teacher in physics and mathematics at the Handelsakademie, a business-oriented secondary school. There he stayed and raised his family until March 1938, when Hitler's troops marched peacefully into Austria and the *Anschluss* with Nazi Germany placed Nohel in grave peril.

Due to anti-Semitic racial laws, he lost his job. Surprisingly, he found another one, teaching at the Zvi Perez Chajes *gymnasium*, a Jewish school that remained open. He was assigned to teach physics again; when the Gestapo arrested the Latin teacher, he took over that subject as well. What little we know about this period comes from an admiring memoir piece written by a physics student he inspired named Walter Kohn. Kohn was one of the lucky ones. After leaving Vienna on the *Kindertransport*, he ended up in England and then Canada, where he was interned for two years as a refugee. After his release he embarked on a stellar career in physics, developing the extremely important density-functional theory for modeling complex quantum mechanical systems, which was eventually recognized by the 1998 Nobel Prize in Chemistry. He continued to harken back gratefully to his teacher.[92]

Nohel was one of the unlucky ones. His son Heinrich (later Yeshayahu) had been sent to Palestine, but Nohel could not bring himself to make that move. He recognized, however, the need to get out of Central Europe if he could, and he began writing widely to employers in the United States offering his services in exchange for an affidavit and a visa to emigrate. Naturally, in describing his qualifications, he mentioned his former position as Einstein's assistant in Prague, and this is how Emil Nohel's name, after several decades, again crossed the physicist's desk in Princeton. Einstein, who signed so many affidavits of his own, responded to Nohel quickly in May 1939 to confirm the latter's story: "I gladly confirm for you that you were my assistant in Prague during 1910–1912 [*sic*]. I would gladly be of assistance you besides this, if you would allow me the opportunity."[93]

Einstein had to attest to Nohel's bona fides to several of his correspondents as well, who were themselves flooded with requests and did not know how to evaluate his story. Arthur Ruark, a physicist at the University of North Carolina at Chapel Hill, wrote Einstein in July 1939 inquiring as to both the veracity of Nohel's account and the nature of an affidavit. Einstein was optimistic about Nohel's chances and told Ruark to wait. "There is some prospect to get an affidavit for Professor Nohel from other sources. If you don't hear from me in the contrary about this you may assume that the matter of the affidavit has been settled," he responded. "The issuing of an affidavit does not involve any real financial obligation. It is rather a matter of form to satisfy the immigration-laws."[94] The same week, he assured Isidor Rabinovitz of R. & M. Industrial Laboratories that "it is perfectly true that Dr. Emil Nohel was mine and Professor Pick's assistant in Prague. He is a fine and reliable person both as personality and in his work. In my opinion it would be highly justified to give him an affidavit." Given immigration restrictions, however, possession of the document would not necessarily allow Nohel to enter the United States, but it might enable him to go elsewhere in Europe. "I should have sen[t] Mr. Nohel my own affidavit if I could have done it without endangering the ones I have already given to other people," Einstein wrote in another letter.[95] He sent two other letters in this vein in August.[96] It seems that Einstein's confidence was unjustified, however, and he lost track of his former assistant.

Nohel left Vienna, hoping to avoid confrontation with the authorities if he resided in the countryside. He moved to Hradec Králové in Eastern Bohemia to live with his sister, Otilie Mahlerová. For roughly three years, Nohel resided in the Nazi-occupied Protectorate of Bohemia and Moravia, keeping a low profile. These were very lonely years for him, although he put on a brave face in the letters he periodically managed to send to his son in Palestine. The letters make for heartbreaking reading, and they are suffused with regret and despair. "You know that, unfortunately, before 1938 I was not an organized Zionist for reasons that were not dishonorable. I managed all the normal achievements of a Zionist, probably to a higher degree than the majority of organized Zionists," he wrote in the summer or fall of 1941, "but I had not yet gone so far that I wanted to dedicate my whole life to this goal—that is, diligently learn Hebrew, make myself familiar with the culture of our people, and if possible myself emigrate to the Land of Israel."[97] When an opportunity to be smuggled out came his way, he demurred out of fear of betrayal and retreated farther into the countryside. "My life, that indeed for the most part was built on false foundations, is already behind me; I would like only to earn my bread through the work of my own hands."[98] Sorting through the rumors that rushed around Bohemia in those days, he understood that deportation almost certainly meant death.

In late 1941, he wrote another letter to his son—trusting that eventually it might reach its destination—explaining that deportations from his region ended up in Theresienstadt (Terezín), a fortress named after Joseph II's mother Maria Theresa, "where a large camp for Jews is erected. No news at all comes from there."[99] He had no illusions about what would happen next: "We must also reckon with the fact that soon we will also be interned. How it will be then, no one knows; but in any case the sole—and for me modest—postal connection with you will be interrupted."[100] Incredibly, he remained free one more year, a year in which he managed to send a few more lengthy letters, chronicling his disappointment in himself for not committing to Bar Kochba or following the guidance of Bergmann, Brod, and Robert Weltsch (all of whom he mentioned by name).[101] The letters convey the sickening sense of an apocalypse rolling across the countryside, the feeling that one day he would no longer be able to escape.

That day was 21 December 1942, when he was transported to Theresienstadt. His mentor Georg Pick had preceded him by a few months and passed away in late July. Nohel stayed at the camp for almost two years. On 16 October 1944 the Nazi guards put him on a train to Oświęcim, where he disembarked at Auschwitz and was murdered.[102] In February 1946, a man from Newark named Fred Schwarz wrote to Albert Einstein in Princeton to pass along the news of Nohel's death. Einstein, not inured to the barbarity of what had transpired, wrote back sadly: "Our efforts to bring him to safety have unfortunately foundered upon the onrush of well-known events."[103]

* * * * *

Einstein seems to have barely ever set foot in Josefov, preferring to take his long walks in the woods surrounding the city. By the time he lived in Prague, the former ghetto had been radically transformed so that it no longer resembled the city of Jews from legend and lore. When reflecting on his complex and erratic identification with Jewishness—whether as a religious faith, a Zionist philosophy, or a series of interpersonal relationships—Einstein never brought up his time in Prague. Yet at every corner, there was a Bohemian tinge to his later ruminations on the subject. When he left Prague, Einstein was comprehended as a German *and* a Jew by certain elites who wanted to score political points against Vienna, and from then on the question of Einstein's Jewishness always bore some mark of his potential Germanness, identifications imposed upon him that he sometimes, fitfully, embraced. The blend of German and Jew was characteristic of a man of his time and place: a former professor of the German University in Prague.

At the very moment that Einstein was beginning to be troubled by the course that Zionism had taken in Palestine, he was also becoming ever more vocal on behalf of Jews being persecuted by Hitler's National Socialist regime that seized power in Germany in 1933. From the most famous German scientist, overnight Einstein had become the world's most prominent Jewish refugee. From his position on the other side of the Atlantic, Einstein issued blistering denunciations of anti-Semitic measures taken within Germany, and it was clear that these attacks further enraged the Nazi hierarchy, who had a virtual mania about Einstein. Back in Berlin, Max Planck wrote his former colleague asking for a moderation of his rhetoric so that those at home opposed to the racial policies would have more room to maneuver. On 6 April 1933, two weeks after the Enabling Act essentially granted Hitler a free hand in running the country, Einstein responded rejecting Planck's position:

> I must now however also recall that . . . especially in recent years I have been systematically persecuted in the right-wing press without anyone having taken the trouble to intercede for me. Now however the war of annihilation against my defenseless Jewish brothers has compelled me to place the influence that I have in the world on the scales for their benefit. In order that you better comprehend this, I ask you to imagine yourself for a moment in the following position. You are

a University professor in Prague. A regime comes to power there that robs the Czech Germans of their means of existence and that simultaneously forcibly prevents them from leaving the country. They have posted guards at the border who are supposed to shoot at those who seek to leave without permission the country whose regime is conducting a bloodless war of annihilation against them. Would you then find it correct to take this silently, not to speak out for them? Isn't the annihilation of German Jews through starvation the official program of the current German regime?[104]

This is a fascinating passage, not least because, despite its subjunctive mood, the circumstances being described were not actually hypothetical, at least not for Einstein. He in fact *had been* a professor at the German University in Prague, and he had been exposed to the anxieties of colleagues like Lampa and Lippich that Czech chauvinism would make life intolerable for the Germans. The Nazi-affiliated Sudeten German Party headed by Konrad Henlein had painted precisely such a fantasy of anti-German discrimination to stoke anti-Czech resentment among the substantial German-identified minority within Czechoslovakia. Einstein was here not only using Henlein-style scenarios to make the situation of the Jews more understandable to Planck, but was also himself filtering his understanding of the persecution of the Jews through his own identification as a German. As we have repeatedly seen, whenever Einstein approached topics or people connected with the Jewish experience in Prague, the notion of Germanness was never far from hand. The conflation was intrinsic to Einstein's understanding of both categories, and it was a conflation born in Bohemia.[105]

A final example brings us back to Martin Buber, whose words opened this chapter and whose speeches helped ignite the active community of Prague Zionists that provided Einstein's first exposure to the movement. In the logs of phone conversations held during his final years with Hanna Fantova, herself originally from Brno but then living in Princeton, she recounts Einstein's reaction to a story she had told him about a Prague family during the Holocaust. "Your story of the young man whose whole family was murdered in Prague has moved me deeply," she noted Einstein as responding. "The German Jews are indeed terrible; they are traveling again to Germany. Even Martin Buber himself traveled to Germany and allowed himself to be celebrated with a Goethe Prize. People are simply vain. I have declined everything and given them a kick in the backside."[106] A story about Prague immediately triggered a refusal to ever return to Germany. He never did.

CHAPTER 7

From Revolution to Normalization

> Also theories have a large influence on the formation of
> modern society; certain scientific systems and certain facts
> have indeed become the leading ideas of their eras, as
> *Galilei*'s science about the orbit of the earth around the sun
> is the landmark of a new era; *Newton* gave character to the
> seventeenth century with his theory of gravity; *Darwin* is the
> characteristic man of the nineteenth century, just as
> *Einstein* is distinctive of today's era.
>
> —*Emanuel Rádl*[1]

While he lived there, Albert Einstein's Prague was unquestionably German. All of his interactions took place in the German language—they could not have really happened in any other—and he repeatedly referred to the German University where he worked as connected with the web of other Germanophone institutions across Germany, Switzerland, and Austria-Hungary. When Czechs appeared in his correspondence, it was always as an aside, and he did not flesh out his superficial impressions of the vast majority of the local population (if he even realized it was so large) with any substantial qualities. Prague was a German space, but not one that he particularly enjoyed.

After he left, however, on the few occasions that he reflected on Prague it was often as a *Czech* metropolis, with his sympathies extended to a marginalized population striving for political independence. In October 1918, during what would turn out to be the final weeks of World War I, the "Washington Declaration"—drafted in the American capital— proclaimed the independence of a new entity, "Czechoslovakia," an amalgamation of the Czech-dominated Bohemia and Moravia with the

Slovak-dominated regions of northern Hungary (and some other parcels of land at the edges) to produce a state where the once-regnant Germans (constituting just shy of a third of the population) and Hungarians were now national minorities. Einstein had a soft spot for Czechoslovakia and for its long-serving president, former Czech University philosophy professor Tomáš Garrigue Masaryk. His political admiration of the new state tended to color his memories of Prague in his later invocations of the country.

In 1923, the Borový publishing house in Prague issued a translation of Einstein's slim 1916 book, *Relativity: The Special and the General Theory*, which had been written with an educated lay audience in mind. It is worth noting that he wrote this *before* the 1919 eclipse expedition had turbocharged demand for such a book; when confirmation came, this volume was near to hand. Translations appeared in a host of languages, and so a Czech one was to be expected. Unlike so many of the new editions, however, this one was not simply a rendering of Einstein's engaging German text in a new language: the physicist penned a new foreword to the book, one that tied the creation of general relativity to his time in Bohemia. The one-page foreword began:

> I am happy that this small booklet, in which the chief thoughts of relativity are presented without their mathematical realization, now appears in the national language of that country in which I found the necessary composure to gradually give a more definite form to the fundamental thoughts of the general relativity theory, which had been gathering already since 1908. In the quiet rooms of the Institute of Theoretical Physics of the German University in Prague on Viničná ulice, I came in 1911 to the discovery that the equivalence principle required an observable degree of bending of light beams by the sun, without knowing that more than a hundred years earlier a similar consequence had been drawn from Newtonian mechanics in connection with Newton's emission theory of light. In Prague I also discovered the result, still not definitely established, of the redshift of spectral lines.[2]

(Directly afterward, he confessed that though he began the work in Prague, he did not hit on the path to a full solution until he returned to

Zurich.) For all his nostalgia about his time in what was now the capital of the First Republic of Czechoslovakia, there is something odd—one might almost say tone-deaf—about this tribute to the Institute, the location of which he gave with its Czech address: it was written, and published, in German. (A Czech rendition immediately followed.) This was probably the press's decision, not the scientist's, but it does point to Einstein not quite understanding the delicate position of the resentful German minority or the feeling of Czech elites toward it.

Einstein was still a reference point for that German minority, a symbol to conjure with when prominent intellectuals proclaimed the continued cultural importance of German-speakers in Prague even as their political significance declined precipitously. After the eclipse expedition, he continued to be cited as a beacon of German intellectual leadership—though admittedly sometimes with tongue in cheek. For example, on 19 December 1919, dramatist and theater critic Heinrich Teweles penned a feuilleton in the *Prager Tagblatt*, the country's leading German daily, which he had edited from 1900 to 1910, recalling Einstein:

> But what do I really know about Einstein? When he taught at the German University in Prague there were rumors: This is a great guy! But nothing came from the rumors. Those that knew something about him thought it was entirely right that the Huns called him to Berlin, where he received an institute and a salary in order to research without the obligation to share his knowledge with young people, who besides had not yet begun to understand a thing.
>
> Einstein intrigued me. Suddenly, during the war, the English dispatched an expedition to study the solar corona in order to test Einstein's thesis or hypothesis and Triumph! Einstein wins. God knows what Clemenceau will make Germany pay for that.[3]

Einstein found the piece so hilarious he wrote Teweles a fan letter.[4] He also penned a note of congratulations on the tenth anniversary in 1927 of the Prague Urania, the German-led astronomy and physics showcase, which was lovingly printed in their newsletter.[5] If a man like Einstein bestowed his favor on these German institutions in a Czechoslovak state, then they might borrow some of his international celebrity to bolster their significance. Einstein was a talisman that could dispel accusations of provinciality and instill in the German community of Bohemia the spirit of cosmopolitanism. Because he had once lived and worked for 16 months in Prague, the Czech metropolis, he was also available for such use by Czech scientists and intellectuals. He deprovincialized Bohemia.

This final chapter turns to a question that has lingered in the background from the outset: What was the *Czech*—understood both as a language and as a personal identification—version of Einstein? Einstein's Prague was roughly 7 percent German-identified and 93 percent Czech-identified, and yet Czech voices have been sparse in the preceding pages. This is not because Prague lacked a significant Czech-speaking and -writing physics community, but rather because the literatures of the two groups were segregated. Czech-identified scientists had to engage with the Germanophone literature—and the French and English literatures, like every other scientist—but the reciprocal relationship did not hold. When a physicist chose to write in Czech, he or she (although female physicists were rare in Czechoslovakia in this era) was making a zero-sum choice about audiences: *local* engagement at the cost of the international community completely ignoring the work. (For writers in the three dominant languages, writing for a local readership did not mean the sacrifice of foreign attention.)[6] Going back to 1911 and Einstein's arrival in Prague, if one reads the Czech periodicals, one sees a world that he lived in but of which he was unaware.

What follows traces the ways in which Einstein and relativity theory were invoked and deployed against the background of the brutal shocks that transformed Czechoslovakia across most of the twentieth century, stopping in 1979, the year of the centenary of Einstein's birth (and 14 years before the country ceased to exist). The dividing line in this story—as with so much of twentieth-century history—is the Second World War. Before that point, Czech writers under the Habsburg Empire and the First Republic (1918–1938) engaged with relativity and its creator as a token of their cosmopolitanism. They were writing in a so-called "minor language" from the capital of a small European country, but their facility with the new physics made them citizens of the world. Theirs was a bilingual vision that faced outward, reaching across the language barrier to incorporate Philipp Frank's German University as well. After the dismemberment, occupation, genocide, and reconstruction of the state, Czechoslovakia (now with slightly different borders) was a changed place intellectually. Brought into the Soviet political orbit and then, after a coup in 1948, under a Communist regime, the new Czechoslovaks continued to discuss the theory of relativity, but this time as a marker of distinction from global trends. Einstein had been domesticated: rather than serving as a ticket to the world, he became a monument to Prague's apartness.

* * * * *

The story of the development of a scientific infrastructure in the Czech language cannot be told without the surrounding context of the largely Germanophone institutions that represented Bohemian distinctness. The Royal Bohemian Society of Sciences, founded in 1784 without the royal imprimatur (which came six years later), formed a major outlet in which scholars who identified as Czechs were able to communicate to a broader European readership. They were required to make a linguistic compromise: the cultural charge of the moment was that the journal not be *in Latin*, and so the Society published its *Abhandlungen* in German, even though many of the early articles needed to be translated into that language to be published.[7] Only in 1873 did the organization's meetings become officially "utraquist," allowing Czech to serve as an official language. By midcentury, the *Abhandlungen* published articles in Czech, but it also published in French, Italian, English, and other Slavic languages. As with the university in Prague, utraquism led to a final division along linguistic lines, with the Royal Bohemian Society splitting in 1890 into the Czech Academy and the Society for the Promotion of German Science, Art, and Literature: "From an utraquist institute with a German character a Czech institute was created."[8]

This pattern was repeated across the cultural landscape. Where there had once been a general "Bohemian" institution, either functioning entirely in German or nominally bilingual or utraquist, parallel organizations cropped up in domain after domain that proclaimed their explicit goal of building institutions that worked exclusively in Czech. Quite successful from their earliest years, these new establishments cast the Bohemian institutions, which conceived of themselves as nonnational in theory if not always in practice, as "German" redoubts. The Union of Czech Mathematicians was created in 1862, motivated by the poor state of instruction in the discipline; by the 1870s, German-speaking teachers had begun to drop out, citing feelings of exclusion. Founded in 1866, the Association of Architects and Engineers in Bohemia, which published a quarterly in both languages, met a similar fate as German-identified members felt forced out for nationalist reasons after only three years. (They formed the German Polytechnic Association instead.) In the mid-1880s the association changed its statutes to become purely Czech.[9]

The various groups that dotted the landscape of civil society presaged the much larger transformation that followed the division of the university in 1882. As the Czech University and Polytechnic were the only Czech-language institutions of higher education in the world, it stands to reason that they established Prague as the epicenter of a Czechophone community of physicists, essentially all of whom passed through their halls at one time or another. When the university had been unified and Ernst Mach—who spoke Czech—had been the major luminary in the science, he had trained a large number of students, overseeing 17 doctoral theses in all. Just about the entire first generation of modern Czech physicists attended his classes.[10]

After the split in 1882, August Seydler was appointed an extraordinary professor of physics in the Czech University and elevated to ordinary professor six years later. When he died in 1891, the position was divided into a chair in astronomy, held by Gustav Gruss, and one in mathematical physics, occupied by František Koláček. Koláček in turn held the chair until his death in 1913, with the exception of a brief sojourn to Brno. The chair then passed to a student of Koláček's, František Záviška, whom we will encounter in greater detail later. (A second chair in theoretical physics opened up in 1922.) In experimental physics, a similar succession pattern of teacher to student took place, with Čeněk Strouhal, Mach's prize student, forming the stem of that tree.[11] Although their pedigrees—and the nationalist historiography—emphasize their training in Prague, many of these leading figures conducted part of their studies in Germany, working in a language they knew but outside the constricting embrace of Habsburg Vienna.

Over the late Habsburg period, Czech-speaking intellectuals focused especially on promoting linguistic purism in scholarly communication and scientific terminology. Although of course these scholars continued to maintain polyglot intellectual contacts across Austria-Hungary and beyond, for political and cultural reasons there was a marked turn after 1880 toward Slavic engagements and away from the German language.[12] For many within the empire and its successor states—and this partly applies even to the Hungarians, who are exceptions to many of these generalizations—German formed the obvious "vehicular language" to reach outside of their national(ist) intelligentsia, one that any educated Slav or Magyar would have been exposed to from an early age.[13] The nature of the dilemma for the Czech scientist was well expressed by Vilém Mathesius, the founder of the resolutely polyglot and internationally acclaimed Prague Linguistic Circle, in 1925:

That very fact carries weight that a really independent and still today to a certain extent self-sufficient Czech scientific world has developed. On the one hand, the Czech scientific worker was more strongly inclined to the obligations of building a Czech cultural and political life. This took from him time and energy needed for the struggle for scientific knowledge. On the other hand, by means of this work threads were unraveled and weakened which had earlier bound every scientifically working Czech to the broad domain of German science. Papers written in Czech were left almost exclusively restricted to Czech readership and Czech scientific knowledge remained as before merely a publication in itself, not a living article of international scientific circulation. Czech scientific life became a cloistered creation of the Czech scientific world. But everything has moved beyond this point, necessary to establish contributions to an independent Czech scientific tradition in the sense that there are independent English, French, German, and Russian scientific traditions. And all those disadvantages and all those obstacles would have been very insignificant for Czech scientific production if it were not the case that the modern Czech scientific atmosphere is not the natural patron for the growth of scientific creators.[14]

With strenuous effort, a Czech physics terminology was built using foreign (largely German) templates, fueling the proliferation of Czech-language scientific publications.[15]

As impressive as this was, it remained the case that very few non-Czech-identified scientists had Mach's facility with the language. The solution was to publish abroad in one of the three dominant vehicular languages of international science of that era. Because of the specific valences of nationalist politics, German was discouraged, though it would have been the easiest choice. Chemist Bohuslav Brauner, son of an important nationalist lawyer and the first chair of inorganic chemistry at the Czech University, chose to run his laboratory in English—the command of which he had perfected while a postdoctoral researcher in Henry Roscoe's laboratory in Manchester—rather than resort to German. (He did publish occasionally in that language when necessary.) Brauner made a point of befriending the most famous Russian chemist of his day, Dmitrii I. Mendeleev, and learning Russian proved another path out of the dominant Germanism of his discipline. The Anglophone tradition would be continued by Jaroslav Heyrovský, the only Czech Nobelist in the sciences (1959), who co-edited the *Collection of Czechoslovak Chemical Communications*—including translations into English and French as well as original articles composed in those languages—as an important outlet during the First Republic.[16]

* * * * *

The First Republic's story is often told through the lens of the Czech majority and its Slovak allies and is particularly embodied in the person of Tomáš Masaryk, who was promoted as the personification of a nation-state instead of what was in reality a multiethnic country: Czechs just barely made up the majority with 51 percent of the population, followed by Germans, Slovaks, Hungarians, Ukrainians (Ruthenians), Jews, Poles, and Roma in significant numbers. Excellent recent scholarship has

dispelled some of the romance of this period, a product of the later identification of Czechoslovakia in the 1930s as a hopeful island that successfully resisted much of the right-wing politics that dominated Central Europe.[17] What was at the time for many both inside and outside the
country a vindication of a kind of cosmopolitan nationalism—a benign
variant of patriotic ethnic self-identification that provided a universal-
izing model for a peaceful path to a post-imperial continent—could seem
to those who identified as Germans inside the country something rather
different.

The formal transition in the status of these Czechoslovak "Germans"
was striking. They had always been a minority in Bohemia and for several generations (at least) a minority in Prague, dwindling by the early
twentieth century to percentages in the single digits, but they had also
been subjects of a Habsburg monarch who ruled over an empire that was
conceptualized as Germanophone and where those who identified as
Germans enjoyed dominance. When the new states of Czechoslovakia
and Austria signed the Treaty of Saint-Germain-en-Laye in 1919, Masaryk's government explicitly committed itself to protecting the rights
of German-identified citizens as *minorities*. The rights of Germans were
protected in the state's constitution in the same way the Jews' rights were.
This was, in terms of status, an amazing victory for the latter and a severe
shock to the former.

The drama played out on a smaller stage at the bifurcated universities
of the capital city. On 19 February 1920, the government proclaimed the
Lex Mareš. Gone was the fiction that both universities, carved in 1882
out of the Charles-Ferdinand University, were equal heirs of its legacy.
Now the Czech University would be called the "Česká Univerzita Karlova," the Czech Charles University, and the German University became
simply the "Deutsche Universität Prag," or German University of Prague.
The law explicitly stated that the German University—denuded of reference to Charles IV in its name—was *new*. It was a school for minorities, just as the Czech University had been conceptualized earlier by the
Habsburgs.[18] Other reforms swept through higher education: a university was established in Brno in late January 1919, while the philosophy
faculty of both universities in Prague, where physics and other natural

sciences had been ensconced and where Einstein had taught, was broken into separate philosophy and natural sciences faculties. A consequence of the former change was to diminish the uniqueness of the Czech University as the only institution teaching in the Czech language; a consequence of the latter was a striking boom in the natural sciences in both institutions.[19]

The status of the German University would remain a major point of dispute during the two decades of the First Republic. Amid claims emanating from newly National Socialist Germany that the German University was discriminated against in favor of the Czech, defenders in 1934 argued that retaining the former at all was "a concession made to the Germans" and reminded observers about the earlier financial asymmetries between the twin institutions that had meant that "for a long time the Czech University was treated as a Cinderella."[20] Looking at the numbers, it seems that independence was a boon for Einstein's former employer. It had become harder for German-identified Czechoslovak citizens to study in Vienna—where the now-foreign university used a different currency in an age of wild monetary instability—and this was the only Germanophone institution in their country. Before the war, the German University had enrolled 2,295 students in the winter semester and 2,067 in the summer, while by 1918 (before the Lex Mareš) the average had grown to 3,000, and by academic year 1930–1931 it had risen higher than 5,000.[21]

The changes were even more striking for the Jews, who did not have to suffer the imposition of the *numerus clausus* imposing low quotas for their admission, common in so many Habsburg successor states. Not that this prevented discrimination in higher education. On 17 February 1922 an anti-Semitic scandal concerning the physiologist Armin Tschermak-Seysseneg rocked the German University, followed by another concerning the election of the Jewish medievalist and papal historian Samuel Steinherz to the position of rector. In 1929 German and Czech students at *both* universities ineffectually protested in favor of the introduction of a *numerus clausus*. The continued open status of the institution led to some educational immigration by Jews from across Central and Eastern Europe who could get an education in Prague in German without suffering an imposed penalty for their ethnicity or confession.[22]

Of course, the Jewish-identified professors at the German University remained in their positions, including Philipp Frank, Einstein's successor as professor of theoretical physics and mainstay of the philosophical movement of logical empiricism identified with the Vienna Circle. He was joined, at various points over this period, by Erwin Finlay Freundlich in astronomy, the first person who had abortively attempted to verify the bending of light rays around the sun during an eclipse as predicted by Einstein's Prague theory of general relativity; Leo Wenzel Pollak in geophysics, who had made the connection between Freundlich and Einstein; and, initially at least, Anton Lampa in experimental physics, the man who had been most responsible for bringing Einstein to Prague in the first place.

In May 1919, before the Lex Mareš, Frank wrote to Einstein with some information and a request for assistance. The information was to assure Einstein that rumors of decay at the now-minority university in Czechoslovakia were entirely false. "The University here works exactly as well and as poorly as it had worked earlier. I have heard that people in Germany often believe that our University is dissolved or transformed into a Czech one or that something of the sort is planned," Frank stated. "That is all entirely incorrect. So far nothing essential has changed in it besides the professors receiving a considerable raise. Though scientific activity in the past months has very much suffered because of the ongoing uncertainty."[23] Things were looking so good that the university was seeking to fill a new position: Lampa's.

Frank had never much enjoyed Lampa's company or esteemed his talents. As he wrote in his postwar biography of Einstein: "There was a considerable gap between [Lampa's] high aspirations and his scientific capacities . . . and as a result he was animated by an ambition he could not satisfy. Since he was a man of high ethical ideals, he consciously sought to suppress this ambition, but the result was that it played an even greater role in his subconscious life."[24] There was also the fact that Lampa—who had been dean of the philosophy faculty in the last year of the Great War—remained a committed German chauvinist and, despite his Budapest birth and Bohemian upbringing, continually suspicious of the Slavs who resided in his city. In 1919, Lampa found he was unable in good conscience to swear loyalty to Czechoslovakia and re-signed. He moved to Vienna, where he served as head of public education.[25]

In that capacity he even maintained some contact with Einstein. In 1920 he wrote his former colleague, now in Berlin, asking for his views about the controversial Viennese physicist Felix Ehrenhaft. He did not mention Prague.[26] In Einstein's supportive response (now lost), he must have recalled their time together at the German University, for Lampa exulted: "That you also think with pleasant feelings about your time in Prague is for me a true satisfaction. Although you belonged to Austria only a short while, yet the cultural history of that country can be proud of the fine fact of having offered you your first full professorship."[27] (Telling here is the geography Lampa uses: Prague is Austria.) A few months later, as an afterthought, Lampa wrote that he regretted they had not de-veloped a working relationship in 1911–1912, a time that, as described earlier, Einstein had found himself rather isolated from his physicist col-leagues. "I am very sorry that I do not have the opportunity to be able to occasionally be with you, one is also not likely to appear. You will now perhaps think that I should have used the opportunity in Prague better. But that was unfortunately not possible," he wrote in May. "The extraor-dinary significance of your ideas was already clear to me then. Your call to Prague is an experimental proof of that. But I was then not yet entirely clear on your thoughts and I wrestled with how to incorporate them into my epistemological thoughts or style of thinking—what perhaps seems almost full of contradictions for someone whose style of thinking is oriented toward Mach. But it is the case."[28] As far as we know, Einstein never followed up. Lampa continued his career within the fledgling nation-state of Austria, serving as president of the Vienna Urania from 1927 to 1934, and dying in January 1938, a few days before his seventieth birthday. That would be a fatal year for Czechoslovakia too; one wonders what Lampa would have made of that.

All this, however, lay in the future as Philipp Frank attempted to pre-serve the high quality of physics instruction and research at the German University. He continued his correspondence with Einstein throughout the First Republic, and when fascism turned Prague into an oasis where refugees could study unmolested, Frank sent some of his best students to Einstein at Princeton's Institute for Advanced Study, including Einstein's future close collaborator Peter Gabriel Bergmann.[29] In 1932, Frank also managed to bring the preeminent philosopher Rudolf Carnap to the German University in a position within the natural sciences faculty, an end-run around Frank's former sparring partner Oskar Kraus, who disdained Carnap's philosophy of language from his perch in the parallel philosophy faculty. (Carnap's appointment was specifically supported by Masaryk.)[30] If physics at the German University muddled through pretty much as it had under the Habsburgs, however, the transformation in science at the Czech University and in the Czechophone public sphere was something else entirely.

* * * * *

Czech-speaking and -writing physicists had been excited about relativity theory years before Einstein arrived in the Bohemian capital. František Záviška gave a summary account of the special theory as a lecture in 1905, the same year it was formulated, and published the text of the lecture two years later.[31] A popularization of the theory intended for high school teachers and the scientifically literate lay public appeared in 1911—the author, Augustin Žáček, noted that Einstein was "now professor of physics at the German university here"[32]—in the periodical *Živa*, a mainstay of Czech-language popular science that had been founded in 1853 by Jan Evangelista Purkyně (1787–1869), the renowned physiologist and Prague professor. Further popularizing articles penned by Prague's leading physicists from the Czech University continued to appear until the demise of the Habsburg Empire.[33]

In the new nation of Czechoslovakia, interest in relativity on both the popular and the research fronts grew. The first full book on relativity theory appeared in 1921 under the byline of František Nachtikal—a surname that indicates Czechified Germanic origins (*Nachtigall* means "nightingale")—a professor of physics at the Czech Polytechnic in Brno, the twin institution to the one where Gustav Jaumann taught. Nachtikal's primer, *The Principle of Relativity*, was steeped both in local

Bohemian and Moravian references and in the current events of the physics world. On the one hand, he embedded several references to Tycho Brahe, "the celebrated Prague astronomer," in the text; on the other, he noted the emergent anti-Einstein movement promoted by Philipp Lenard and the controversy it caused at the Bad Nauheim physics meeting.[34] The book could be read, even today, as a serviceable popular introduction to the theory, as could its successor, Arnošt Dittrich's *On the Principle of Relativity*, published in Slovakia the following year. Dittrich was a careful writer, and he took pains to distance relativity theory from philosophical and moral relativism on the very first pages before explaining the Galilean relativity principle using the example of a boat trip up the Vltava. Striking in Dittrich's account is the emphasis on Hermann Minkowski's spacetime formalism; Einstein only becomes the protagonist halfway through the book.[35]

Both books preceded the Czech edition of Einstein's own popularization and thus conditioned the reception of the theory among Czech readers who were unable to access the German version. Since they had thus not been exposed to the buzzing confusion of the antirelativity writings by individuals such as Oskar Kraus that saturated the Germanophone press, monolingual Czechs were blessed with a fair and thorough introduction to Einstein's work. It is thus not surprising that Prague philosophers like Emanuel Rádl would in 1926 write intelligently about relativity, in Rádl's case with a judicious weighing of both the German and the Czech sources.[36] Almost all Czech physicists were also bullish on general relativity. As Nachtikal noted in a short letter to a physics journal in 1924 about the solar spectrum: "Today one can now summarize the judgment of the general theory of relativity: the theory is warranted up to those levels where it has been generally accessible to experimental examination. The viewpoint of physicists is now clear: *Natura locuta, lis finita* [Nature has spoken, the matter is closed]."[37]

The most dedicated exponent of Einsteinian relativity, as well as one of the most talented theoretical physicists working in Czechoslovakia, was František Záviška. Born on 18 November 1879—eight months and four days after Albert Einstein—in the Moravian town of Velké Meziříčí (Großmeseritsch), he was the eldest of three children of a small farmer who died just before Záviška turned 12. (His mother, who remarried, died in 1911.) Educated at the state *gymnasium* in Třebíč (Trebitsch) until fifth grade, and then at the Czech higher *gymnasium* in Brno until eighth grade, he passed his secondary school completion examination on 12 July 1898 and subsequently enrolled as an ordinary student at the Czech University in mathematics and physics. (This was something of a disappointment for his mother and stepfather, who had hoped he would be a priest.) There he fell under the sway of František Koláček, a professor of theoretical physics. Koláček smoothed the path for his protégé in all respects.

In 1906 Záviška spent a year at the Cavendish Laboratory in Cambridge, England, as a research student working on the influence of X-rays on the condensation of water droplets in cloud chambers. It was an excellent education, but the pull of theoretical physics and Prague proved stronger. At age 27 he habilitated at the Prague Polytechnic in theoretical physics with a dissertation "On the Polarization of Limiting Lines of Total Reflection"; he had wanted to habilitate at Vienna, but Ludwig Boltzmann's suicide the previous year scotched that plan. Like all of his publications, including his important articles on electromagnetic waves, his dissertation was written in Czech, which accounts for his total obscurity outside of Czechoslovakia.[38] In 1908 he was appointed an assistant in theoretical physics at the Czech University, assuming his mentor Koláček's position when the latter fell ill in 1910 and receiving a permanent appointment in 1913 upon Koláček's death. This meant that Záviška was Einstein's precise counterpart at the Czech University during his time in Prague. They never met.[39]

The absence of a personal connection did not deter Záviška from pursuing research on the electromagnetic aspects of relativity and popularizing the theory in the decades to come. In fact, in the winter semester of 1910, before Einstein's arrival at the German University, Záviška had already directed a course "On the Principle of Relativity." He kept teaching about relativity, both special and general, in 1914 and 1916, in courses that were in many ways similar to what Philipp Frank was offering at the German University.[40] These two scientists in parallel faculties did develop a professional relationship. Záviška translated some of Frank's articles into

Czech, and Frank was thanked in Záviška's 1925 book on relativity for helpful conversations.[41] Given that Frank, despite his long commitment to the city, never seems to have learned much Czech, one must presume that the dialogue happened in German (although both did speak English).

Záviška's most widely referenced work on relativity falls easily under the rubric of "popularization," which is indeed how I have so far referred to it. Today, this label is often considered somewhat derogatory, as indicating a lesser scientific genre. In the case of relativity, however, it is important to keep in mind how difficult the theory was for practicing scientists in the early twentieth century. The most obvious audience for books like Záviška's—or even Einstein's—was other scientists who were looking for a guide to orient themselves before delving into the more specialized literature.[42] In both his articles and his book-length publications, Záviška served as a mediator for this cosmopolitan new theory from a cosmopolitan scientist; by providing an explanation of relativity in Czech, he helped bring an otherwise marginal, nationally focused readership into a transnational intellectual conversation.

He also policed countervailing efforts to undercut relativity theory. In 1920 a 65-page booklet was published by Jindřich Skokan entitled *Einstein's Principle of Relativity: A Critical Study*. Záviška's scathing review of this work began with a punch and did not relent: "In this critical study there is rather a lot of criticism but rather little study."[43] But rather than simply dismissing the book as not worth reading, he took the opportunity to explain inertial reference frames and the significance of Einstein's insights about clock coordination. In his conclusion, though, he made it clear that readers should not pay attention to the critiques from this sort of author:

It is impossible to enumerate all the mistakes and illogic which the author has amassed in these few pages of his study. About the principle of relativity the reader learns very little, and what he or she does learn is perhaps all bad. The author does not know even the simplest things about Einstein's theory and his polemic gives the impression in many places of Don Quixote's battle against the windmills; that

which the author smears with such vehemence *is not* Einstein's principle of relativity, but something totally different.[44]

Skokan was easy; Bohuslav Hostinský was another matter altogether.
Hostinský was a professor of mathematics at the Czech University, and
thus Záviška's close colleague, and a specialist in differential geometry
(a field in which he published roughly 140 works over his career, typically in Czech)—the counterpart to Einstein's former colleague Georg
Pick at the German University. Hostinský's first exposure to relativity was
reading Einstein's 1916 popularization when it came out in German. He
was unimpressed. Especially with regard to general relativity, he believed
that until the conceptual details were expanded upon, "it will not be possible to assume that the general principle of relativity was exactly formulated or clearly affirmed, that it was established with reliable observations."[45] He continued his attack when Einstein published his 1921
Princeton lectures as *The Meaning of Relativity*. No one could accuse Hostinský of selecting a simplified popular presentation for his criticism:
these lectures were for physicists, and again he found their reasoning
sloppy. "Briefly said: the concept of kinetic energy does not lend itself
to relativization. Likewise, the difference between real and apparent
forces do not lend themselves to be cleared away by meditations about
tensors, invariants, equivalent systems of coordinates, light rays in four-
dimensional space time etc.," he fumed. "Mach and Einstein's wish to
understand centrifugal forces as an effect of all matter in the universe remains a wish and the physical significance of inertial systems remains
invariant toward all relativistic efforts."[46]

Záviška pushed back against his senior colleague. He cited recent findings from the Mount Wilson Observatory in California that provided
the strongest evidence to date for general relativity, and he also pointed
out that arguments similar to Hostinský's had been produced by members of the *völkisch* anti-Einstein crowd like Lenard and Ernst Gehrcke.[47]
Given the anti-German tenor of Czechoslovak public discourse at this
time, these critiques smarted. Hostinský defended himself in print, only
to be rebutted rather decisively again by Záviška, after which Hostinský,
bruised, left the field.[48] As a capstone to these efforts, Záviška published

a clear and straightforward introduction to relativity in 1925 that synthe-sized the expositions and defenses he had been developing over the previous 20 years.[49]

After Záviška's book, debates over relativity seem to have faded in Czechoslovakia, which again puts it on a similar timeline to that found in the rest of Europe—except for the dwindling number of diehards we encountered in an earlier chapter, relativity was by the 1920s simply an accepted part of physics. Záviška continued to research and teach in Prague, always in Czech. This came at significant cost to his professional standing outside the country, but it was a choice that many academics in his position also favored. His own story did not end as happily as the relativity debates. Under the Nazi occupation of Prague, he was suspected of being part of the resistance and arrested on the evening of 21 Janu-ary 1944 at his apartment. He died from illness and maltreatment just before the end of the war and was buried on 17 April 1945 at a cemetery in Gifhorn, Germany; at the time of his death, he had been on his way back home from a liberated concentration camp. In the 1960s, his body was exhumed and cremated.[50] As this melancholy aftermath suggests, Záviška, like the rest of his countrymen, was swept up in the nihilistic maelstrom that engulfed Masaryk's First Republic. In its wake everything about Czechoslovakia was changed.

* * * * *

The war arrived in Prague long before it started. In 1933, democracy ended in both Germany and Austria as Adolf Hitler and Engelbert Dollfuss abolished legislatures and tightened dictatorial control over their respective countries. Dollfuss did not last long: he was assassi-nated the following year as Nazi agents attempted to enact a coup; his successor Kurt Schuschnigg intensified what some would later call the "Austrofascist" regime until 1938. In that year, the dilemma Czechoslo-vakia faced in being sandwiched between two oppressive dictatorships ended on 12 March when Austria was annexed by the German Reich in the *Anschluss*. Now there was just the one fascist state. More than ever, Czechoslovakia was isolated and vulnerable in the heart of Europe. For most of the 1930s, it bordered on Germany, Austria, Hungary, Romania, and Poland—all of them right-wing dictatorships.

The refugees had been flooding in for years. After the Bolshevik take-over of the Russian state in November 1917 and the ensuing civil war, thousands of displaced Russians and Ukrainians, including many intel-lectuals, were welcomed by Masaryk into Czechoslovakia, where the gov-ernment subsidized their existence in the hopes that the communist fever sweeping the eastern reaches of the continent would burn itself out and then these individuals could return to rebuild the political, legal, and academic infrastructures of their states. Months stretched to years stretched to decades, and these refugees were soon joined by those flee-ing oppressive regimes, especially that of the National Socialists in Germany. Jews, antifascist leaders, and artists all swarmed into the First Republic. There were 1,100 by the middle of 1933, and the numbers only grew.[51] As literature and culture were heavily censored in the states they fled, refugees like Bertolt Brecht and Heinrich Mann participated in the flourishing of Germanophone letters in Prague.[52] Philipp Frank tired himself out helping displaced scientists find positions in the city.[53]

In 1936, despite the worrying encirclement of Czechoslovakia, Frank remained optimistic. Masaryk, who had maintained largely unquestioned leadership of the country from independence until 1935, retired, but his deputy Edvard Beneš easily slipped into control of the First Republic. "In Prague it is still relatively pleasant," Frank wrote to Einstein. "The presi-dential election has happened peaceably. Masaryk has installed his fa-vorite Beneš as his successor. This was truly an effect of Hitler. For this election was only possible because for the first time the Catholics and the Communists joined together to prevent the election of a Nazi-friendly candidate."[54] Frank had continued his cordial—and exceptional—relations with his counterparts in both physics and philosophy at the Czech University, including the philosopher Albína Dratvová, one of the first women to habilitate in the new Czechoslovakia.[55] He kept her in his thoughts as he prepared to leave Prague for a scheduled lecture series in the United States. He left for London in August to work a bit on his English before undertaking the long oceanic passage. His trip to America was supposed to last from 10 October to 10 December. The year was 1938.

Frank was fortuitously absent during Czechoslovakia's fateful hour. He kept close watch on events. "But the Czechoslovak republic is now monstrously popular," he wrote to Dratvová on 1 September. "In a tea house my waitress asked where I came from. I said 'from Czechoslovakia.' She said 'I admire Czechoslovakia.' I ask 'Why?' She answers: 'Because they resist Hitler.'"[56] The government in Prague had for some time been caught in a new iteration of a demagogic movement centered around the grievances of the German-identified citizens of the republic, who had adopted the designation "Sudeten Germans." These individuals comprised almost a quarter of the population of the country, concentrated heavily in the regions bordering Germany and Austria, and in 1933 an enterprising gymnastics teacher named Konrad Henlein had founded the Sudeten German Party to give them a voice. The timing was not a coincidence. Capitalizing on Hitler's success in Germany, Henlein stirred up right-wing sentiment amid allegations of systematic discrimination against the Sudetenlanders, demanding autonomy.

Seeing an opportunity to capitalize on the movement in order to gain access to the industry- and resource-rich border regions, Hitler—in some ways manipulating Henlein, in others manipulated by him—intervened diplomatically, threatening to invade Czechoslovakia in the spring of 1938 (in the wake of the *Anschluss*) to liberate the Sudeten Germans from ostensible Czechoslovak oppression.[57] Eager to avoid (or at least postpone) a military engagement with Germany, the leaders of France and Britain met with Hitler in Munich to broker a settlement, which was inked on 29 September 1938. Now an infamous synonym for "appeasement" of aggression, the deal permitted Nazi Germany to annex the so-called "Sudetenland" to the expanding Reich in exchange for guarantees to preserve the sovereignty of the rest of Beneš's country, now known as the Second Republic or Czecho-Slovakia. As a result, the land lost a third of its population, and the demographic balance of its ethnicities shifted dramatically. Almost all heavy industry was incorporated into Germany, and unemployment skyrocketed. More refugees—this time citizens of the very state they were fleeing to—poured in.[58] It was an unquestioned victory for Hitler and a catastrophe for everyone else.

Einstein, watching the situation as closely as he could from Princeton, was outraged. "You have confidence in the English and in particular in Chamberlain? *O sancta simpl* . . . !," he wrote to his closest friend Michele Besso, who resided in Switzerland. "He is sacrificing Eastern Europe in the hope that Hitler will tire himself out against Russia. But we will see that here too cleverness has short legs. . . . Now he has at the last moment saved Hitler in that he has placed the wreath of peace-lovingness and brought France into betraying the Czechs. . . . I no longer give a farthing for Europe's future."[59] He warned another friend from his Bern days, Maurice Solovine, that this concession to Germany would not assuage Hitler's appetite for long. "It is awful that France has betrayed Spain and Czechoslovakia. The worst thing about it is that it will be bitterly avenged."[60] War was inevitable.

Philipp Frank and his wife were on board a steamer in the middle of the Atlantic when the news of the agreement broke. He reported to Dratvová from Chicago in late October, after the carving up of Czechoslovakia had already taken place, that the French and British passengers had celebrated: there would be no war. The Czechs on board, by contrast, had been crestfallen. So had the Franks. "For us ourselves, my wife and I, we were suddenly also personally put into a difficult position," he wrote. "First difficult in a financial sense and second even more difficult in a moral sense."[61] He still intended to return to Prague in either December or January, but he was sure "I will find an entirely changed position. Our university, at which I worked for 25 years, will scarcely exist any longer and I do not at all know for sure how I will then start over. One naturally learns here very little about all the details of what happened."[62] He wrote much the same to Jaroslav Heyrovský, the leading chemist at the Czech University: "If there were any possible work, I would gladly remain in Prague and my wife especially is very attached to Prague. I can however not imagine in what a difficult position public instruction is in Prague and I am quite certain that it must today do many things that it does not do gladly."[63] He never returned. His colleagues Reinhold Fürth and Leo Pollak emigrated to Britain and Ireland in early 1939. The generation of Einstein's successors had abandoned the German University.

More to the point, it had abandoned them. As geopolitics brought Germany and Czechoslovakia into conflict, something analogous happened on a miniature scale between the German University and the Czech Charles University. Many of the faculty of the former (Frank was an exception) had never entirely reconciled with the Lex Mareš's ostensible diminution of its status as the—or at least *a*—legitimate heir of the institution Charles IV had founded in 1348. Students at the German University, many from the Sudetenland, pushed the local politics of the institution rightward, beginning with an explosion of student disorder in 1934 in response to the Nazi takeover in Germany. In reaction, the state transferred the university insignia from the German University to the Czech one, severing the last symbolic connection to the medieval ancestor.[64] In December 1934 the minister of education insisted that "the crown of Bohemia, however, never belonged to Germany and the university founded in Prague was not, therefore, established in Germany. The Charles University is the oldest university in central Europe but it is not a university of Germany. The simple fact is that there was a university in Bohemia at a time when the German lands did not possess such a seat of learning."[65] The status of the two universities would become a bellwether of the changing fortunes of Czechoslovakia. Running up to the Munich negotiations, a series of histories were published that argued, in contrast to the minister's stance, that the university had always been German.[66]

It would certainly become so. On Einstein's sixtieth birthday, 14 March 1939, Slovakia broke away from the Second Republic to form a separate client state of the Third Reich; the following day, Hitler's troops violated the Munich Accords and invaded. Two days later, Hugo Bergmann, writing in Jerusalem, recorded in his diary a meeting with Prague émigré and Bar Kochba stalwart Robert Weltsch: "Hitler in Prague yesterday evening. I was earlier at Robert's and found him entirely broken. He cannot imagine the swastika flags on the Hradschin."[67] Germany created the Protectorate of Bohemia and Moravia. Not only did the invasion make European war now inevitable, but it marked a turning point for the Reich by incorporating millions of non-Germans under its rule for the first time, creating a host of problems for Nazi racial laws as well

as linguistic problems related to managing the region.[68] The Protectorate proved to be a testing ground for Hitler's empire and a storehouse of
conscripted labor to power the war machine.

On 29 April, Hitler gave a speech in Prague declaring the city an an-
cient core of the German Empire—a standard German translation of
the "Holy Roman Empire"—and asserting that its university had been
the first university in Europe where Germans had been able to study in
German.[69] (This was of course not true; Charles IV's university had functioned in Latin for over four centuries.) If the university had not originally been "German" in a Nazi sense, Protectorate policies quickly made
it so. As a 1943 propaganda volume about Prague intoned:

> What German spirit and German industry has built over centuries in
> Bohemia was smashed in a few decades, and a dismal red cloud of em-
> bers, issued from Prague, hovered over German history. Prague Uni-
> versity indeed was made desolate, decayed, and became for a long time
> entirely insignificant for Europe because it stood under the spell of a
> confining, Czech nationalist intolerance. Only in our times is it suc-
> cessfully again in the full possession of its rights and its value. After
> its renewal as the Charles-Ferdinand University during the time of the
> Counterreformation and after the tragic interlude of its gagging by the
> Czechoslovak Republic the events of 1939 have given it back its free-
> dom and its old honorable name. It is now again called by the name
> its founder gave it—Charles University—and it has arisen to a new
> flourishing as a bulwark of the German spirit and German science.[70]

The Czech University had been affected even before the invasion: under
the presidency of Emil Hácha (who had succeeded the exiled Beneš), the
state had purged Jews from the employment rolls on 27 January 1939,
a process that had begun at the university in the fall and winter of the
previous year, when 77 teachers had been furloughed or dismissed. (Since
most of them had been part of the medical faculty, this produced a crisis
in health care.)[71] It was not long for this brave new world. On 28 October demonstrations burst out at the Czech University and were brutally
suppressed leading to some student deaths. At the funeral of one of them
on 15 November, protests resumed, and the Protectorate responded two

days later by closing down all Czech institutions of higher education, including the universities in Prague and Brno, and sending 1,200 students to prison camps.[72] Two weeks earlier it had renamed the German University the "German Charles University." The Lex Mareš was no more, and there was only one university in Prague.[73] Packed with professors from other Reich universities, it was explicitly a *German* university.[74]

The tragedies that followed struck down so many lives: of soldiers, of slaves, of victims of starvation and state brutality, of Jews. The Nazis murdered more than 78,000 "Jews" (as defined by the Nuremburg Laws) in the Protectorate; only 14,045 survived. Prewar Bohemia and Moravia had been home to 6,500 Roma; a mere 583 returned at war's end. By this time, all of the populations of Europe had been devastated. That of a reconstituted Czechoslovakia—reincorporating the wartime puppet state of Slovakia into a polity with somewhat changed borders (at the insistence of the Soviet Union, which annexed the far eastern portion by the Carpathians)—was among the most dramatically different. The Jews and Roma had been devastated. The Ukrainians now lived in the Soviet Union. And the Germans . . . well, the Germans were gone. Before 1939 roughly a third of Czechoslovakia's population had considered themselves German and somewhat over half had identified as Czechs and Slovaks; in the 1950 census the Czechs and Slovaks comprised 94 percent of the postwar country.[75]

What happened to the Germans has become one of the most persistent unresolved legacies of World War II. What was at first euphemistically labeled a "transfer" (*odsun* in Czech) has now generally been more accurately labeled an "expulsion" (*vyhnání* in Czech, *Vertreibung* in German). Millions of German-identified citizens of Poland and Czechoslovakia were deported to occupied Germany in the years immediately following the war, predominantly to the American sector in the case of the latter. First proposed by Beneš and then vigorously supported by the Soviets and the local Communists, this plan to remove "Germans" (and, to a lesser extent, Hungarians) from Czechoslovakia as a form of retribution for the role Sudeten Germans had played in the onset of the conflict was condoned by the Allies. In the first quasi-spontaneous waves of the "wild transfer," thousands of citizens of Czechoslovakia who had been identified as Germans were murdered; in the end, millions were displaced and forced to make a new life outside of the country where their families had lived for centuries. The population of Czechoslovakia after the expulsion was 12.4 million. It would take over 30 years for it to return to its 1935 level.[76] The new postwar Czechoslovakia faced a different reality than it had ever confronted: essentially, a land without Germans. A symbolic marker of the change was the abolition of the German University on 18 October 1945 and the reopening of the Czech institution as the only Charles University.[77] After a millennium, Bohemia was gone.

In its aftermath, Czechoslovak politics was reoriented away from what had always been its polestar of nationality-baiting and confronted a new organizing principle: engagement with global communism. The Czechoslovak Communist Party had been founded in 1921, and at the time was the only party in the country that had both Germans and Czechs in its ranks.[78] During its few months of existence, the Second Republic criminalized the Communist Party, proscribing its publications in one of its first acts in October 1938 and then banning it altogether in December. That, and the dominant role the Red Army played in liberating the Protectorate, boosted the popularity of the party in 1945, making it a force to contend with at the polls. In May 1945, there were roughly 50,000 Communists in the country, who worked hard building a network of cells; two months later the party boasted 597,000 members. That size was extraordinary, even compared to Russia. By January 1948, despite a purge the previous year to thin out the ranks, membership had reached 1.31 million, making it the largest communist party in the world on a per capita basis.[79] In the 1946 elections, the Communist Party won 38 percent of the vote.[80] Edvard Beneš, who was again serving as president, was forced to move in the Communists' political landscape.

The coming of Soviet-style Marxism-Leninism to Czechoslovakia would change just about everything in Czechoslovak life and culture. Some of those transformations had to do with the party and the state it transformed. Others were consequences of the trauma of the war and the expulsions. Rather than the ebullient cosmopolitanism that had accompanied the optimistic Masaryk, the coming years would see an inward turn in many areas of Czechoslovak culture. One way to track the reverberations of this new world is to return to the image of the physicist who had once taught at a now-defunct university in the capital city in 1911–1912, a Habsburg moment that seemed so far distant as to be almost in the Pleistocene, although it was recent enough that there were living individuals who remembered Einstein and tried to invoke and shape his legacy for the Marxist future. Rather than reaching outward to the world of learning, this Communist-era invocation of Einstein and his relativity theory faced inward, closing off Prague from the non-Soviet world. Its main propagandist was a native of the city named Arnošt Kolman.

* * * * *

In 1892, Jaromír Kolman, a nonobservant Jew who worked as a postal official, and his wife Julie had a son they named Arnošt. Precocious and good at school, he advanced quickly through his courses—as it happened, at the same *gymnasium* that Oskar Kraus had attended two decades earlier—and in 1910 enrolled in the Czech Polytechnic to study electrical engineering. Drawn to politics, his first love was Zionism, which he would later claim was triggered by outrage at the anti-Semitic Hilsner affair.[81] He studied Hebrew in courses offered by the Bar Kochba student group, and he was the translator into Czech of Martin Buber's *Three Lectures on Jewry* that had so galvanized Hugo Bergmann and the rest of the Prague Zionists.[82] Over his parents' objections (they wanted him to have a practical education) he transferred to the Czech University in 1911 to study mathematics. Upon graduation, he took a job as a calculator at the observatory.

Kolman enters our story because of an incident he related in his memoirs, originally a 720-page typescript written in Moscow beginning in 1970 and only published posthumously, first in an abridged German translation, and then a few years later in a longer, but still abridged, Russian version.[83] The story takes place in 1911, when Kolman's friend Růžek, a fellow math student, persuaded him to attend lectures being given by the new professor at the German University, Albert Einstein:

As is customary at the inaugural lecture of a new professor, the rector introduced Einstein. Einstein was a medium-sized, rather sturdy, still very young man with wildly curly hair, somewhat carelessly dressed for a professor and for this ceremonial moment. And without beating about the bush he began at a rapid pace to expound the principles of the special theory of relativity. That was supposed to be only the introduction to his lectures about general relativity, on which he was then working, especially on its application to cosmology. Already in this first lecture it was obvious to see that the majority of the audience were hardly in a position to follow the special "dry" physical-mathematical presentation of the speaker, and that Einstein—at least this time—would not offer those generally philosophical reflections on behalf of which the audience most of all came—when they weren't there just to see this Coryphaeus, or because it was good form to be here.

Already after the second lecture the auditorium began to noticeably thin out. My friend Růžek and I remained, and it happened in a decreasing geometric progression until it had shriveled up into a stable little cluster of something more than ten enthusiasts. Einstein was not particularly struck by this. Apparently he held to the same principle as one of my mathematics professors, Sobotka, who liked to say: "That doesn't matter, *tres faciunt collegium*—three make a society—God the Father, the Son, and the Holy Ghost are always present, and therefore I will also lecture when none of you, gentlemen, shows up."

Einstein came to the following lectures dressed informally, sometimes in a sweater, often without a tie. Once though he came in two ties! At that time people wore knickers, the legs of which were fastened underneath with ribbons. These ribbons were in his case often open, hanging about loosely, and were soiled from the mud of the Prague streets. He understood that we students counted Einstein among the oddballs. But the odd thing was something entirely different—the content of his lectures, the form of his presentation and his relation to the audience. Almost every one of his lectures was comparable to a creative act. He talked about what was occupying him at that

moment, partly arguing with himself and encouraging us to participate in this confrontation. He wrote an equation on the board, and
sometimes he did this with gigantic symbols, often however in small,
almost illegibly small, script. And then he suddenly stopped, thought
about it, and said completely excitedly: "Gentlemen, why didn't you
say anything? This is all wrong, this is a mistake!" He wiped everything
off and wrote something else on the board.

Sometimes he would postpone continuing and explain: "We must
first think more about this." What a difference between Einstein and
the infallible, godlike politicians who want to recognize nothing in the
world as a mistake which they have committed! Einstein liked it when
he was interrupted—that would have been unthinkable with other
professors—when one posed the most devious questions to him. He
loved long discussions with the audience. Often—even during rainy
weather, which is often the case in Prague—a small crowd of students
would accompany him on the walk home after the lecture. And when
the interesting discussion had not yet come to an end, then he would
be—notwithstanding his world fame—quite capable of turning
around and accompanying his accompaniers back to the university
dining hall.

But Einstein's popularity also already then had its flip side. Nationalist, anti-Semitic German students who belonged to various dueling
societies tried to spoil his time in Prague using every means. Perhaps
then there awoke in Einstein—partly as a form of internal protest—
the consciousness of his belonging to the Jewish people, or at least it
was intensified by this. He began to associate with the Prague Jewish-
nationalist and Zionist intelligentsia. And although he was not an
Orthodox Jew, he began to play his beloved violin in the synagogue
on the Jewish high holy days.

As you know, Einstein went from Prague to Berlin [sic], where he
was offered extraordinarily favorable conditions for his work, until the
horrifying wave of National Socialism flooded Germany.

Without any embarrassment I confess that it was very difficult for
me to understand Einstein's lectures. One had to learn a new apparatus of special tensor analysis for this. But I persisted, together with

Růžek, to whom I owe my meeting with the greatest physicist of our time, the first of the many remarkable people I have met in my life (which has not in other parts been all sunny). This seems to me a small compensation for that. I remained one of the few audience members until the end.[84]

This could never have happened. First, Einstein had not yet developed his theory of general relativity while he was in Prague, so he could not have lectured on it in 1911. Second, the only public lecture he *did* give, on special relativity, took place on one evening in the spring and was devoid of mathematical detail. Third, Kolman's name does not show up on the list of auditors of Einstein's courses. (The business about him playing violin in the synagogue is a complete fabrication.) However vivid the description—and the prose is characteristic of Kolman's memoirs— the account of this episode is unreliable.

Unreliability is something of a theme of Kolman's career. Despite the apocryphal nature of this first encounter, it is worth following Kolman's relationship with Einstein closely; even if it was a conscious fabrication (which is by no means clear), Kolman's fascination with relativity theory is amply confirmable, and he seems to have believed his own story. Commentators often referred to him as a "one-time student of Albert Einstein."[85] Perhaps nobody played a larger role in shaping the popular understanding of special relativity in the Soviet bloc, which represented at its height the largest scientific community in the world. From these inauspicious beginnings, Kolman's career encapsulated in miniature the tumultuous career of Czech communism, and his obsession with Einstein runs like a red thread through it.[86]

First, he had to leave Zionism and find communism. When the Great War broke out in 1914, Kolman was immediately drafted into the Habsburg army. He was not a lucky soldier. Sent to the eastern front, he was wounded, and after recuperation he was sent back and wounded again, this time falling into Russian hands and interned in a prisoner of war camp. Gifted in languages, he worked as a translator, and when the Bolsheviks arrived in 1918 to free the prisoners and declare the advent of the Marxist revolution, Kolman—who had dabbled in social-democratic politics back in Prague—was all for it.[87] He joined the Communist Party and threw himself into a variety of roles: property expropriation, undercover work in Germany, organizing in Moscow, and more. According to his memoirs, while on a Comintern mission to Düsseldorf, he ran into Einstein again and asked the physicist to give a talk to raise funds to help provide relief for a famine in the Soviet Union. "He greeted me, and it even seemed to me as though he still dimly remembered me from Prague," he recalled.[88]

By the beginning of the 1930s, after a motley collection of experiences, Kolman had returned to his earlier love of mathematics and science and resolved to become a philosopher of science, earning his doctorate in the subject in 1934 after working for several years at the leading ideological journal of the Soviet Union, *Under the Banner of Marxism*. It was during this decade and in this periodical that he began his career as a prominent exponent of relativity theory within the framework of dialectical materialism, the Soviet Union's official philosophy of science. This turns out to have been a crowded and quite cutthroat field.

As the Bolsheviks began to reconstruct the scientific system they had inherited from imperial Russia, they both needed new cadres to replace those that had either fled or been removed from their positions and wanted to enforce ideological conformity to the philosophy of science developed by Vladimir Lenin in *Materialism and Empirio-Criticism* and earlier by Friedrich Engels in *Anti-Dühring* and *The Dialectics of Nature*. The last had been posthumously published in 1925 in both Russian and German editions, a process that Kolman was intimately involved with (he would also help prepare the Czech translation in 1952 and write the preface for the Czech translation of *Anti-Dühring* in 1947).[89] Strangely enough, Einstein himself was consulted in 1924 about whether *Dialectics of Nature* should be published at all and provided this assessment: "If this manuscript came from an author who was not interesting as a historical personality, I would not advise publication because the content is not of especial interest either from the standpoint of today's physics nor also for the history of physics. On the other hand I can imagine that this text would come into consideration for publication insofar as it forms an interesting contribution to the illumination of Engels's intellectual personality."[90] Obviously, the curators of the Marx-Engels Institute in Moscow went ahead with it.

During the 1920s, as with so much else in early Soviet culture, there was a wide degree of experimentation in dialectical materialism, but by 1930 Joseph Stalin had intervened and forced both of the warring camps of so-called "mechanists" and "Deborinites" into line. The emphasis would no longer be interpreting the philosophy, but deciding which sciences were compatible with it—and so much the worse for those that were not. All cutting-edge developments in modern science were up for discussion: psychoanalysis, genetics, quantum theory, and relativity.[91]

The positions on relativity ranged from those that denied that Einstein's theories (special or general) had any validity at all and demanded a return to Newtonian physics to those—espoused by a subset of the physicists—that considered the philosophy irrelevant to the science. Arnošt Kolman appeared in the middle of these pitched debates, widely considered both at the time and by later commentators as one of the few philosophers who fully understood the technical issues and tried to find a way to preserve the theory in the face of the hardening Stalinist line.[92] Throughout the rest of his career, which displayed what can charitably only be called significant ideological flexibility, Kolman continually wrestled with how to interpret both relativity and its creator.

Kolman maintained throughout his life two main positions on these questions: one about Einstein and another about relativity. He did not have much patience for Einstein's philosophical views, which he contended "stand on Machian and simply idealist positions," positions that Lenin had devoted the bulk of his writings on science to excoriating. That did not mean that a good Marxist should dismiss him: "However he is not just a bad philosopher, but is still above all a great physicist. He cannot therefore reconcile himself to the rotting of physics through such a theoretical direction, which in effect would mean the liquidation of the science."[93] Not only was his philosophy bad, but his politics were also disappointing: "Einstein always was in politics the same kind of unreliable fellow traveler as in philosophy, and together with all of the 'left'-radical bourgeois intelligentsia of Western Europe preached a mix of pacifism, Zionism, humanism, and other cheap democratic liberal nonsense."[94] He wasn't anti-Soviet, but he wasn't pro-Soviet either. This need not be a problem, Kolman contended: by separating the science Einstein did from his philosophical interpretation of that science, Soviet physicists could continue to work on relativity without fear of ideological apostasy.

And a good thing, too, since relativity—especially the gravitational theory of general relativity—was a boon to dialectical materialism. As Kolman told a group of Columbia University students visiting Moscow State University on 5 August 1935: "Relativity theory has proven that spacetime is one form of existence of matter, that our relative knowledge is knowledge of the absolute material world, it has affirmed once again the Engels and Lenin thesis of the asymptotic, spiraling approach of knowledge to reality."[95] Once one neutralized the allegations of philosophical relativism, which was distinctly un-Marxist, there was nothing that demanded censure. Kolman used this position, buttressed by his continued reading of the state of the art in gravitational physics, to defend the theory against even more hard-line Stalinists—he himself was no saint and had orchestrated purges within the Moscow mathematical community—who were on the warpath.[96] To maintain this balance, he repeatedly attacked Philipp Frank as a Machian shill who attempted to pervert the theory with bourgeois window-dressing.[97]

During World War II, Kolman headed the department of dialectical materialism at the Institute of Philosophy at the Soviet Academy of Sciences. His brusque manner ruffled some feathers, and the leadership soon needed to find something else for him to do. Fortunately, the perfect opportunity arose, and they decided to send him to Prague to advise the Czechoslovak Communist Party. Several Soviet professors were also being sent to the Charles University to teach subjects like Soviet literature and history, and Kolman was assigned to teach Soviet philosophy. (There was a dearth of trained philosophers among Communist Party cadres in Prague.) There were some difficulties in obtaining the post (Kolman had not habilitated and retained his Soviet citizenship), but they were smoothed over thanks to the enormous good feeling toward the Soviets. Also, Kolman, unlike several of these new professors, spoke fluent Czech, although it was by now studded with Russianisms, which actually gave him an additional appeal.[98] He was quite popular among students. In the meantime, he ran agitation and propaganda efforts for the Communists.[99]

His work was a success, from the party's point of view. Since the 1920s, the Communist Party had been a fixture in Czechoslovak politics, operating with a certain degree of distance from Moscow. After the Second Republic banned the Czechoslovak Communist Party and the Protectorate police hounded its members, the leadership fled to the Soviet Union, where linkages between the two groups became extremely tight. As the Communists won parliamentary elections in the postwar republic, the restored president Edvard Beneš was forced to deal with their leader, Klement Gottwald. With the onset of a frigid Cold War between Stalin and the Western allies and the defeat of Communist forces at the polls in Italy, Moscow pushed for a nonparliamentary path to power. Fearing they would lose in the upcoming May 1948 elections, Gottwald and his deputy, Rudolf Slánský, maneuvered Beneš into a corner, and he capitulated on 25 February 1948. The coup was complete, and the Communists began over four decades of rule.

Kolman, buoyed by this victory, continued to propagandize without clearing his views with Gottwald and especially Slánský. In September 1948, the philosopher published an article in the weekly journal *Tvorba* that warned that Czechoslovakia ran the risk of straying from the Stalinist line and leading the country into a revisionist nationalism analogous to that installed in Yugoslavia by Josip Broz (aka Tito).[100] Slánský was not amused. He called Moscow, and the secret police arrested Kolman in Prague on 26 September. He was whisked back to Moscow, where he spent over three years imprisoned in the Lubianka, the fearsome NKVD headquarters. Western Prague-watchers noted Kolman's disappearance but did not know what had happened to him.[101] His family feared the worst. And then, just like that, in 1952 things changed. Stalin decided to purge the Czechoslovak Party and make an example out of Slánský for being . . . a revisionist nationalist like Tito. Slánský was hanged in December 1952, and Kolman was free.

He was also unemployed. He soon got a job teaching mathematics at an automechanical institute, but he wanted more: a position at the Academy of Sciences working on dialectical materialism or a return to Prague. In April 1953, a month after Stalin's death, Kolman turned to his former boss in Moscow Communist politics: Nikita Khrushchev, now chair of the Central Committee. The problem, as Mikhail Suslov, Khrushchev's advisor, told the general secretary, was that nobody in Prague wanted him back. (The philosophers at the Academy were equally adamant.) Instead, Khrushchev found another sinecure for him, as a historian and philosopher of mathematics at the Institute of the History of Science in Moscow.[102] From this position, he wrote biographies of Nikolai Lobachevskii, one of the creators of non-Euclidean geometry, and of Bernard Bolzano, the Bohemian philosopher and mathematician, and spearheaded a successful campaign to rehabilitate the new field of cybernetics from its purdah as a bourgeois pseudoscience incompatible with dialectical materialism.[103] He also continued to publish, as always, about relativity and Einstein, hitting the same themes he had worked out during the Stalinist 1930s.[104]

After once again airing his views about relativity for six years, Kolman got his wish: a return ticket to Prague. He was appointed director of the Institute of Philosophy of the Czechoslovak Academy of Sciences, as well as the editor-in-chief of the new journal *Filosofický časopis*. Here, though, he was not insulated from critiques of his views—including his interpretation of relative motion—to the same degree he had been in Moscow.[105] He encountered resistance both from the Stalinist old guard, who resented the returned prodigal, and from younger Marxists, who were experimenting with what would come to be called "revisionism," which blossomed in the 1960s into a reformist trend within Czechoslovak communism. However, Kolman's views concerning politics and Stalin had adapted to the changing winds of the moment, and he was closer to this new generation than its members had anticipated. (It seemed only his positions on relativity held firm.) In December 1962, he gave a lecture to the Union of Writers in Prague, ostensibly about current developments in science and technology and how they necessitated a reevaluation of older views, but he quickly shifted to a criticism of bureaucratism:

> Despotism, tyranny as the catalyst of the mold of bureaucratism penetrated into all social institutions in art, in literature, the same as in science. Bureaucratism manifests primarily in that bureaucratic centralism established control of the positions of democratic centralism,

that it created a closed, privileged civil service which strove for the heartless regimentation of cultural life. The bureaucratization of culture, science, and art had also as a consequence the disarming of an array of cultural workers by careerists, people who are alien to the interests of literature, art, and science.[106]

These statements were immediately denounced, and Kolman was summoned back to the Soviet Union.[107] This time he was not placed under arrest. He left full of bitterness that he, an actual comrade of Lenin, had been so slighted by the Czech *apparat*. As he put it in his memoirs: "So I left Prague and yet did not have any illusions that it would be better for me in Moscow."[108]

Indeed it was not. The comments he had made at the writers' meeting obviously percolated back to the Soviet leadership, and he was tarred as a "revisionist" himself, an identification that was reinforced by the involvement of his daughter Ada and his son-in-law, the physicist František Janouch, who were living in Prague with Kolman's grandchildren, in the moderate socialism of what was coming to be called the "Prague Spring." He continued to publish articles in Czech that invoked special relativity as an example of proper philosophy, railing against the "dogmatism" of the older guard.[109] He was occasionally allowed to travel back to Prague, but in 1968 that became much more difficult.

On the night of 20–21 August 1968, the Soviet leader Leonid Brezhnev—who had succeeded Kolman's patron Khrushchev in a 1964 coup—sent Warsaw Pact tanks into Prague to end the reforms pursued by the regime of Alexander Dubček. The rollback, led by the increasingly hardline Gustáv Husák, was called "normalization" (*normalizace*).[110] Within a few years, Ada and František Janouch would emigrate to Stockholm. Before then, Kolman began to notice that his right to travel within Eastern Europe had become restricted. He last traveled to Prague in May 1971, a privilege he attributed to an oversight by the regime.[111] After he was denied permission to visit his grandchildren in Sweden, he and his daughter petitioned the regime aggressively to let him go, corralling a pantheon of foreign supporters who had met Kolman during his decades of philosophical jetsetting.[112] Intriguingly, one of his most frequently cited complaints about the Soviet Union was that it refused to publish an article he had written about relativity theory.[113] A plea to Brezhnev from Sweden's socialist prime minister Olof Palme—about Kolman's desire to visit his family, not Einsteinian physics—eventuated in an exit visa for Kolman and his wife.

If the Soviet regime expected the Kolmans to return, it was rudely disappointed. Kolman resigned from the party (after 58 years), filed for political asylum, and announced his apostasy in an open letter to Brezhnev reprinted around the world, including the 13 October 1976 op-ed page of the *New York Times*. For a man who had built his entire life around the interpretation of universal science, the letter to Brezhnev emphasized his Czechness. "However, 1968 was the real turning point for me, when I had occasion to observe the 'Prague Spring' and see with my own eyes with what enthusiasm the united people of Czechoslovakia backed the strivings of the party to rekindle the Socialist ideals and the fight for Socialism with a human face," he wrote. "When your tanks and armies occupied Czechoslovakia, subjecting it to your political *diktat* and merciless economic exploitation—in short, into your colony—I lost any illusions I may have had about the nature of your regime."[114] That said, he remained a Marxist, a determined critic of religion and also of capitalism.[115] He continued to write, almost obsessively, about relativity.

He was panicked about the fact that the article that he had written about relativity—the one that the Soviet journals had censored—remained unpublished. He wrote again and again to his longtime friend Dirk Struik, a mathematician (and former Dutch communist activist) at the Massachusetts Institute of Technology, and Robert S. Cohen, a Marxist philosopher of science at Boston University, asking them to publish his piece: "During the last six years I have troubles with the publishing of all what I write, here and in my native land. This article was accepted by the editorial board of 'Voprossy filossofii' two years ago, three times it was inserted in the current number and three times taken off by an intervention from outward."[116] In November 1976, Cohen wrote him that the paper would be forthcoming in 1977 in a series he edited, because there was simply no other place that might take it: the English was bad (Cohen tried his best to fix it), it was too focused on science to be of interest to general leftist magazines, and the philosophy in it was too elementary for professional venues.[117] Even while in the hospital in 1978, however, Kolman had his wife inquire about whether the piece had appeared and whether he could get an offprint.[118]

It is not clear that he ever saw it. On 22 January 1979, Arnošt Kolman died in Stockholm. Obituaries for him, focusing on his transformation into a dissident, did not neglect to emphasize one of his own favorite ways of describing himself: as a "confidant of Lenin and pupil of Einstein."[119] A long life that had been intertwined with the history of twentieth-century Czechoslovakia, international Marxism, and the philosophical interpretation of relativity theory had come to a close. He was also one of the last people who could have claimed (however dubiously) to have met Einstein in Prague.

* * * * *

Almost exactly two months after Kolman's death, the world would celebrate the centenary of Albert Einstein's birth. This would turn into an important moment for Husák's regime to commemorate Einstein's time in Prague, but it was a commemoration very different from the celebrations of cosmopolitanism found in the tributes by František Záviška or even—after a fashion—Kolman. This would be a commemoration of a local Einstein for a city mired in the doldrums of normalization.

The state had shut down the cultural and political efflorescence of the Prague Spring of the 1960s in a process that swept through every sector of Czechoslovak life, muzzling intellectuals, freezing their children out of education and opportunity, and attempting to tamp down any signs of dissent. These harsh measures were supplemented by officially sanctioned celebrations, and anniversaries fit the bill nicely. You could always find one. In 1970, Lenin's centenary was commemorated everywhere. The following year, the Czechoslovak Communist Party marked its fiftieth birthday. In 1973, the twenty-fifth anniversary of the Communist takeover of Czechoslovakia was observed, attended by Brezhnev—who had ordered the invasion only five years earlier. Then followed the thirtieth anniversary of liberation from the Nazis in 1975, what would have been Gottwald's eightieth birthday in 1976, and so on.[120]

On 18 April 1969, nine months after tanks had rolled through Wenceslaus Square in the center of Prague, an article appeared on the front page of the youth newspaper *Mladá fronta* entitled "Prague Forgets Einstein." Although there had been a few obituaries and historical articles in Czech physics journals since the scientist's death in 1955, the author argued, there was insufficient acknowledgment of his time in Prague: "It is not necessary to discuss here his contribution to science and to humanity. As yet there is in Prague no memorial plaque to this scientist; to honor his memory with the issuing of a single sign is not an extreme demand."[121] There were streets named after Einstein around the world, as well as countless memorial plaques. Why not in Prague, where he had lived from April 1911 to August 1912?

Articles about Einstein in Prague began to appear in popular science journals throughout the 1970s, including one that included a rare interview with Einstein's eldest son, Hans Albert.[122] Yet it was not entirely true that there was no memorial to the physicist in the city. In 1974 Hugo Bergmann, then in Israel, was consulted about a commemorative plaque to Einstein that had been unveiled in 1966 at the physics institute on Viničná ulice, which the authorities now proposed to move from an inside wall to the outside of the building so it would be visible to the public. (Bergmann told none other than Arnošt Kolman, with whom he had reconnected after several decades, about it in a letter.)[123] The impending arrival of the hundredth anniversary of Einstein's birth on 14 March 1979 provided another opportunity to fête the physicist.

On the eve of that date, a statue was unveiled on the outside of Einstein's former apartment on Lesnická ulice (figure 4). J. Kilián, of the mayor's office, and Jaroslav Koževník, chair of the Czechoslovak Academy of Sciences, presented the statue designed by academic sculptor Milan Benda and engineer Ivan Hněvkovský.[124] The event capped a series of celebrations that had begun in February when Koževník and Zdeněk Česka, rector of the university, presided over a symposium at the main hall of the Carolinum, where Einstein had taught. The featured speakers were John Archibald Wheeler, a distinguished physicist from Princeton University who had known Einstein personally, Einstein's former assistant and collaborator at the Institute for Advanced Study (and

FIGURE 4. The memorial bust of Einstein outside his apartment at Lesnická ulice 7, in the Prague neighborhood of Smíchov a few steps from the Vltava River. The lettering reads: "Here lived and worked in the years 1911 and 1912 Albert Einstein." *Source:* Author's photograph.

onetime Frank student at the German University) Peter Gabriel Berg-mann, and Czechoslovak gravitational physicist Jiří Bičák.[125] Bičák also published a slim, 60-page booklet entitled *Einstein and Prague*, which in-cluded his own historical introduction, heavily shaped by Philipp Frank's biographical treatment of Einstein's time in the capital, and the German original and Czech translation of the paper on starlight bend-ing around the sun, the chief scientific fruit of Einstein's Bohemian year.[126]

It is hard to overstate how unusual this monument was in Czechoslo-vakia, a country where monuments were, both before and after the Communist coup, taken very seriously as ways of narrating public his-tory.[127] It appears that this monument from 1979 was the only public bust erected that commemorated a professor from the *German* University. (Later, monuments to Ernst Mach and others would be unveiled.) Ein-stein was, in many ways, an excellent subject. He was world famous and so would lend some luster to Prague and its scientific legacy. True, he had not been a Czech, but he also had not been a "German," not really. He certainly had not been a Prague German, a member of the city's long-standing Germanophone community, which had been largely expelled in the wake of World War II. He had also not been a Habsburg German imported from Vienna. He had barely even been a German German: a Swiss citizen, he had renounced his affiliation with the *Kaiserreich* and was widely celebrated as one of the great opponents of the Third Reich. His Jewishness also mattered: lionizing Einstein was another way to castigate the bygone political menace of the Nazis, always useful in public propaganda.

These were all internal justifications within the idiom of Czechoslovak communism. Focusing on them, as the discourse around the plaque outside his apartment did, required muffling the other symbolic resonances that surrounded the physicist, including the many that had emanated from his time in Prague and that have been explored in the earlier chapters of this book. The motto underneath the bust was a prime example of this closed-off vision of Einstein: "Here lived and worked in the years 1911 and 1912 Albert Einstein." That should definitely not be the last word.

Princeton, Tel Aviv, Prague

My circle of Prague attractions is different, indeed rather
difficult to define—I want to say that I love something in
Prague that occupies the middle between "historic" and
"everyday," an unmodern Prague: things that are neither
especially worth seeing nor especially new—somewhat
half-old things, that also when they were still new were not
exactly worth seeing—that yet even in this modesty and
unobtrusiveness, yes, unpretentiousness of their entire
essence convey so much more of their own spirit of times
past as the historically important parade pieces. A past in
plainclothes, in a weekday suit . . .

—*Max Brod*[1]

By 1979, the year of the Einstein centenary, just about everyone who had
known Albert Einstein during his 16 months in Bohemia was dead. Many
of them had passed away before Einstein did on 18 April 1955, which only
stands to reason given that he had been appointed ordinary professor at
the German University at an extremely young age. Those who had stayed
in Prague had been culled by two world wars, the Nazi occupation and
genocide during the Protectorate, the economic and political ravages of
communism, the Soviet-led invasion of 1968, and normalization—as well
as the passage of 67 years. Most of Einstein's interlocutors whom we have
followed in the preceding pages did not die in Prague but outside of it,
displaced by those same events. Einstein watched many of them go: Ernst
Mach in 1916, Bertha Fanta in 1918, Max Abraham in 1922, Gustav Jaumann
in 1924, Anton Lampa in 1938, Georg Pick and Oskar Kraus in 1942,

Mileva Marić-Einstein (whom I have called, during her Prague period, Einsteinová) in 1948, and Gerhard Kowalewski in 1950. Those who have featured most prominently in this book died after him: Philipp Frank in 1966, Max Brod in 1968, and Hugo Bergmann (then called Schmuel Hugo Bergman) in 1975. Arnošt Kolman passed away in January 1979, a few months before the centenary celebrations, which he most certainly would have enjoyed. The eyewitnesses were gone, but the birth of the Einstein industry has made it possible to preserve their traces and follow the ramifications of Einstein's Bohemian moment.

The year 1979 was also a cusp for Prague, if not for the Bohemia that had passed into oblivion decades before. The changes were hard to see in Czechoslovakia—blocking change was the whole point of Gustáv Husák's policy of normalization—but they were there. In January 1977 a group of artists, writers, philosophers, and citizens released a document they called Charter 77, drafted the previous year and signed in December. It called for respecting human rights and political freedoms; its promulgation was considered a political crime by the regime. At first, the repression seemed to work. In May 1979 one of the most visible of what had come to be called (in particular by the foreign press) "dissidents," a blacklisted playwright named Václav Havel, was once again imprisoned, this time for his longest term. (He was released in February 1983.) Cracks were showing in the Eastern Bloc, ranging from the Soviet invasion of Afghanistan in 1979, to the revolving funeral cortege of geriatric leaders of the Soviet Union, to economic and political protests in Poland and elsewhere. By 1989 the lid had blown off the region, and former prisoner Havel had become the front man for a sweeping political movement. On 29 December 1989, he was unanimously elected president of Czechoslovakia by the Federal Assembly. In 1993, Czechoslovakia was dead, split into two states, the Czech Republic and Slovakia.[2] Prague was still the capital of the former, but it had traveled quite a long way from the Habsburg Empire.

The two trajectories—of Einstein and of Bohemia—had sharply diverged, though each would supply ample material for posters in college dormitories, especially as tourists began to discover the charms of the now-accessible Prague. Einstein stood in more than ever for the quintessential genius, an increasingly disembodied icon who represented abstract ideas ostensibly deracinated from physical location and historical epochs, even as they were fundamentally about space and time. Prague became a synecdoche, first for political liberation and then for artistic ferment and experimentation. The links that had bound individual and city together, forged as they had been by personal connections, had evaporated when the people who had sustained them were no more.

It is the goal of history to make such bonds visible anew. The Einstein who has emerged on these pages, viewed through the filter of his ties to Habsburg Prague, does not appear as precisely the same individual who can be found in his many biographies, the dominant genre in which we have treated this most charismatic of scientists. Our Einstein was an intensely social being, was not always in the right politically, philosophically, or scientifically, and was not always the main character of the story. (One intrinsic feature of biographies as a genre is that the titular character rarely yields the stage.) He was a human being, defined as all human beings are by how he interacted with the world around him. Prague, too, is different: a city shot through with a scientific sensibility, one where Germanophones and the Czech-identified repeatedly interacted over intellectual topics, and one where the memory of the past was a constant resource for the present. Those aspects of Habsburg Prague, of the capital of what was once known as Bohemia, have been altered beyond recognition with the transformations of the city. (To be clear: there is still plenty of scientific activity in Prague; its links with the international system have, however, been entirely reformatted.)

While the emphasis on the city of Prague and Einstein himself has enabled us to see all of these resonances, it has also shunted to the side those stories about Einstein and Bohemia that unfolded much later and far outside the geographic bounds of the Czech lands. In conclusion, I will briefly visit two of those other far-flung locations—Princeton, New Jersey, and Tel Aviv, Israel—before returning briefly to the metropolis that has animated this book.

* * * * *

In the Rare Books and Special Collections Department of Firestone Library at Princeton University, you can call up a surprising coffee-table book. The volume itself is not that extraordinary: a German-language work by Karel Plicka from 1953, entitled *Prague: A Photographic Picture Book*, it contains gorgeous black-and-white photographs of the classic vistas of the city on the Vltava. There are many such books. What makes this one special is a German inscription inside:

> To dear Hanne
>> for mutual homesickness
>>> A. Einstein. 54.[3]

The book was one of two inscribed by Einstein and given to Hanna Fantova—the other being (somewhat vainly) a copy of Carl Seelig's biography of the physicist, *Albert Einstein und die Schweiz*—both of which she donated to Firestone Library, where she worked for many years as the first librarian dedicated to collecting maps.[4] That Einstein gave Fantova gifts is not that peculiar: they had formed a close relationship in the several years before his death, speaking on the phone almost every evening (Fantova kept an invaluable log paraphrasing these discussions, also deposited in the archives at Firestone), and taking frequent sailing trips (figure 5). She was younger and vivacious, clearly admired the distinguished scientist, and was someone with whom he could speak German without triggering the anger he felt toward the German people.[5] The inscription seems a bit of a puzzle. "Homesickness" for Prague? That, too, makes sense when we trace Fantova's life to the point at which she got to stand on a sailboat in Lake Carnegie.

Hanna Fantova was born Johanna Bobasch in Brünn (today Brno), the capital city of Moravia, the Habsburg province just to the east of Bohemia, on either 6 or 9 April 1901.[6] She was 10 years old when Einstein arrived in Prague, and they did not meet then. She went to *gymnasium* in Varnsdorf, at the very northern edge of what would later be called the Sudetenland on the border with Germany, and at some point settled in Prague, where she met the man she would eventually marry: Otto Fanta. After the creation of Czechoslovakia, it became commonplace for citizenship documents for Germanophone citizens to include their name written in the Czech style, so Johanna Bobasch became Johanna Fantová (somewhere in her travels, the accent was lost). It was through Otto Fanta

FIGURE 5. Albert Einstein and Hanna Fantova preparing to sail on Lake Carnegie, an activity
they undertook often (though she seems overdressed on this particular occasion).
Source: Hanna Fantova Collection of Albert Einstein, Rare Books and Special Collections,
C0703, Princeton University Library, Box 2.

that she came into Einstein's orbit. Given his personal connections to the physicist, it would have been hard for her not to.

Otto Fanta was the son of Bertha Fanta, the woman who ran the salon where Einstein made so many connections to persons, like Brod and Bergmann, who would be prominent representatives of Prague in his later life (as well as those, like Franz Kafka, whom he would forget). Otto was born on 26 October 1890 in Prague, which meant he was of university age during Einstein's tenure at the German University. He audited a physics course during the physicist's last semester in summer 1912, though Fanta was officially enrolled at the University of Berlin.[7] (Later, Philipp Frank would serve on Fanta's doctoral committee.) He also met with Einstein frequently during the discussion evenings at his mother's house, where he became especially close to the mathematician Gerhard Kowalewski, himself an astute observer of Einstein.[8] In October 1915, Otto Fanta began teaching chemistry at a *Staatsrealgymnasium* in Prague, retiring from that position in April 1930. In subsequent years, he moonlighted as an unpaid handwriting analyst for the Prague Police Department.[9]

Just before leaving his teaching job, Otto took his young bride to Berlin, where he wanted to take in the theatrical season. He renewed his acquaintance with Albert Einstein, now at the pinnacle of his career in the German capital. Einstein definitely remembered Otto Fanta, although perhaps not from the salon or from a course audited almost 20 years earlier. In the early 1920s, Fanta had collaborated with Otto Buek, Rudolf Laemmel, and Georg Nicolai on a screenplay for a movie entitled *The Foundations of Einstein's Theory of Relativity* that had been intended to popularize the relativity of motion and simultaneity for a mass audience using clever animations. Einstein had been aware of the venture— when he had traveled to Prague in 1921 for his visit with Philipp Frank and debate with Oskar Kraus, he had shared his train ride with what he called "the filmmaker Fanta."[10] When the movie had finally been produced by the Colonna Filmgesellschaft in Berlin, he had registered his displeasure at the media campaign surrounding it: "The reason for this lies in the fact that people called the film the 'Einstein Film' rather than the 'Relativity Film.' I would like herewith to urgently request you to do

FIGURE 6. Poster for the popular film about relativity theory for which Otto Fanta co-wrote
the screenplay. As Einstein himself had objected, the film was misleadingly advertised as
being about him, and not about the theory. *Source:* "Quellen zur Filmgeschichte 1922:
Daten zum Einstein-Film," https://www.kinematographie.de/einstein.htm.

me the favor to select in the future such a title for your film which obviates such misconceptions"[11] (figure 6). Indeed, Ernst Gehrcke specifically attacked the film as part of his anti-Einstein campaign in these years.[12] A review in *Die Umschau* in 1922 mostly complained that two hours was far too long for sustained thought by the average viewer.[13]

This, then, was the person who showed up at Einstein's apartment in 1929 with his new wife. Hanna and Albert hit it off. As she recalled the meeting in the preface to her record of Einstein's phone conversations: "My husband had displayed great interest in the scientist's unique collection of books, and on his suggestion, and with Einstein's consent, I began to classify the library that was rather chaotically scattered throughout the house. I gave the catalog to Einstein on his 50th birthday."[14] They returned to Prague, and she enrolled at the German University, earning her bachelor of arts degree there in 1934. This interaction fixed her in his memory when politics in Europe turned dark. In late October 1938, Einstein sent her a letter, written in English:

Dear Frau Fanta:

I am glad to hear that there is a prospect that you will come to America. I hereby extend you a friendly invitation to visit me in Princeton.

My daughter, who has recently undergone a serious operation, will be especially glad to be able to chat about old times with you.

Very cordially yours,

A. Einstein[15]

A month and a half later, he sent a supplementary testimonial: "I hereby confirm that a few years ago Frau Hanne Fanta performed for me the great service of composing a catalogue of my library and to put it in thorough order."[16] For unknown reasons, the couple delayed their exit (accompanied by Otto's sister Else, now divorced from Hugo Bergmann) until March 1939, almost the last feasible moment.

They made it to Great Britain, but that was for some time the only good news in their lives. By this time they had lost all their money. In the hopes of being able to transfer some assets out of Czecho-Slovakia, Fanta sold his house in Prague and then gave the money to someone who claimed to be selling the couple an apartment in New York City. The apartment did not exist; the confidence scheme annihilated their savings.[17] Then Otto Fanta was interned by British authorities, a common enough occurrence for refugees fleeing from Central Europe. While still in custody of the state, he died in a hospital in Aylesbury, County of Buckingham. Fantova had to leave for the United States alone.

The first person she turned to was Einstein. As she recounted their conversation, he said to her: "I am so glad you are here, but what are you going to do now? You know, you'll have to work in America."[18] Together, they hit on an idea that recalled their previous meeting in Berlin. She enrolled in the Library School of the University of North Carolina–Chapel Hill in 1939 and graduated in 1941. Knowledgeable in English, French, German, Dutch, Czech, Polish, and Italian, she was highly employable in the war years, taking a job as senior cataloguer at Oberlin College from February 1942 until the end of August 1944. Then the position of inaugural curator of maps at the Princeton University Library opened up, and she was hired. Upon her retirement, university president Robert F. Goheen noted that she had expanded the maps collection 18-fold, and also "recall[ed] [her] helpful assistance to Einstein in the early 1930's in cataloguing his personal library."[19]

Einstein's relationship with Fantova was important for both of them, as their almost daily sailing trips attested. Fantova recalled that Einstein "quite frequently revived old mutual memories of Prague."[20] They were certainly old memories, but they were not mutual. The link between them, Otto Fanta, had died in Britain, and the memories they now shared were largely of Berlin and Princeton. So what was this "mutual homesickness" about? It was not a concept that Einstein had a high opinion of in general. In 1938, he raged in a letter to his close Swiss friend Michele Besso about the state of European politics, digressing with a commentary about the situation in New Jersey:

> However I like it here quite well, and one rarely finds someone who would rather go back to refined Europe. I know that you have an incurable weakness for your Italy, like most German Jews have for Germany. This form of sentimental weakness comes from our longing to be led back to a solid home on this unstable Earth, in which we fall prey to the fallacious illusion that the goyim have such a home and we don't. But I think that a home in which a rational person cannot open his or her mouth is not a proper home. A German jurist and goy, married to a Jew, who survived here only with terrible trouble, answered my question about whether he ever suffered homesickness: "But I am not a Jew!" The man got it right.[21]

The feeling of homesickness he invoked with Fantova may have partly been a recognition that, as discussed in the introduction to this book, the notion of belonging to a place was not something that Einstein fully grasped. His sickness was not having a home in the first place. Perhaps it also matters that he was writing about Prague, and not only when trying to charm his lady friend.[22] (She too, recall, was not from there.) The city was still shaping his everyday life, even when he was half a world and 40 years distant from it.

* * * * *

The story in Tel Aviv is briefer, in large part because every character is familiar. Hugo Bergmann, as we have seen, emigrated to Palestine, eventually assuming a prominent position at the Hebrew University of Jerusalem as a philosopher and academic administrator. Shuttling back and forth from the Israeli capital to the city on the coast, he was also a link for the community of those displaced from Prague and the Bohemia of the Habsburg era. His letters and diaries, especially from the 1960s and early 1970s, read like a reprise of the account of characters in this book. It was almost as though Bergmann had reconstituted the Fanta salon on the eastern shores of the Mediterranean. More to the point, he had unwittingly re-created *Einstein's* Bohemia.

Bergmann traveled back to Prague occasionally, including immediately after World War II, although the journey was surely depressing. Prague itself was not badly damaged by the conflict that had left so many other cities in ruins, but the people who had brought that city to life, the people he had known, were gone. The lucky ones he still saw regularly—they had also made it to Palestine during the interwar period. The unlucky ones had been murdered, with many now lying in unmarked graves at extermination camps. And then there were those who had fallen between lucky and unlucky.

On 8 November 1946, the eve of the anniversary of Kristallnacht, Bergmann was in Prague again, and he ran into someone he had not thought about in years: "Remarkably, the Communist professor of philosophy here, who is the preacher of materialism in the Czech philosophical movement, is an old Bar Kochba member from my days, Kolmann [*sic*], who returned here from Russia."[23] Arnošt Kolman was back in Prague as a Stalinist philosopher of science, a great conceptual distance from his role as translator of Martin Buber's Prague lectures into Czech. He and Bergmann enjoyed reconnecting, and they hoped to stay in touch. Then

Kolman dropped out of contact. In 1949 Bergmann heard a rumor that the philosopher had been recalled to the Soviet Union.[24] These were Kolman's years in the Lubianka.

After his release, their correspondence resumed. Kolman sent the biography he had written of the early-nineteenth-century Bohemian philosopher Bernard Bolzano—about whom Bergmann had written his dissertation—to Israel, adorned with a dedication in Hebrew (to Bergmann's astonishment).[25] Bergmann continued his own work, including an article extolling the philosophical writings of Philipp Frank, which was published in November 1966. He only later found out that the subject of the piece had passed away before it appeared.[26] (In 1969, Bergmann was even reading a work by Frank's old nemesis Oskar Kraus again: his biography of their mutual teacher Franz Brentano, of course.)[27] Throughout this decade he continued to publish variations of his reminiscences about Albert Einstein, a topic in increasing demand both in Israel and abroad. In the midst of all this work, out of the blue, he received a note in 1967 from Kolman stating that he was in Cyprus and planned to pop over to Israel. Here is how Bergmann described the reunion to Robert Weltsch, another shipwreck from Bar Kochba and the Fanta circle: "It is a very beautiful and uncommonly cordial reunion. When we saw each other in Prague in 1946 it was more polite than cordial, as when two men from different sides of the barricade reach out a hand to each other. This time it is entirely cordial; perhaps the air of the Land of Israel is a factor, perhaps also his cordial goyish wife, who is an author for children. She speaks only Russian, but she understands Czech and some German."[28] Not everyone was so delighted. Max Brod was incensed that Bergmann kept in touch with Kolman, resenting everything from his politics to his ideas about art and literature.[29]

Bergmann's Tel Aviv was another resonance of Einstein and Bohemia. Of course it was not *merely* that, but the conversations that occurred among Brod, Bergmann, Kolman, and the other Prague émigrés would again and again drift back to Einstein, a national hero in the new Jewish state. Their image of Einstein was different from the ones current in Berlin, London, Paris, or New York. They had all known him very well for a brief period of time, and every time they heard of him again, the news

would be filtered through the lens of those Prague memories. Brod, Bergmann, and Kolman, each in his own way, as Philipp Frank had in his,
contributed a Bohemian tinge to the Einstein that was passed down to
later generations. Bohemia was still present for them.

* * * * *

What, in the end, about Prague itself? As the personal connections between Einstein and the city began to weaken—whether because people
had died or because they no longer lived in Prague—broader consciousness of the physicist's former presence in the city diminished as well.
Before the run-up to the centenary discussed in the last chapter, Einstein's
name only appeared as a universal signifier, not as a plausible local
memory.

One prominent such moment was the publication of Josef Nesvad-
ba's story *Einstein's Brain* (*Einsteinův mozek*) in 1960, a collection of science fiction tales that begins with a short story of that title. You might
expect that it would have something to do with relativity, or quantum
theory, or even perhaps the actual saga of Einstein's brain, which had been
purloined by a coroner and secreted in the trunk of his car against the
physicist's express wishes that he be cremated. (That last one, sadly non-
fictional, is a great story, but it doesn't have any connection with
Prague.)[30] Rather, Nesvadba's tale is about a quest for an artificial biological brain that will be able to divine the meaning of life, something ever
more powerful computers are not able to do. (It does and decides to
starve itself to death.) There is nothing here about Einstein except for
metaphor. If anything, the book that its title pays homage to, Jakub Arbes's
Newton's Brain (*Newtonův mozek*)—published in 1877, it is often considered the first Czech science fiction work—is more like it. Arbes's book
imagines a con man, supposed to have died in the Battle of Königgrätz
between Austria and Prussia in 1866, who receives a transplanted brain
from Isaac Newton. Using his new powers of genius, he builds a space-
ship that enables him to travel faster than the speed of light and therefore take photographs of the past. This was almost three decades before
the formation of the theory of special relativity (and almost two before

H. G. Wells's *Time Machine*)! For Nesvadba, Einstein was a universal signifier of intelligence, not a Bohemian icon. The closest we get to the
actual person is the epigraph Nesvadba chose: Einstein's famous slogan
"Imagination is more important than knowledge."[31]

You can still find traces of Einstein in the city of Prague if you know
where to look.[32] There are the memorial plaques inside his office building and outside his home from 1911–1912, erected during the centenary
year of 1979. Those are tied to real places where the physicist once wandered. There is now a third plaque too, on Old Town Square, outside
the home where Bertha Fanta once held her salon. It depicts a very aged
Einstein, much too old to relate to the historical moment it describes,
and the text narrates the connections Einstein made in this house with
Franz Kafka. This, too, is not very faithful to the historical record. The
plaque is also in English. The audience for this memorial, unlike the
other two, is not the local population but the masses of tourists who
stream through the beautiful city of Prague. For them, the city offers a
new Einstein, one who is more human in appearance than Nesvadba's,
but no less fictional.

These final Prague stories have less to do with Einstein himself than
with his legacy, and that is only to be expected. Bohemia has long since
vanished, at least from the real cartography of the world. It has moved
instead to the land of myth. There you can also find Einstein. He still lives
in that Bohemia—at least a part of him.

Acknowledgments

Although it seems as though I have been reading Einstein biographies for as long as I can remember, I first became fully aware that he had taught for three semesters in Prague sometime during my first three semesters of graduate school. Being interested in both Einstein (who isn't?) and science in the Slavic world, I filed the notion away as an intriguing topic for a future article. One thing led to another, and after about 20 years of marinating, that article idea grew into the book you have before you. Since *Einstein in Bohemia* is an experiment in thinking historically, proper acknowledgment of everyone who has shaped how I approach these matters would require a separate book in itself. So let me start by thanking those of you I cannot mention here. You may find traces of yourself somewhere in these pages, and you can be sure that I am grateful for all the conversations that led to them.

I conducted the vast majority of the research for this book in 2015–2016 while on a sabbatical fellowship at the Wissenschaftskolleg zu Berlin, which was a captivating environment to think about both displacement and belonging. Rector Luca Giuliani, Secretary Thorsten Wilhelmy, and Academic Coordinator Daniel Schönpflug generated a welcoming scholarly environment that enabled a kind of concentrated focus that I had never before experienced and that I miss with regularity. (My fellow Fellows, one of whom has since become the current rector, had a lot to do with that as well.) I am especially grateful to Sonja Grund and her team at the library, who undertook absolutely heroic efforts to assemble materials I requested from across the continent. Upon my return to

Princeton, the Interlibrary Loan Office at Firestone Library continued those good services. For the space, the time, and the content, many thanks.

Several archives and libraries were very generous with offering me access to their holdings. I began my investigations by poring over the Albert Einstein Duplicate Archive in the Department of Rare Books and Special Collections in Firestone Library at Princeton. The staff was uncommonly helpful in navigating that rather complex collection, as well as the papers of Hanna Fantova. Diana Buchwald, editor of the Einstein Papers Project, has been unfailingly supportive of this work, granting access to several specific documents from the Albert Einstein Archives at the Hebrew University in Jerusalem. (Her masterful leadership of the *Collected Papers of Albert Einstein* has shaped the book on almost every page, as the endnotes will attest.) Thomas Binder was kind enough to send me materials from the Nachlass Oskar Kraus at the Franz Brentano Archive in Graz, Austria, as was Valery Merlin from the Yad Vashem Archives in Jerusalem. Jan Musil performed research about both Oskar Kraus and Arnošt Kolman in the archives of the Czech Academy of Sciences and of Charles University in Prague—this project would be much poorer without his efficient assistance. Finally, I would like to thank Tomáš Herrmann of Charles University, who not only put me in touch with Jan but offered encouragement at a very early stage of my thoughts about this work.

Several friends and colleagues patiently read chapters and offered numerous suggestions that have improved the book both empirically and conceptually. Gary B. Cohen, Stanley Corngold, David Kaiser, Matthew Stanley, and a group of sharp scholars at the Berlin Center for the History of Knowledge read selected chapters, and Elizbeth Baker and Patrick McCray selflessly commented on the entire manuscript. I also presented early versions of parts of this project at the Wissenschaftskolleg zu Berlin, the Max Planck Institute for the History of Science in Berlin, the Works-in-Progress series at the Princeton History Department, the Princeton Society of Fellows, the Hamilton Colloquium Series in the Princeton Physics Department, Yale University, Queen's University, the Ludwig-Maximilians-Universität in Munich, the University of

New Hampshire, Oregon State University, the University of Minnesota, and Central European University. I am especially grateful to Jan Maršálek, Jiří Podolský, and Jiří Bičák, for conversations about this material during a memorable visit to Prague. Last, but also most definitely first, Erika Lorraine Milam has endured countless ramblings about Einstein, Prague, and other topics for many years now, and despite that *still* read the whole manuscript when we both knew she didn't really have the time to do so. I am always grateful, for everything.

Penultimate thanks go to Princeton University Press, which has from the beginning been the most natural home for this book. Al Bertrand encouraged me when the concept was still embryonic, and Eric Crahan and Pamela Weidman expertly shepherded the manuscript to realization. Derek Sayer and Diana Buchwald served as the anonymous reviewers for Princeton University Press (thankfully anonymous no longer), and they provided generous and very helpful correctives to a number of factual points and (most gratifyingly) really understood and welcomed the enterprise. Mark Bellis was a superb production editor, and Sarah Vogelsong's copyediting improved the manuscript in numerous places. My thanks to them all.

My final debt is to Peter Galison, who was the first historian of science I ever met. Back in 1992, Peter was also the first person to make both relativity and Einstein *real* to me, and he has never ceased to surprise me by making those topics (and just about everything else he addresses) more new and more real. He has been a constant presence and inspiration in his person and his writings ever since. This book is dedicated to him.

Notes

INTRODUCTION: A SPACETIME INTERVAL

1. Leopold Infeld, *Albert Einstein: His Work and Its Influence on Our World* (New York: Charles Scribner's Sons, 1950), 125.

2. "Interlude": Carl Seelig, *Albert Einstein: A Documentary Biography*, tr. Mervyn Savill (London: Staples Press Limited, 1956), 119; "sojourn": Anton Reiser [Rudolf Kayser], *Albert Einstein: A Biographical Portrait* (New York: Albert & Charles Boni, 1930), 87, and Ze'ev Rosenkranz, *Einstein Before Israel: Zionist Icon or Iconoclast?* (Princeton, NJ: Princeton University Press, 2011), 33 and 254; "detour": Albrecht Fölsing, *Albert Einstein: A Biography*, tr. and abridged Ewald Osers (New York: Penguin Books, 1997 [1993]), 322; "way station": Jürgen Neffe, *Einstein: A Biography*, tr. Shelley Frisch (New York: Farrar, Straus and Giroux, 2007 [2005]), 161; "intermezzo": Giuseppe Castagnetti et al., *Einstein in Berlin: Wissenschaft zwischen Grundlagenkrise und Politik* (Berlin: Max-Planck-Institut für Bildungsforschung, [1994]), 10; Dieter Hoffmann, *Einsteins Berlin: Auf den Spuren eines Genies* (Weinheim: Wiley-VCH, 2006), 2.

3. Arnold Sommerfeld, "Zum Siebzigsten Geburtstag Albert Einsteins," *Deutsche Beiträge* 2 (1949): 141–146, on 143; Lewis Pyenson, *The Young Einstein: The Advent of Relativity* (Bristol: Adam Hilger Ltd., 1985), 61 (although Einstein's secretary Helen Dukas definitively rejected the idea on p. 64: he was "anything but 'bohemian,'" from a letter of Dukas to Pyenson of 16 September 1974); Fölsing, *Albert Einstein*, 114; Jean Eisenstaedt, *The Curious History of Relativity: How Einstein's Theory of Gravity Was Lost and Found Again*, tr. Arturo Sangalli (Princeton, NJ: Princeton University Press, 2006 [2003]), 112; Roger Highfield and Paul Carter, *The Private Lives of Albert Einstein* (New York: St. Martin's Press, 1993), 133.

4. See especially Richard Miller, *Bohemia: The Protoculture Then and Now* (Chicago: Nelson-Hall, 1977); and Jiří Kořalka, *Tschechen im Habsburgerreich und in Europa 1815–1914: Sozialgeschichtliche Zusammenhänge der neuzeitlichen Nationsbildung und der Nationalitätenfrage in den böhmischen Ländern* (Vienna: Verlag für Geschichte und Politik, 1991), 62.

5. Starting in the ninth century foreign sources began to refer to the region using a variety of names: Beheim, Bohemia, Beimi, Boemani, and Beheimare. The term is derived from the name of a Celtic tribe, the Boii, who were the last occupants of the region in the pre-Christian era, during the Roman Empire. They did not last long. At the end of the fifth century the Germanic Langobards moved in, followed in the sixth century by the Slavis (Sklavinoi in the west and Antoi in the east), the ancestors of the Slavs who presently dominate the area. Jiří Sláma, "Boiohaemum-Čechy," in Mikuláš Teich, ed., *Bohemia in History* (Cambridge: Cambridge University Press, 1998), 23–38.

6. Thun quoted in Stanley Z. Pech, *The Czech Revolution of 1848* (Chapel Hill: University of North Carolina Press, 1969), 30. For more on the adjectival distinction, see Tilman Berger,

"Böhmisch oder Tschechisch?: Der Streit über die adequate Benennung der Landessprache der böhmischen Länder zu Anfang des 20. Jahrhunderts," in Marek Nekula, Ingrid Fleischmann, and Albrecht Greule, eds., *Franz Kafka im sprachnationalen Kontext seiner Zeit: Sprache und nationale Identität in öffentlichen Institutionen der böhmischen Länder* (Köln: Böhlau, 2007), 167–182.

7. See, for a small selection: Anne Jamison, *Kafka's Other Prague: Writings from the Czechoslovak Republic* (Evanston, IL: Northwestern University Press, 2018); Scott Spector, *Prague Territories: National Conflict and Cultural Innovation in Franz Kafka's Fin de Siècle* (Berkeley: University of California Press, 2000); Pavel Eisner, *Franz Kafka and Prague*, tr. Lowry Nelson and René Wellek (New York: Golden Griffin Books, 1950); and Christoph Stölzl, *Kafkas böses Böhmen: Zur Sozialgeschichte eines Prager Juden* (Munich: Edition Text + Kritik, 1975).

8. Reiner Stach, *Kafka: Die frühen Jahre* (Frankfurt am Main: S. Fischer, 2014), 469; Margarita Pazi, "Franz Kafka, Max Brod und der 'Prager Kreis,'" in Karl Erich Grözinger, Stéphane Mosès, and Hans Dieter Zimmermann, eds., *Franz Kafka und das Judentum* (Frankfurt am Main: Jüdischer Verlag bei Athenäum, 1987), 71–92, on 88; Hartmut Binder, "Der Prager Fanta-Kreis: Kafkas Interesse an Rudolf Steiner," *Sudetenland*, no. 2 (1996): 106–150, on 136.

9. Quoted in Binder, "Der Prager Fanta-Kreis," 136. It is true that Kafka also did not mention in his letters, diaries, and notebooks individuals such as Friedrich Nietzsche whom we are certain he was engaged with, which has led some to hold out hope that an intellectual connection to Einsteinian physics persisted beneath the surface. Franz Kuna, "Rage for Verification: Kafka and Einstein," in Franz Kuna, ed., *On Kafka: Semi-Centenary Perspectives* (London: Paul Elek, 1976), 83–111. Granted that absence of evidence is not evidence of absence, it is hardly evidence of presence either.

10. On the Berlin visits, see Hubert Goenner, *Einstein in Berlin, 1914–1933* (Munich: C. H. Beck, 2005), 211; John Forrester, "Die Geschichte zweier Ikonen: 'The Jews All over the World Boast of My Name, Pairing Me with Einstein' (Freud, 1926)," tr. Bettina Engels, in Michael Hagner, ed., *Einstein on the Beach: Der Physiker als Phänomen* (Frankfurt am Main: Fischer Taschenbuch, 2005), 96–123, on 305n13. On "perversity," see the description of the ostensible incident in R. W. Stallman, "A Hunger-Artist," in Angel Flores and Homer Swander, eds., *Franz Kafka Today* (Madison: University of Wisconsin Press, 1958), 61–70, on 61.

11. Bergmann to Paul Amann, 28 April 1955, reproduced in Schmuel Hugo Bergman, *Tagebücher & Briefe*, 2 vols., ed. Miriam Sambursky (Königstein: Athenäum, 1985), 2:196. See also Hugo Bergman, "Personal Remembrances of Albert Einstein," in Robert S. Cohen and Marx W. Wartofsky, eds., *Logical and Epistemological Studies in Contemporary Physics* (Dordrecht: D. Reidel, 1974), 388–394, on 390, where he repeats the claim.

12. Franz Halla, "Anläßlich des Todes von Albert Einstein," *Mitteilungen aus der anthroposophischen Arbeit in Deutschland* 32 (June 1955): 74–75.

13. On Steiner's visits to Prague in this period, see Zdeněk Váňa, "Rudolf Steiner in Prag: Zur Geschichte der tschechischen anthroposophischen Bewegung," *Beiträge zur Rudolf Steiner Gesamtausgabe*, no. 109 (1992): 1–64; Binder, "Der Prager Fanta-Kreis," 132; and idem, "Rudolf Steiners Prager Vortragsreise im Jahr 1911: Berichtigungen und Ergänzungen zu der Kritischen Ausgabe der Tagebücher Kafkas," *Editio* 9 (1995): 214–233. It is conceivable that Einstein was

dragged to Steiner's later lectures in Prague on 28 and 30 April 1912, but those were on different topics and the scientist never mentioned it. Claims that Einstein was devoted to the teachings of the theosophist H. P. Blavatsky (Sylvia Cranston, *HPB: The Extraordinary Life and Influence of Helena Blavatsky, Founder of the Modern Theosophical Movement* [New York: Tarcher/Putnam, 1993], xx, 434, and 557n11) are also scarcely credible once one chases down the sources.

14. Franz Kafka, *Tagebücher: Bd. 1: 1909–1912* (Frankfurt am Main: Fischer, 2008), 29–31. See also June O. Leavitt, *The Mystical Life of Franz Kafka: Theosophy, Cabala, and the Modern Spiritual Revival* (New York: Oxford University Press, 2012).

15. Res Jost, "Einstein und Zürich, Zürich und Einstein," *Vierteljahrsschrift der Naturforschenden Gesellschaft in Zürich* 124 (1979): 7–23; Max Flückiger, *Albert Einstein in Bern: Das Ringen um ein neues Weltbild* (Bern: Verlag Paul Haupt, 1974); Paul Cohen and Brenda Cohen, "The Einstein House in Berne, Switzerland: Visiting the Home of One of the Century's Great Minds," *Journal of College Science Teaching* 25, no. 6 (May 1996): 440–441; Carl Seelig, *Albert Einstein und die Schweiz* (Zurich: Europa-Verlag, 1952); L. Kollros, "Albert Einstein en Suisse: Souvenirs," *Helvetica Physica Acta* 29 (1956): 271–281; Michel Biezunski, *Einstein à Paris: Le temps n'est plus . . .* (Paris: Presses Universitaires de Vincennes, 1991); Thomas Levenson, *Einstein in Berlin* (New York: Bantam Books, 2003); Castagnetti et al., *Einstein in Berlin*; Goenner, *Einstein in Berlin*; Hoffmann, *Einsteins Berlin*; Christa Kirsten and Hans-Jürgen Treder, eds., *Albert Einstein in Berlin, 1913–1933*, 2 vols. (Berlin: Akademie-Verlag, 1979); Hans Eugen Specker, ed., *Einstein und Ulm: Festakt, Schülerwettbewerb und Ausstellung zum 100. Geburtstag von Albert Einstein* (Ulm: W. Kohlhammer, 1979). The list can be extended. It is surprising that there is no "Einstein in Princeton," given that he lived in that town longer than he did anywhere else.

16. Jiří Bičák, ed., *Einstein a Praha: K stému výročí narození Alberta Einsteina* (Prague: Jednota československých matematiků a fyziků, 1979). On Bičák's distinguished contributions to his science, see Oldřich Semerák, "Preface," in O. Semerák, J. Podolský, and M. Žofka, eds., *Gravitation, Following the Prague Inspiration: A Volume in Celebration of the 60th Birthday of Jiří Bičák* (River Edge, NJ: World Scientific, 2002), v–vii.

17. Especially useful studies include József Illy, "Albert Einstein in Prague," *Isis* 70 (1979): 76–84, in Czech translation as Illy, "Albert Einstein a Praha," *Dějiny věd a techniky* 12, no. 2 (1979): 65–79; Martin Šolc, "Poznámky k pobytu Alberta Einsteina v Praze," *Bulletin Plus*, no. 3 (2003), accessible at http://wwwold.nkp.cz/bp/bp2003_3/12.htm, last accessed on 7 April 2016; Miroslav Brdička, "Einstein a Praha: Česká einsteinovská pohlednice," *Československý časopis fyziky* A29 (1979): 269–275; Rudolf Kolomý, "Albert Einstein v Praze," *Vesmír* 53, no. 4 (1974): 112–115; Dieter Hoffmann, "Einstein in Prag," *Physik in unserer Zeit* 35, no. 5 (2004): 244; and Eva Rozsívalová, "Albert Einstein v Praze," *Pokroky matematiky, fyziky a astronomie* 4 (1959): 352–354. Less scholarly accounts, often with significant errors and mythologization, include Rudolf Kolomý, "Albert Einstein a jeho vztah v Praze," *Pokroky matematiky, fyziky a astronomie* 17 (1972): 265–272; and the disappointing website "Albert Einstein's Years in Prague, 1911–1912," accessible at http://www.einstein-website.de/z_biography/prague.html, last accessed on 2 May 2016. Understandably, pre–World War II biographies of Einstein tended to place a lot more emphasis on Prague than those produced later, for the simple reason that Einstein's life was at that point shorter and thus the Prague moment loomed larger: David Reichinstein, *Albert Einstein: A Picture of His Life and His Conception of the World* (Prague: Stella Publishing House, 1934), 23; H. Gordon

Garbedian, *Albert Einstein: Maker of Universes* (New York: Funk & Wagnalls, 1939), 92–95, 98–103; and Dimitri Marianoff with Palma Wayne, *Einstein: An Intimate Study of a Great Man* (Garden City, NY: Doubleday, Doran and Co., 1944), 49. They are, however, unreliable as to facts, being based mostly on impressionistic and partial interviews.

18. Release from Württemberg Citizenship, 28 January 1896, *CPAE* 1:16, on 20. See also Ronald W. Clark, *Einstein: The Life and Times* (New York: Thomas Y. Crowell, 1971), 27–28.

19. A. Einstein, "Meine Meinung über den Krieg," reproduced in *CPAE* 6:20, on 212.

20. Goenner, *Einstein in Berlin*, 65.

21. Einstein to Berlin-Schöneberg Office of Taxation, 10 February 1920, *CPAE* 9:306, on 420.

22. Fritz Haber to Einstein, 9 March 1921, *CPAE* 12:87, on 125. Emphasis in original.

23. Einstein to Haber, 9 March 1921, *CPAE* 12:88, on 128.

24. See, for example, the arrangements for his trip to East Asia: Einstein to Swiss Embassy in Berlin, 18 September 1922, *CPAE* 13:361, on 514–515.

25. Einstein to Gilbert Murray, 13 July 1922, *CPAE* 13:286, on 408.

26. Nonetheless, after arguing unpersuasively for a "Habsburg Einstein," the usually astute Péter Hanák makes the enthusiastically ahistorical declaration: "I vote for late Austro-Hungarian citizenship for the late Albert Einstein." Péter Hanák, *The Garden and the Workshop: Essays on the Cultural History of Vienna and Budapest* (Princeton, NJ: Princeton University Press, 1998), 154 (quotation), 160.

27. For the resolution of the dispute, see Heinrich Lüders to Einstein, 15 February 1923, *CPAE* 13:431, on 719; and Einstein, "Note on Prussian Citizenship," 7 February 1924, *CPAE* 14:209, on 329. Yet in 1924 Einstein continued to demand that the Swiss government provide him and his wife with a diplomatic passport for their international travels: Einstein to the Swiss Foreign Ministry, 10 July 1924, *CPAE* 14:283a, in vol. 15 on 31; Einstein to the Swiss Foreign Ministry, 28 July 1924, *CPAE* 14:296a, in vol. 15 on 32.

28. Banesh Hoffmann with Helen Dukas, *Albert Einstein: Creator and Rebel* (New York: Viking, 1972), 234; Alice Calaprice, ed., *The New Quotable Einstein* (Princeton, NJ: Princeton University Press, 2005), 337.

29. Fred Jerome, *The Einstein File: J. Edgar Hoover's Secret War Against the World's Most Famous Scientist* (New York: St. Martin's Griffin, 2002), ch. 16.

30. The classic piece is Hans Tramer, "Prague—City of Three Peoples," *Leo Baeck Institute Year Book* 9 (1964): 305–339. See also Oskar Wiener, ed., *Deutsche Dichter aus Prag* (Vienna: Ed. Strache, 1919), 6–7.

31. This has become the basis for a healthy pushback in recent scholarship in favor of those previously untold histories of people, often peasants, who were "indifferent" to nationalist intellectuals' talking points. See especially Tara Zahra, "Imagined Noncommunities: National Indifference as a Category of Analysis," *Slavic Review* 69, no. 1 (Spring 2010): 93–119.

32. Rudolf Hilf, *Deutsche und Tschechen: Bedeutung und Wandlungen einer Nachbarschaft in Mitteleuropa* (Opladen: Leske, 1973); Johann Wolfgang Brügel, *Tschechen und Deutsche, 1918–1938* (Munich: Nymphenberger, 1967); idem, *Tschechen und Deutsche, 1939–1946* (Munich: Nymphenburger, 1974); Ferdinand Seibt, *Deutschland und die Tschechen: Geschichte einer Nachbarschaft in der Mitte Europas* (Munich: Piper, 1993).

33. Rogers Brubaker and Frederick Cooper, "Beyond 'Identity,'" *Theory and Society* 29, no. 1 (February 2000): 1–47, on 14.

34. James J. Sheehan, "What Is German History?: Reflections on the Role of the Nation in German History and Historiography," *Journal of Modern History* 53, no. 1 (March 1981): 1–23; Peter J. Katzenstein, *Disjoined Partners: Austria and Germany since 1815* (Berkeley: University of California Press, 1976); and, especially for Bohemia, Ronald M. Smelser, *The Sudeten Problem, 1933–1938: Volkstumspolitik and the Formulation of Nazi Foreign Policy* (Middletown, CT: Wesleyan University Press, 1975).

35. Katherine Arens, "For Want of a Word . . . : The Case for Germanophone," *Die Unterrichtspraxis* 32, no. 2 (1999): 130–142. On the implications of Bohemia for Germanophone intellectual history, see David S. Luft, "Austrian Intellectual History and Bohemia," *Austrian History Yearbook* 38 (2007): 108–121; and William M. Johnston, *The Austrian Mind: An Intellectual and Social History, 1848–1938* (Berkeley: University of California Press, 1972).

36. The ur-text here is Angelo Maria Ripellino, *Magic Prague*, tr. David Newton Marinelli, ed. Michael Henry Heim (London: Macmillan, 1994 [1973]). For others that follow his lead, see Joseph Wechsberg, *Prague: The Mystical City* (New York: Macmillan, 1971); Jiří Kuchař, *Esoteric Prague: A Guide to the City's Secret History* (Prague: Eminent, 2002); and Bernard Michel, *Prague, Belle Époque* (Paris: Aubier, 2008). Magical Prague appears in the Einstein literature as well. In Lewis S. Feuer, *Einstein and the Generations of Science*, 2nd. ed. (New Brunswick, NJ: Transaction, 2010 [1974]), the author wildly exaggerates both the mysticism of the local Zionists and their effect on the physicist.

37. Derek Sayer, *The Coasts of Bohemia: A Czech History* (Princeton, NJ: Princeton University Press, 1998); Peter Demetz, *Prague in Black and Gold: Scenes in the Life of a European City* (New York: Hill and Wang, 1997).

38. As Larry Woolf has demonstrated, that imaginary boundary has its origins much further back than the late 1940s: *Inventing Eastern Europe: The Map of Civilization on the Mind of the Enlightenment* (Stanford, CA: Stanford University Press, 1994).

39. Soňa Štrbáňová, "Patriotism, Nationalism and Internationalism in Czech Science: Chemists in the Czech Revival," in Mitchell G. Ash and Jan Surman, eds., *The Nationalization of Scientific Knowledge in the Habsburg Empire, 1848–1918* (Basingstoke: Palgrave Macmillan, 2012), 138–156; Luboš Nový, ed., *Dějiny exaktních věd v českých zemích do konce 19. století* (Prague: Nakl. ČSAV, 1961); Adalb. Wraný, *Geschichte der Chemie und der auf chemischer Grundlage beruhenden Betriebe in Böhmen bis zur Mitte des 19. Jahrhunderts* (Prague: Fr. Řivnáč, 1902).

40. Important models for me have been Derek Sayer, *Prague, Capital of the Twentieth Century: A Surrealist History* (Princeton, NJ: Princeton University Press, 2013); Brigitte Hamann, *Hitler's Vienna: A Portrait of the Tyrant as a Young Man* (New York: Tauris Parke, 2010 [1999]); and Philippe Sands, *East West Street: On the Origins of "Genocide" and "Crimes Against Humanity"* (New York: Knopf, 2016).

CHAPTER 1: FIRST AND SECOND PLACE

1. Quoted in Oskar Kraus, *Franz Brentano* (Munich: C. H. Beck, 1919), 139.

2. Tara Zahra, "Imagined Noncommunities: National Indifference as a Category of Analysis," *Slavic Review* 69, no. 1 (Spring 2010): 93–119. See also idem, *Kidnapped Souls: National Indifference and the Battle for Children in the Bohemian Lands, 1900–1948* (Ithaca, NY: Cornell University Press, 2008).

3. Einstein to Pauline Einstein, [28 April 1910], *CPAE* 5:204, on 238.

4. On the evolution of this system and the interconnections among universities across different sovereign states during the nineteenth century, see Christa Jungnickel and Russell McCommach, *The Intellectual Mastery of Nature: Theoretical Physics from Ohm to Einstein*, 2 vols. (Chicago: University of Chicago Press, 1986).

5. On Lippich, see Jan Havránek, "Ke jmenování Alberta Einsteina profesorem v Praze," *Acta Universitatis Carolinae—Historia Universitatis Carolinae Pragensis* 17, no. 2 (1977): 105–130, on 109; on Lecher, see Andreas Kleinert, "Anton Lampa und Albert Einstein: Die Neubesetzung der physikalischen Lehrstühle an der deutschen Universität Prag 1909 und 1910," *Gesnerus* 32, no. 3/4 (1975): 285–292, on 285. On Prague's less than coveted status, which produced on average a younger professoriate as the venerable lions decamped for more elite locales, see Josef Petráň, "The Philosophical Faculty 1848–1882," in František Kavka and Josef Petráň, eds., *A History of Charles University*, 2 vols. (Prague: Karolinium, 2001): 2:109–122, on 119; and idem, *Nástin dějin filozofické fakulty Univerzity Karlovy v Praze (do roku 1948)* (Prague: Univerzita Karlova, 1983), 184.

6. The documents and background for the empanelment of the commission and its deliberations are provided in Havránek, "Ke jmenování Alberta Einsteina profesorem v Praze."

7. Proposal of the committee nominating for the vacant position in mathematical physics, 27 January 1910, reproduced in Havránek, "Ke jmenování Alberta Einsteina profesorem v Praze," 121.

8. Ibid., 122.

9. Ibid., 122.

10. Max Planck, quoted in ibid., 122–123.

11. Ibid., 123.

12. Ibid., 123.

13. Ibid., 125–126.

14. Einstein to Arnold Sommerfeld, July [19]10, *CPAE* 5:211, on 246.

15. For examples, see Wolfgang Wolfram von Wolmar, *Prag: Die älteste Universität des Reiches* ([Munich]: Arbeitsgemeinschaft Prager und Brünner Korporationen, 1998); Eugen Lemberg, "Die Prager Universität und das Schicksal Mitteleuropas," in *Die deutsche Universität in Prag: Ein Gedenken anläßlich der 600 Jahrfeier der Karls-Universität in Prag* (Munich: Edmund Gans, 1948): 12–38; Adolf Hauffen, "Zur Geschichte der deutschen Universität in Prag: Mit einem bibliographischem Anhang," *Mittheilungen des Vereines für Geschichte der Deutschen in Böhmen* 38 (1900): 110–127, on 115. On the university as essentially "Czech/Bohemian," see Ernest Denis, "University of Prague," *Czechoslovak Review*, November 1919, 316–319, 374–377. For a moderate view from the leading Czech philosopher active in the interwar period, see the comments declaring it a regional "Bohemian" university, as opposed to an ethnic university, in Emanuel Rádl, *Der Kampf zwischen Tschechen und Deutschen*, tr. Richard Brandeis (Reichenberg [Liberec]: Gebrüder Stiepel, 1928), 35.

16. Lisa Wolverton, *Hastening Toward Prague: Power and Society in the Medieval Czech Lands* (Philadelphia: University of Pennsylvania Press, 2001); Ferdinand Seibt, *Deutschland und die Tschechen: Geschichte einer Nachbarschaft in der Mitte Europas* (Munich: Piper, 1993), 76.

17. Michal Svatoš, "The Studium Generale (1347/8–1419)," in Kavka and Petráň, *History of Charles University*, 1:23–88. For another informative study, with a slightly more German nationalist flavor, see Peter Moraw, "Die Universität Prag im Mittelalter: Grundzüge ihrer Geschichte

im europäischen Zusammenhang," in *Die Universität zu Prag*, vol. 7 of *Schriften der Sudeten-deutschen Akademie der Wissenschaften und Künste* (Munich: Verlaghaus Sudetenland, 1986), 9–134.

18. V. Patzak, "The Caroline University of Prague," *Slavonic and East European Review* 19, no. 53/54 (1939–1940): 83–95, on 88; Renate Dix, "Frühgeschichte der Prager Universität: Gründung, Aufbau und Organisation, 1348–1409" (Ph.D. diss., Rheinischen Friedrich-Wilhelms-Universität zu Bonn, 1988), on 133; and especially Sabine Schumann, "Die 'nationes' an den Universitäten Prag, Leipzig und Wien: Ein Beitrag zur älteren Universitätsgeschichte" (Ph.D. diss., Freie Universität Berlin, 1974).

19. Schumann, "Die 'nationes' an den Universitäten Prag, Leipzig und Wien," 102. On the proportions, see Svatoš, "Studium Generale," 74.

20. Malcolm Lambert, *Medieval Heresy: Popular Movements from the Gregorian Reform to the Reformation*, 3rd. ed. (Cambridge, MA: Blackwell, 2002 [1977]), 323; Moraw, "Die Universität Prag im Mittelalter," 26; Thomas A. Fudge, *The Trial of Jan Hus: Medieval Heresy and Criminal Procedure* (New York: Oxford University Press, 2013), 125–126.

21. Seibt, *Deutschland und die Tschechen*, 160; Schumann, "Die 'nationes' an den Universitäten Prag, Leipzig und Wien," 41, 205, 233; and František Šmahel, "The Kuttenberg Decree and the Withdrawal of the German Students from Prague in 1409: A Discussion," *History of Universities* 4 (1984): 153–166.

22. R. W. Seton-Watson, *A History of the Czechs and Slovaks* (Hamden, CT: Archon, 1965 [1943]), 46.

23. Wolfgang Wolfram von Wolmar, *Prag und das Reich: 600 Jahre Kampf deutscher Studenten* (Dresden: Franz Müller Verlag, 1943), 59.

24. On the theological questions, see Hieromonk Patapios, "*Sub utraque specie*: The Arguments of John Hus and Jacoubek of Stříbro in Defence of Giving Communion to the Laity under Both Kinds," *Journal of Theological Studies* 53, no. 2 (October 2002): 503–522. On the functioning of the Utraquist Church over the next two centuries, see Zdeněk V. David, *Finding the Middle Way: The Utraquists' Liberal Challenge to Rome and Luther* (Washington, DC: Woodrow Wilson Center Press, 2003).

25. For classic works, see Howard Kaminsky, *A History of the Hussite Revolution* (Berkeley: University of California Press, 1967); František Šmahel, "The Hussite Movement: An Anomaly of European History?," in Mikuláš Teich, ed., *Bohemia in History* (Cambridge: Cambridge University Press, 1998), 79–97. On the university as a hotbed of the Hussite Reformation, see Thomas A. Fudge, *The Magnificent Ride: The First Reformation in Hussite Bohemia* (Aldershot: Ashgate, 1998). On the university during the war years, see Howard Kaminsky, "The University of Prague in the Hussite Revolution: The Role of the Masters," in John W. Baldwin and Richard A. Goldthwaite, eds., *Universities in Politics: Case Studies from the Late Middle Ages and Early Modern Period* (Baltimore: Johns Hopkins Press, 1972): 79–106.

26. Petráň, *Nástin dějin filozofické fakulty Univerzity Karlovy v Praze*, 52, 79. On the two institutions, see Petr Lozoviuk, *Interethnik im Wissenschaftsprozess: Deutschsprachige Volkskunde in Böhmen und ihre gesellschaftlichen Auswirkungen* (Leipzig: Leipziger Universitätsverlag, 2008), 93; Michal Svatoš, "The Utraquist University (1419–1556)," in Kavka and Petráň, *History of Charles University*, 1:187–197; and Ivana Čornejová, "The Jesuit Academy up to 1622," in Kavka and Petráň, *History of Charles University*, 1:217–234.

27. Josef Petráň and Lydia Petráňová, "The White Mountain as a Symbol in Modern Czech History," in Teich, *Bohemia in History*, 143–163; R.J.W. Evans, "The Significance of the White Mountain for the Culture of the Czech Lands," *Bulletin of the Institute of Historical Research* 44 (1971): 34–54.

28. Howard Louthan, *Converting Bohemia: Force and Persuasion in the Catholic Reformation* (Cambridge: Cambridge University Press, 2009).

29. Käthe Spiegel, "Die Prager Universitätsunion (1618–1654)," *Mitteilungen des Vereines für Geschichte der Deutschen in Böhmen* 62 (1924): 5–94; Patzak, "Caroline University in Prague," 90–91; Jan Krčmář, *The Prague Universities: Compiled According to the Sources and Records* (Prague: Orbis, 1934), 20; Ivana Čornejová, "The Administrative and Institutional Development of Prague University (1622–1802)," in Kavka and Petráň, *History of Charles University*, 1:261–297, on 275.

30. S. Harrison Thomson, "The Czechs as Integrating and Disintegrating Factors in the Habsburg Empire," *Austrian History Yearbook* 3, pt. 2 (1967): 203–222, on 209; Hans Lemberg, "Universität oder Universitäten in Prag—und der Wandel der Lehrsprache," in Lemberg, ed., *Universitäten in nationaler Konkurrenz: Zur Geschichte der Prager Universitäten im 19. und 20. Jahrhundert* (Munich: R. Oldenbourg, 2003), 19–32, on 25.

31. On language, see Peter Burian, "The State Language Problem in Old Austria (1848–1918)," *Austrian History Yearbook* 6–7 (1970–1971): 81–103. On the *Toleranzpatent* and the university, see Joseph F. Zacek, "The Czech Enlightenment and the Czech National Revival," *Canadian Review of Studies in Nationalism* 10 (Spring 1983): 17–28, on 20; Lemberg, "Universität oder Universitäten in Prag," 23. On the subsequent transformations of the position of the Jews from Joseph II's era through the nineteenth century, see the comprehensive account in Hillel J. Kieval, *The Making of Czech Jewry: National Conflict and Jewish Society in Bohemia, 1870–1918* (New York: Oxford University Press, 1988).

32. Irena Seidlerová, "Science in a Bilingual Country," in Teich, *Bohemia in History*, 229–243, on 230; Peter Bugge, "Czech Nation-Building, National Self-Perception and Politics, 1780–1914" (Ph.D. diss., University of Aarhus, 1994), on 18; *Die Deutsche Karl-Ferdinands Universität in Prag unter der Regierung seiner Majestät des Kaisers Franz Josef I.* (Prague: Verlag der J. G. Clave'schen K. u. K. Hof- u. Universitätsbuchhandlung, 1899), 21–22; Jan Havránek, "The University: Organization, Administration, Students (1802–1848)," in Kavka and Petráň, *History of Charles University*, 2:25–34, on 30.

33. Hugh LeCaine Agnew, *Origins of the Czech National Renascence* (Pittsburgh: University of Pittsburgh Press, 1993); John F. N. Bradley, *Czech Nationalism in the Nineteenth Century* (Boulder, CO: East European Monographs, 1984); Joseph Frederick Zacek, *Palacký: The Historian as Scholar and Nationalist* (The Hague: Mouton, 1970).

34. Jaroslav Goll, *Rozdělení Pražské university Karlo-Ferdinandovy roku 1882 a počátek samostatné university české* (Prague: Nakl. Klubu historického, 1908), 3.

35. On "utraquist" in this usage, see Ingrid Stöhr, *Zweisprachigkeit in Böhmen: Deutsche Volksschulen und Gymnasien im Prag der Kafka-Zeit* (Köln: Böhlau, 2010), 87–88, 93. See also *Die Karl Ferdinands-Universität in Prag und die Čechen: Ein Beitrag zur Geschichte dieser Universität in den letzten hundert Jahren (1784–1885)* (Leipzig: Duncker & Humblot, 1886), 11 and 28.

36. *Die Deutsche Karl-Ferdinands-Universität in Prag*, 24. The use of Czech grew in every area of the university, despite occasional ministerial reactions, such as the abolition of its use in oral

examinations in 1862. Jan Havránek, "The University, Its Organization, Administration and Students (1848–1882)," in Kavka and Petráň, *History of Charles University*, 2:71–88, on 86.

37. Havránek, "University, Its Organization, Administration and Students (1848–1882)," 81–83.

38. Rudolf Hemmerle, "Die Deutsche Technische Hochschule in Prag," in Franz Böhm, ed., *Alma mater Pragensis: Ein Dank an Prag und seine hohen Schulen* (Erlangen: Verlag Karl Müller, 1959), 78–87. For more on technical education in Prague, see David F. Good, *The Economic Rise of the Habsburg Empire, 1750–1914* (Berkeley: University of California Press, 1984), 62.

39. Gary B. Cohen, *The Politics of Ethnic Survival: Germans in Prague, 1861–1914*, 2nd. rev. ed. (West Lafayette, IN: Purdue University Press, 2006 [1981]), 100.

40. Peter Burian, "Die Teilung der Prager Universität und die Österreichische Hochschulpolitik," in *Die Teilung der Prager Universität 1882 und die intellektuelle Desintegration in den böhmischen Ländern: Vorträge der Tagung des Collegium Carolinum in Bad Wiessee vom 26. bis 28. November 1982* (Munich: R. Oldenbourg, 1984), 25–36, on 28.

41. Emil Brix, *Die Umgangssprachen in Altösterreich zwischen Agitation und Assimilation: Die Sprachenstatistik in den zisleithanischen Volkszählungen 1880 bis 1910* (Vienna: Hermann Böhlaus, 1982).

42. For an excellent example of how this worked in practice over the decades before and after 1880, see Jeremy King, *Budweisers into Czechs and Germans: A Local History of Bohemian Politics, 1848–1948* (Princeton, NJ: Princeton University Press, 2002).

43. Stöhr, *Zweisprachigkeit in Böhmen*, 110–111, 122–123; Josef Erben, ed., *Statistika královského hlavního města Prahy*, vol. 1 (Prague: Obecní statistická kommisse královského hlavního města Prahy, 1871). The essential reference on the demographics of the German population remains Cohen, *Politics of Ethnic Survival*.

44. Cynthia Paces, *Prague Panoramas: National Memory and Sacred Space in the Twentieth Century* (Pittsburgh: University of Pittsburgh Press, 2009), 19.

45. See chapter 6. On the reasons for the population inversion, see Jan Havránek, "The Development of Czech Nationalism," *Austrian History Yearbook* 3, pt. 2 (1967): 223–260.

46. Peter Horwath, "The Erosion of 'Gemeinschaft': German Writers of Prague, 1890–1924," *German Studies Review* 4, no. 1 (February 1981): 9–37, on 12.

47. Pieter M. Judson, "'Whether Race or Conviction Should Be the Standard': National Identity and Liberal Politics in Nineteenth-Century Austria," *Austrian History Yearbook* 22 (1991): 76–95. See also Gerald Stourzh, "Ethnic Attribution in Late Imperial Austria: Good Intentions, Evil Consequences," in Ritchie Robertson and Edward Timms, eds., *The Habsburg Legacy: National Identity in Historical Perspective* (Edinburgh: Edinburgh University Press, 1994), 67–83.

48. For examples of studiously balanced literature on the national conflict, see Elizabeth Wiskemann, *Czechs and Germans: A Study of the Struggle in the Historic Provinces of Bohemia and Moravia*, 2nd. ed. (London: Macmillan, 1967 [1938]); Johann Wolfgang Brügel, *Tschechen und Deutsche, 1918–1938* (Munich: Nymphenberger, 1967); Seibt, *Deutschland und die Tschechen*; Jan Křen, *Die Konfliktgemeinschaft: Tschechen und Deutsche, 1780–1918* (Munich: R. Oldenbourg, 1996); Bugge, "Czech Nation-Building"; Derek Sayer, "The Language of Nationality and the Nationality of Language: Prague 1780–1920," *Past & Present*, no. 153 (November 1996): 164–210.

49. Emil Skála, "Die Entwicklung des Bilinguismus in der Tschechoslowakei vom 13.-18. Jahrhundert," *Beiträge zur Geschichte der deutschen Sprache und Literatur* 86 (1964): 69–106; Pavel Trost, "Der tschechisch-deutsche Makkaronismus," *Wiener slawistischer Almanach* 6 (1980): 273–278.

50. On the mechanics of the division of the university, see H. Gordon Skilling, "The Partition of the University in Prague," *Slavonic and East European Review* 27, no. 69 (May 1949): 430–449. Goll, who himself served as rector in 1907–1908, published a more polemical version as his inaugural address from 19 November 1907: Goll, *Rozdělení Pražské university*. This volume contains many valuable appendices reprinting the primary documents in the original German. On Mach's central role, see Dieter Hoffmann, "Ernst Mach und die Teilung der Prager Universität," in Lemberg, *Universitäten in nationaler Konkurrenz*, 33–61; idem, "Ernst Mach and the Conflict of Nations," in John Blackmore, ed., *Ernst Mach—A Deeper Look: Documents and New Perspectives* (Dordrecht: Kluwer Academic, 1992), 29–55; and idem, "Ernst Mach in Prag," in Dieter Hoffmann and Hubert Laitko, eds., *Ernst Mach: Studien und Dokumente zu Leben und Werk* (Berlin: Deutscher Verlag der Wissenschaften, 1991), 141–178, on 151–155.

51. Erich Schmied, "Die Altösterreichische Gesetzgebung zur Prager Universität: Ein Beitrag zur Geschichte der Prager Universität bis 1918," in *Die Teilung der Prager Universität 1882 und die intellektuelle Desintegration in den böhmischen Ländern*, 11–23.

52. Goll, *Rozdělení Pražské university*, 41; Jiří Pešek, "Die Prager Universitäten im ersten Drittel des 20. Jahrhunderts: Versuch eines Vergleichs," in Lemberg, *Universitäten in nationaler Konkurrenz*, 145–166, on 149.

53. Jan Havránek, "The Czech University, 1882–1918," in Kavka and Petráň, *History of Charles University*, 2:123–131. On the corpses, see Schmied, "Die Altösterreichische Gesetzgebung zur Prager Universität," 20. On the doors, see Seibt, *Deutschland und die Tschechen*, 291.

54. Jaroslav Goll, *Der Hass der Völker und die österreichischen Universitäten* (Prague: Bursík & Kohout, 1902), 26.

55. Wiskemann, *Czechs and Germans*, 63. On German-speaking Bohemians opting for foreign universities, see Pešek, "Die Prager Universitäten im ersten Drittel des 20. Jahrhunderts," 155; idem, "Les étudiants des pays tchèques entre Prague et Vienne: Comparaison du rôle des trois universités en 1884," tr. Frédéric Bègue, in *Allemands, juifs et tchèques à Prague/Deutsche, Juden und Tschechen in Prag, 1890–1924: Actes du colloque international de Montpellier, 8–10 décembre 1994* (Montpellier: Université Paul-Valéry-Montpellier III, 1996): 101–113.

56. Ludmila Hlaváčková, Alena Míšková, and Jiří Pešek, "The German University in Prague, 1882–1918," in Kavka and Petráň, *History of Charles University*, 2:163–174; Pešek, "Die Prager Universitäten im ersten Drittel des 20. Jahrhunderts," 149.

57. Seibt, *Deutschland und die Tschechen*, 291; Cohen, *Politics of Ethnic Survival*, 100.

58. Ortsrat Prag des deutschen Volksrates für Böhmen, *Prag als deutsche Hochschulstadt* (Prague: Carl Bellmann, 1909).

59. Havránek, "Ke jmenování Alberta Einsteina profesorem v Praze," 109.

60. Kleinert, "Anton Lampa und Albert Einstein," 285–288.

61. Emilie Těšínská, "Profilování teoretické fyziky na pražské univerzitě a vazby s pražským působením A. Einsteina před 100 lety," *Pokroky matematiky, fyziky a astronomie* 57, no. 2 (2012): 146–168, on 153.

62. Biographical details are taken from E. Lohr, "Gustav Jaumann," *Physikalische Zeitschrift* 26, no. 4 (15 February 1925): 189–198.

63. Philipp Frank, "Die Bedeutung der physikalischen Erkenntnistheorie Machs für das Geistesleben der Gegenwart," *Die Naturwissenschaften* 5, no. 5 (2 February 1917): 65–72, on 69. See also idem, *Einstein: His Life and Times*, tr. George Rosen, ed. and rev. Shuichi Kusaka (Cambridge, MA: Da Capo Press, 2002 [1947]), 136.

64. John Blackmore, R. Itagaki, and S. Tanaka, eds., *Ernst Mach's Vienna, 1895–1930: Or Phenomenalism as Philosophy of Science* (Dordrecht: Kluwer, 2001), 23. Mach also composed a middle-school science textbook with a Czech, J. Odstrčil, which was published in 1887 but only approved for implementation in the classroom by the state in 1893. Adolf Hohenester, "Ernst Mach als Didaktiker, Lehrbuch- und Lehrplanverfasser," in Rudolf Haller and Friedrich Stadler, eds., *Ernst Mach—Werk und Wirkung* (Vienna: Hölder-Pichler-Tempsky, 1988), 138–163, on 144.

65. Ernst Mach, "Prager Testament," reproduced in Joachim Thiele, *Wissenschaftliche Kommunikation: Die Korrespondenz Ernst Machs* (Kastellaun: Henn, 1978), 229.

66. Gustav Jaumann to Ernst Mach, 27 October 1906, reproduced in Thiele, *Wissenschaftliche Kommunikation*, 227.

67. Minister of Religion and Education Count Karl Stürgkh to the emperor, 16 December 1910, reproduced in Havránek, "Ke jmenování Alberta Einsteina profesorem v Praze," 128. On the search process in general, and the frequent favoritism shown to Austrians in reordering the list, see Tatjana Buklijas, "The Politics of *Fin-de-siècle* Anatomy," in Mitchell G. Ash and Jan Surman, eds., *The Nationalization of Scientific Knowledge in the Habsburg Empire, 1848–1918* (Basingstoke: Palgrave Macmillan, 2012): 209–244.

68. Einstein to Jakob Laub, [27 August 1910], *CPAE* 5:224, on 253. Most biographers, such as Frank (*Einstein: His Life and Times*, 78), side with the citizenship explanation, which was later backed up by documentation. Some, however, believe Einstein had a point: Carl Seelig, *Albert Einstein: Leben und Werk eines Genies unserer Zeit* (Zurich: Bertelsmann Lesering, 1960), 117.

69. Hoffmann, "Ernst Mach in Prag," 142.

70. Peter L. Galison, "The Assassin of Relativity," in Galison, Gerald Holton, and Silvan S. Schweber, eds., *Einstein for the 21st Century: His Legacy in Science, Art, and Modern Culture* (Princeton, NJ: Princeton University Press, 2008), 185–204. Coincidentally, Friedrich's father, Victor Adler, founder of the Social Democratic Workers' Party in Austria and a major figure in Viennese politics, was born in Prague. On Adler *père*, see William M. Johnston, *The Austrian Mind: An Intellectual and Social History, 1848–1938* (Berkeley: University of California Press, 1972), 99, 101.

71. Frank, *Einstein: His Life and Times*, 78. The German version reads: "Wenn man Einstein in dem Vorschlag vor mich setzt und glaubt, daß er größere Verdienste hat als ich, so will ich mit einer Universität nichts zu tun haben, die der Modernität nachjagt und das wahre Verdienst verkennt." Frank, *Einstein: Sein Leben und seine Zeit* (Munich: Paul List, 1949), 136.

72. Albrecht Fölsing, *Albert Einstein: A Biography*, tr. and abridged Ewald Osers (New York: Penguin Books, 1997 [1993]), 273; John T. Blackmore, Ryoichi Itagaki, and Setsuko Tanaka, eds., *Ernst Mach's Science: Its Character and Influence on Einstein and Others* (Kanagawa: Tokai University Press, 2006), 134.

73. Max Hussarek von Heinlein (section head at the Habsburg Ministry of Education) to Einstein, 27 September 1910, *CPAE* 5:225 on 255.

74. Count Karl von Stürgkh to Einstein, 13 January 1911, *CPAE* 5:245, on 272–273.

75. Two of his biographers disagree, saying Einstein took Austrian citizenship, but their accounts are so close to the story in Berlin that it is possible there has been some conflation of the two cases: Seelig, *Albert Einstein*, 126; idem, *Albert Einstein und die Schweiz* (Zurich: Europa-Verlag, 1952), 117; Walter Isaacson, *Einstein: His Life and Universe* (New York: Simon & Schuster, 2007), 164.

76. Student petition to retain Einstein to the Faculty of the University of Zurich, 23 July 1910, *CPAE* 5:210, on 243.

77. For the official offer letter, see Count Karl von Stürgkh to Einstein, 15 December 1910, *CPAE* 5:238, on 266. Einstein's resignation is available at Einstein to Department of Education, Canton of Zurich, 20 January 1911, *CPAE* 5:247, on 274.

78. Abraham Pais, *"Subtle Is the Lord . . .": The Science and Life of Albert Einstein* (New York: Oxford University Press, 1982), 193.

79. Mileva Einstein to Helene Savić, [January 1911], reproduced in Milan Popović, ed., *In Albert's Shadow: The Life and Letters of Mileva Marić, Einstein's First Wife* (Baltimore: Johns Hopkins University Press, 2003), 104–105.

80. Otto Brechler, "Das geistige Leben Prags vor Hundert Jahren," *Deutsche Arbeit* 10, no. 6 (March 1911): 329–343.

81. Karl Baedeker, *Österreich: Handbuch für Reisende*, 28th ed. (Leipzig: Baedeker, 1910), 282–292, on 286.

82. Fölsing, *Albert Einstein*, 280. On the new construction in the area, see Zdeněk Guth, "Einsteinovi a Smíchov," *Neděle s Lidová demokracie* 31, no. 15 (12 April 1975): 9–10.

CHAPTER 2: THE SPEED OF LIGHT

1. Max Abraham, "Eine neue Gravitationstheorie," *Archiv der Mathematik und Physik* 20 (1913): 193–209, on 209.

2. Ernest Solvay to Einstein, 9 June 1911, *CPAE* 5:408, on 483.

3. On the Geneva events, see Ronald W. Clark, *Einstein: The Life and Times* (New York: Thomas Y. Crowell, 1971), 124.

4. This is the argument in Richard Staley, *Einstein's Generation: The Origins of the Relativity Revolution* (Chicago: University of Chicago Press, 2008), ch. 10. The classic reference on the 1911 meeting remains Diana Kormos Barkan, "The Witches' Sabbath: The First International Solvay Congress in Physics," in Mara Beller, Jürgen Renn, and Robert S. Cohen, eds., *Einstein in Context, Science in Context* 6, no. 1 (1993): 59–82. For the further trajectory of meetings, see Jagdish Mehra, *The Solvay Conferences on Physics: Aspects of the Development of Physics since 1911* (Dordrecht: D. Reidel, 1975).

5. "On the Present State of Specific Heats," *CPAE* 3:26.

6. Michele Besso to Einstein, 23 October 1911, *CPAE* 5:299, on 343.

7. Einstein to Hopf, [after 20 February 1912], *CPAE* 5:364, on 419.

8. Einstein to Zangger, 15 November [1911], *CPAE* 5:305, on 349.

9. Einstein to Ernest Solvay, 22 November 1911, *CPAE* 5:312, on 358.

10. P. Langevin and M. de Broglie, eds., *La théorie du rayonnement et les quanta: Rapports et discussions de la Réunion tenue à Bruxelles, du 30 octobre au 3 novembre 1911 sous les auspices de M. E. Solvay* (Paris: Gauthier-Villars, 1912), front matter. On "Olympic internationalism" in the European sciences of this period, see Geert J. Somsen, "A History of Universalism: Conceptions of the Internationality of Science from the Enlightenment to the Cold War," *Minerva* 46 (September 2008): 361–379.

11. Einstein to Michele Besso, 13 May 1911, *CPAE* 5:267, on 295; A. Euken to Einstein, 23 January 1912, AEDA 9–257.

12. See especially the magisterial Jürgen Renn, ed., *The Genesis of General Relativity*, 4 vols. (Dordrecht: Springer, 2007). See also John D. Norton, "How Einstein Found His Field Equations, 1912–1915," in Don Howard and John Stachel, eds., *Einstein and the History of General Relativity* (Boston: Birkhäuser, 1989), 101–159.

13. József Illy, "Albert Einstein in Prague," *Isis* 70 (1979): 76–84, on 80; Jan Havránek, "Materiály k Einsteinovu Pražskému působení z Archivu Univerzity Karlovy," *Acta universitatis Carolinae—Historia universitatis Carolinae Pragensis* 22, no. 1 (1980): 109–134, on 112–114. See also "Einstein's Academic Courses," *CPAE* 3, appendix B, on 598–599.

14. Quoted in Rudolf Kolomý, "Albert Einstein v Praze," *Vesmír* 53, no. 4 (1974): 112–115, on 113.

15. Einstein to Heinrich Zangger, [7 April 1911], *CPAE* 5:263, on 290.

16. Editorial note, "Einstein's Teaching Notes," *CPAE* 3, on 3–4.

17. Einstein quoted in Hugo Bergman, "Personal Remembrances of Albert Einstein," in Robert S. Cohen and Marx W. Wartofsky, eds., *Logical and Epistemological Studies in Contemporary Physics* (Dordrecht: D. Reidel, 1974), 388–394, on 390.

18. Reinhold Fürth quoted in G. J. Whitrow, *Einstein: The Man and His Achievement* (New York: Dover, 1967), 48.

19. Otto Stern quoted in Res Jost, "Einstein und Zürich, Zürich und Einstein," *Vierteljahrsschrift der Naturforschenden Gesellschaft in Zürich* 124 (1979): 7–23, on 18.

20. Einstein to Ludwig Hopf, 12 June [1912], *CPAE* 5:269, on 298.

21. Maximilian Pinl with Auguste Dick, "Kollegen in einer dunklen Zeit. Schluß," *Jahresbericht der Deutschen Mathematiker-Vereinigung* 75 (1974): 166–208, on 178–180.

22. Einstein to Marcel Grossmann, [27 April 1911], *CPAE* 5:266, on 294.

23. On the institute, see Dieter Hoffmann, "Ernst Mach in Prag," in Dieter Hoffmann and Hubert Laitko, eds., *Ernst Mach: Studien und Dokumente zu Leben und Werk* (Berlin: Deutscher Verlag der Wissenschaften, 1991), 141–178, on 143; and Emilie Těšínská, "Profilování teoretické fyziky na pražské univerzitě a vazby s pražským působením A. Einsteina před 100 lety," *Pokroky matematiky, fyziky a astronomie* 57, no. 2 (2012): 146–168, on 148.

24. Einstein to Michele Besso, 13 May 1911, *CPAE* 5:267, on 295.

25. On Minkowski, see Peter Louis Galison, "Minkowski's Space-Time: From Visual Thinking to the Absolute World," *Historical Studies in the Physical Sciences* 10 (1979): 85–121.

26. A. Einstein, "Über das Relativitätsprinzip und die aus demselben gezogenen Folgerungen," *Jahrbuch der Radioaktivität und Elektronik* 4 (1907): 411–462, reproduced in *CPAE* 2:47, on 465–466. For more context on this crucial paper, see Arthur I. Miller, "Albert Einstein's 1907

Jahrbuch Paper: The First Step from SRT to GRT," in Jean Eisenstaedt and A. J. Kox, eds., *Studies in the History of General Relativity* (Boston: Birkhäuser, 1992), 319–335. The equivalence principle still has a place in contemporary physics, but it is understood to apply only infinitesimally at each location, which was certainly not how Einstein initially understood it. On the nuances of Einstein's conception, see John D. Norton, "What Was Einstein's Principle of Equivalence?," in Howard and Stachel, *Einstein and the History*, 5–47.

27. In 1918, after he had successfully completed his general theory, Einstein wrote to Eötvös to thank him for his experimental work that had demonstrated the equivalence principle. Einstein to Roland von Eötvös, 31 January 1918, *CPAE* 8:450, on 624.

28. Max von Laue to Einstein, 27 December 1907, *CPAE* 5:70, on 83.

29. Einstein to Arnold Sommerfeld, 5 January 1908, *CPAE* 5:72, on 86.

30. Max von Laue, *Das Relativitätsprinzip* (Braunschweig: Vieweg und Sohn, 1911), 187.

31. Jürgen Renn and Matthias Schemmel, "Theories of Gravitation in the Twilight of Classical Physics," in Christoph Lehner, Jürgen Renn, and Matthias Schemmel, eds., *Einstein and the Changing Worldviews of Physics* (New York: Springer, 2012), 3–22; Frans Herbert van Lunteren, *Framing Hypotheses: Conceptions of Gravity in the 18th and 19th Centuries* (Ph.D. diss., University of Utrecht, 1991); Matthew R. Edwards, ed., *Pushing Gravity: New Perspectives on Le Sage's Theory of Gravitation* (Montreal: Apeiron, 2002); and Scott Walter, "Breaking in the 4-Vectors: The Four-Dimensional Movement in Gravitation, 1905–1910," in Renn, *Genesis of General Relativity*, 3:193–252.

32. Shaul Katzir, "Poincaré's Relativistic Theory of Gravitation," in A. J. Kox and Jean Eisenstaedt, eds., *The Universe of General Relativity* (Boston: Birkhäuser, 2005), 15–37. See also the 1905 attempt to reconcile the Lorentz transformations and gravity in Richard Gans, "Gravitation und Elektromagnetismus," *Physikalische Zeitschrift* 6, no. 23 (1905): 803–805.

33. Albert Einstein, "Autobiographical Notes," tr. Paul Arthur Schilpp, in Schilpp, ed., *Albert Einstein: Philosopher-Scientist*, 2 vols. (New York: Harper & Row, 1951 [1949]), 1:1–95, on 66. For historians' explanations of the ostensible "delay," see Abraham Pais, *"Subtle Is the Lord . . .": The Science and Life of Albert Einstein* (New York: Oxford University Press, 1982), 187–188; V. P. Vizgin, *Reliativistskaia teoriia tiagoteniia: Istoki i formirovanie, 1900–1915* (Moscow: Nauka, 1981), 121.

34. Einstein to Jakob Laub, 10 August 1911, *CPAE* 5:275, on 309. See also the later letter of Einstein to Ehrenfest, [before 20 June 1912], *CPAE* 5:409, on 485.

35. For secondary discussions of these papers, see Jiří Bičák, "Einstein's Prague Articles on Gravitation," in D. G. Blair and M. J. Buckingham, eds., *The Fifth Marcel Grossmann Meeting on Recent Developments in Theoretical and Experimental General Relativity, Gravitation and Relativistic Field Theories: Proceedings of the Meeting Held at the University of Western Australia, 8–13 August 1988* (Singapore: World Scientific, 1989), 1325–1333; idem, "Einstein's Prague Ideas on Gravitation: The History and the Present," in *Trends in Physics, 1984: Proceedings of the 6th General Conference of the European Physical Society, August 1984, Prague, Czechoslovakia* (Prague: Prometheus, 1985), 65–75; Editorial Comment, "Einstein on Gravitation and Relativity: The Static Field," *CPAE* 3, on 122–127; Alexander S. Blum, Jürgen Renn, Donald C. Salisbury, Matthias Schemmel, and Kurt Sundermeyer, "1912: A Turning Point on Einstein's Way to General Relativity," *Annalen der Physik* 524, no 1. (2012): A11–A13; Pais, *"Subtle Is the Lord . . ."*, ch. 11; Vizgin, *Reliativistskaia teoriia tiagoteniia*; and Jagdish Mehra, *Einstein, Hilbert,*

and the Theory of Gravitation: Historical Origins of General Relativity Theory (Dordrecht: D. Reidel, 1974), 3–7.

36. A. Einstein, "Zur Elektrodynamik bewegter Körper," *Annalen der Physik* 17 (1905): 891–921, reproduced in *CPAE* 2:23, on 277.

37. Einstein to Willem Julius, 24 August 1911, *CPAE* 5:278, on 312–313.

38. Albert Einstein, "Über den Einfluß der Schwerkraft auf die Ausbreitung des Lichtes," *Annalen der Physik* 35 (1911): 898–908, reproduced in *CPAE* 3:23, on 494.

39. Einstein, "Über den Einfluß der Schwerkraft auf die Ausbreitung des Lichtes," reproduced in *CPAE* 3:23, on 486.

40. Ibid., 496. Later general relativity would double the size of this effect.

41. It should be noted that the quantitative estimate of the amount of deflection in the static theory was different from the later full general theory's accurate prediction measured by Eddington. On Eddington and his expedition, see especially Matthew Stanley, "'An Expedition to Heal the Wounds of War': The 1919 Eclipse and Eddington as Quaker Adventurer," *Isis* 94, no. 1 (March 2003): 57–89; and idem, *Practical Mystic: Religion, Science, and A. S. Eddington* (Chicago: University of Chicago Press, 2007). On Eddington's advance strategy to publicize the results, see Alistair Sponsel, "Constructing a 'Revolution in Science': The Campaign to Promote a Favourable Reception for the 1919 Solar Eclipse Experiments," *British Journal for the History of Science* 35, no. 4 (2002): 439–467. On further American testing in the 1920s, which proved decisive, see Jeffrey Crelinstein, *Einstein's Jury: The Race to Test Relativity* (Princeton, NJ: Princeton University Press, 2006). On the German-nationalist attempt to find a precursor to Einstein, see Stanley L. Jaki, "Johann Georg von Soldner and the Gravitational Bending of Light, with an English Translation of His Essay on It Published in 1801," *Foundations of Physics* 8, nos. 11/12 (1978): 927–950. After Soldner's long-forgotten paper came to light, fanatical Einstein nemesis and distinguished German experimentalist Philipp Lenard pushed for Soldner's priority.

42. Leo Wenzel Pollak to E. F. Freundlich, 24 August 1911, AEA 11–181.

43. Einstein to Erwin Freundlich, 1 September 1911, *CPAE* 5:281, on 317.

44. On the challenges of this measurement, see H. von Klüber, "The Determination of Einstein's Light-Deflection in the Gravitational Field of the Sun," *Vistas in Astronomy* 3, no. 1 (1960): 47–77.

45. Einstein to Erwin Freundlich, 8 January 1912, *CPAE* 5:336, on 387.

46. Max von Laue to Albert Einstein, 27 December [1911], *CPAE* 5:333, on 385.

47. Einstein to Ernst Mach, 25 June 1913, *CPAE* 5:448, on 531.

48. Erwin Freundlich, "Bedeckung des Sternes BD + 12°2138 durch den Mond während der totalen Sonnenfinsternis am 21. August 1914," *Astronomische Nachrichten* 197 (1914): 335–336.

49. Einstein's foreword in Freundlich, *Die Grundlagen der Einsteinschen Gravitationstheorie*, 3rd. ed. (Berlin: Julius Springer, 1920), i.

50. Klaus Hentschel, "Erwin Finlay Freundlich and Testing Einstein's Theory of Relativity," *Archive for History of Exact Sciences* 47, no. 2 (June 1994): 143–201; Klaus Hentschel, *The Einstein Tower: An Intertexture of Dynamic Construction, Relativity Theory, and Astronomy*, tr. Ann M. Hentschel (Stanford, CA: Stanford University Press, 1997).

51. Einstein to Hopf, [after 20 February 1912], *CPAE* 5:364, on 418.

52. Einstein to Zangger, [before 29 February 1912], *CPAE* 5:366, on 420–421.

53. Einstein to Zangger, 17 March [1912], *CPAE* 5:374a, in vol. 10, on 20.

54. Einstein to Ehrenfest, [10 March 1912], *CPAE* 5:369, on 428.

55. Einstein to Lorentz, 18 February 1912, *CPAE* 5:360, on 413.

56. Einstein to Wilhelm Wien, 24 February [1912], *CPAE* 5:365, on 420.

57. Einstein to Wien, 11 March 1912, *CPAE* 5:371, on 429.

58. Einstein to Wien, 20 March [1912], *CPAE* 5:375, on 433.

59. Einstein to Hopf, [after 20 February 1912], *CPAE* 5:364, on 418.

60. Max Abraham, "Zur Theorie der Gravitation," *Physikalische Zeitschrift* 13, no. 1 (1 January 1912): 1–4. See also the appendix, published as idem, "Das Elementargesetz der Gravitation," *Physikalische Zeitschrift* 13, no. 1 (1 January 1912): 4–5. On Abraham's gravity theory, see the careful analysis in Jürgen Renn, "The Summit Almost Scaled: Max Abraham as a Pioneer of a Relativistic Theory of Gravitation," in Renn, *Genesis of General Relativity*, 3:305–330. On Abraham's context in Milan, see Barbara J. Reeves, "Einstein Politicized: The Early Reception of Relativity in Italy," in Thomas F. Glick, ed., *The Comparative Reception of Relativity* (Dordrecht: D. Reidel, 1987), 189–229.

61. Einstein to Marian von Smoluchowski, 24 March [1912], *CPAE* 5:376, on 434. See also Einstein to Besso, 26 March [1912], *CPAE* 5:377, on 435.

62. Max Abraham, "Relativität und Gravitation: Erwiderung auf eine Bemerkung des Hrn. A. Einstein," *Annalen der Physik* 38 (1912): 1056–1058, on 1056.

63. A. Einstein, "Lichtgeschwindigkeit und Statik des Gravitationfeldes," *Annalen der Physik* 38 (1912): 355–369, reproduced in *CPAE* 4:3, on 130. Nonetheless in 1912 another physicist treated Abraham's and Einstein's theories as essentially equivalent: Jun Ishiwara, "Zur Theorie der Gravitation," *Physikalische Zeitschrift* 13 (1912): 1189–1193.

64. A. Einstein, "Zur Theorie des statischen Gravitationsfeldes," *Annalen der Physik* 38 (1912): 443–458, reproduced in *CPAE* 4:4, on 159. See the thorough discussion of his reasoning in Einstein to Ehrenfest, [before 20 June 1912], *CPAE* 5:409, on 485; as well as the analysis in Vizgin, *Reliativistskaia teoriia tiagoteniia*, 179.

65. Max Abraham, "Nochmals Relativität und Gravitation: Bemerkungen zu A. Einsteins Wiederung," *Annalen der Physik* 39 (1912): 444–448.

66. Einstein to Jakob Laub, 16 March 1910, *CPAE* 5:199, on 231. Einstein continued the following year to polemicize with him at arm's length through Laub: Einstein to Jakob Laub, 10 August 1911, *CPAE* 5:275, on 308–309.

67. Einstein to Zangger, [27 January 1912], *CPAE* 5:344, on 395.

68. Einstein to Besso, 26 March [1912], *CPAE* 5:377, on 436–437. "Untenable (*unhaltbar*)" was a favorite phrase when it came to Abraham's work. See also Einstein to Ehrenfest, 12 February [19]12, *CPAE* 5:357, on 408.

69. Jürgen Renn, "Classical Physics in Disarray: The Emergence of the Riddle of Gravitation," in Renn, *Genesis of General Relativity*, 1:21–80.

70. Einstein to Alfred Kleiner, 3 April 1912, *CPAE* 5:382, on 447–448.

71. A. Einstein, "Relativität und Gravitation: Erwiderung auf eine Bemerkung von M. Abraham," *Annalen der Physik* 38 (1912): 1059–1064, reproduced in *CPAE* 4:8, on 181.

72. Ibid., 184.

73. Ibid., 185.

74. A. Einstein, "Bemerkung zu Abrahams vorangehender Auseinandersetzung 'Nochmals Relativität und Gravitation,'" *Annalen der Physik* 39 (1912): 704, reproduced in *CPAE* 4:9, on 190.

75. Abraham, "Eine neue Gravitationstheorie," 208. In a later essay, he suggested that gravitational waves would be longitudinal. Idem, "Die neue Mechanik," *Scientia* 15 (1914): 8–27, on 22. See also idem, "Das Gravitationsfeld," *Physikalische Zeitschrift* 13, no. 17 (1 September 1912): 793–797.

76. Einstein to Sommerfeld, 29 October [1912], *CPAE* 5:421, on 505. See also the passing reference to Abraham in A. Einstein, "Zum gegenwärtigen Stande des Gravitationsproblems," *Physikalische Zeitschrift* 14 (1913) 1249–1262, reproduced in *CPAE* 4:17, on 488.

77. Max Abraham, "Neuere Gravitationstheorien," *Jahrbuch der Radioaktivität und Elektronik* 11 (1914): 470–520, on 510.

78. Abraham, "Die neue Mechanik," 26.

79. Einstein to Besso, [after 1 January 1914], *CPAE* 5:499, on 588.

80. Theodor von Kármán to Einstein, 22 February 1922, *CPAE* 13:61, on 149.

81. Einstein to Erwin Freundlich, [mid-August 1913], *CPAE* 5:468, on 550.

82. Eva Isaksson, "Der finnische Physiker Gunnar Nordström und sein Beitrag zur Entstehung der allgemeinen Relativitätstheorie Albert Einsteins," *NTM-Schriftenreihe für Geschichte der Naturwissenschaften, Technik und Medizin* 22, no. 1 (1985): 29–52; John D. Norton, "Einstein, Nordström and the Early Demise of Scalar, Lorentz-Covariant Theories of Gravitation," *Archive for the History of Exact Sciences* 45 (1992): 17–94; idem, "Einstein and Nordström: Some Lesser-Known Thought Experiments in Gravitation," in John Earman, Michel Janssen, and John D. Norton, eds., *The Attraction of Gravitation: New Studies in the History of General Relativity* (Boston: Birkhäuser, 1993), 3–29.

83. Gunnar Nordström, "Relativitätsprinzip und Gravitation," *Physikalische Zeitschrift* 13 (1912): 1126–1129; A. Einstein, "Zum gegenwärtigen Stande des Gravitationsproblems," *Physikalische Zeitschrift* 14 (1913): 1249–1262, reproduced in *CPAE* 4:17, on 492.

84. Cornelia and Gunnar Nordström to Einstein, 31 January 1918, *CPAE* 8:452, on 626; see also Einstein to Max Born, 24 June 1918, in Albert Einstein, Hedwig Born, and Max Born, *Briefwechsel, 1916–1955* (Frankfurt: Edition Erbrich, 1982 [1969]), 24; and Einstein to Gunnar Nordström, [28 May 1916], *CPAE* 8:222a, in vol. 15 on 23.

85. Einstein to Ehrenfest, [23 August 1915], *CPAE* 8:112, on 165.

86. Gustav Jaumann, "Theorie der Gravitation," *Sitzungsberichte der Akademie der Wissenschaften in Wien, Mathematisch-Naturwissenschaftliche Klasse, Abt. IIa* 121 (1912): 95–182. Gravity is also treated as a subset of his more general continuum physics in idem, "Geschlossenes System physikalischer und chemischer Differentialgesetze," *Sitzungsberichte der Akademie der Wissenschaften in Wien, Mathematisch-Naturwissenschaftliche Klasse, Abt. IIa* 120 (1911): 385–530, sec. XV. Peter Havas, in a rare mention in the secondary literature, calls this a "serious attempt at a (scalar) field theory fitting in his general continuum theory." Havas, "Einstein, Relativity and Gravitation Research in Vienna Before 1938," in Hubert Goenner, Jürgen Renn, Jim Ritter, and Tilman Sauer, eds., *The Expanding Worlds of General Relativity* (Boston: Birkhäuser, 1999), 161–206, on 173.

87. Gustav Jaumann, "Feststellung einer Priorität in der Gravitationstheorie," *Physikalische Zeitschrift* 15 (1914): 159–160, on 160.

88. Abraham, "Neurere Gravitationstheorien," 484n.

89. E. Lohr, "Gustav Jaumann," *Physikalische Zeitschrift* 26, no. 4 (15 February 1925): 189–198, on 196.

90. A. Einstein and M. Grossmann, *Entwurf einer verallgemeinerten Relativitätstheorie und einer Theorie der Gravitation* (Leipzig: Teubner, 1913), reproduced in *CPAE* 4:13, on 306.

91. Pais, *"Subtle Is the Lord . . .",* 210–212.

92. Jeroen Van Dongen, *Einstein's Unification* (Cambridge: Cambridge University Press, 2010); Michel Janssen, "'No Success Like Failure . . .': Einstein's Quest for General Relativity, 1902–1920," in Janssen and Christoph Lehner, eds., *The Cambridge Companion to Einstein* (Cambridge: Cambridge University Press, 2014), 167–227.

93. Einstein to Willem Julius, 24 August 1911, *CPAE* 5:278, on 312. For the original inquiry, see Julius to Einstein, 20 August 1911, *CPAE* 5:277, on 311.

94. Einstein to Zangger, 20 September 1911, *CPAE* 5:286, on 324–325. For the second approach, see Julius to Einstein, 17 September [1911], *CPAE* 5:284, on 323.

95. Einstein to Julius, 15 November 1911, *CPAE* 5:304, on 347. For more of the context of Einstein's negotiation with Dutch universities in this period, see Maria Rooseboom, "Albert Einstein und die niederländischen Universitäten," *Janus* 47 (1958): 198–201.

96. Einstein to Julius, 22 November 1911, *CPAE* 5:288, on 327.

97. Heinrich Zangger to Ludwig Forrer, 9 October 1911, *CPAE* 5:291, on 332.

98. Robert Gnehm to Einstein, 8 December 1911, *CPAE* 5:317, on 365; Marcel Grossmann to Einstein, 12 December 1911, *CPAE* 5:321, on 368. The offer came in Robert Gnehm to Einstein, 23 January 1912, *CPAE* 5:342, on 393.

99. Marie Curie to Pierre Weiss (Swiss Federal Board of Education), 17 November 1911, AEDA 09-C17, Box 9, Folder "C-Misc."

CHAPTER 3: ANTI-PRAGUE

1. Max Brod, "Der Wert der Reiseeindrücke [1911]," in Brod, *Über die Schönheit häßlicher Bilder: Essays zu Kunst und Ästhetik* (Göttingen: Wallstein Verlag, 2014), 245–246.

2. For details on the bridge, see Kateřina Bečková, *Prague: The City and Its River*, tr. Derek Paton and Marzia Paton (Prague: Karolinum, 2016), 10A–E; Marek Nekula, "The Divided City: Prague's Public Space and Franz Kafka's Readings of Prague," in Nekula, Ingrid Fleischmann, and Albrecht Greule, eds., *Franz Kafka im sprachnationalen Kontext seiner Zeit: Sprache und nationale Identität in öffentlichen Institutionen der böhmischen Länder* (Köln: Böhlau, 2007), 87–108, on 101–102.

3. For his biography, see Joseph Frederick Zacek, *Palacký: The Historian as Scholar and Nationalist* (The Hague: Mouton, 1970). On his role in 1848 and beyond, see Stanley Z. Pech, *The Czech Revolution of 1848* (Chapel Hill: University of North Carolina Press, 1969).

4. Hermann Münch, *Böhmische Tragödie: Das Schicksal Mitteleuropas im Lichte der tschechischen Frage* (Braunschweig: Georg Westermann, 1949), 452; Paul Vyšný, *Neo-Slavism and the Czechs, 1898–1914* (Cambridge: Cambridge University Press, 1977), 19.

5. Cynthia Paces, *Prague Panoramas: National Memory and Sacred Space in the Twentieth Century* (Pittsburgh: University of Pittsburgh Press, 2009), 56, 62–69. For Sucharda's own account

of the production and ceremonial unveiling of the statue, see Stanislav Sucharda, *Historie pomníku Fr. Palackého v Praze: K slavnosti odhalení* (Prague: Eduard Grégr a syn, [1912]); and idem, *Pomník Frant. Palackého v Praze: Jeho vznik a význam* (Prague: Eduard Grégr a syn, [1912]).

6. Nancy M. Wingfield, *Flag Wars and Stone Saints: How the Bohemian Lands Became Czech* (Cambridge, MA: Harvard University Press, 2007).

7. Ladislav Holy, *The Little Czech and the Great Czech Nation: National Identity and the Post-Communist Transformation of Society* (Cambridge: Cambridge University Press, 1996), 38.

8. Paces, *Prague Panoramas*, 52–53; Holy, *Little Czech*, 37.

9. Karl Baedeker, *Österreich, ohne Galizien, Dalmatien, Ungarn und Bosnien: Handbuch für Reisende*, 29th ed. (Leipzig: Baedeker, 1913), 295.

10. Jiří Kořalka, *Tschechen im Habsburgerreich und in Europa 1815–1914: Sozialgeschichtliche Zusammenhänge der neuzeitlichen Nationsbildung und der Nationalitätenfrage in den böhmischen Ländern* (Vienna: Verlag für Geschichte und Politik, 1991), 291; idem, "The Czech Question in International Relations at the Beginning of the 20th Century," *Slavonic and East European Review* 48, no. 111 (April 1970): 248–260, on 259.

11. Claire E. Nolte, *The Sokol in the Czech Lands to 1914: Training for the Nation* (New York: Palgrave Macmillan, 2002), 174–175.

12. Anton Reiser [Rudolf Kayser], *Albert Einstein: A Biographical Portrait* (New York: Albert & Charles Boni, 1930), 85. On Einstein's reservations about this book, see Albrecht Fölsing, *Albert Einstein: A Biography*, tr. and abridged Ewald Osers (New York: Penguin Books, 1997 [1993]), 85.

13. Philipp Frank, *Einstein: Sein Leben und seine Zeit* (Munich: Paul List, 1949), 255.

14. Hans Albert Einstein quoted in Zdeněk Guth, "Einsteinovi a Smíchov," *Neděle s Lidová demokracie* 31, no. 15 (12 April 1975): 9–10, on 10.

15. Einstein to Lucien Chavan, 5 April [1911], *CPAE* 5:262, on 289.

16. Einstein to Heinrich Zangger, [7 April 1911], *CPAE* 5:263, on 289. On the contemporary boycott movements among Czech-identified and German-identified Bohemians, see Catherine Albrecht, "The Rhetoric of Economic Nationalism in the Bohemian Boycott Campaigns of the Late Habsburg Monarchy," *Austrian History Yearbook* 22 (2001): 47–67.

17. Einstein to Heinrich Zangger, 7 November [1911], *CPAE* 5:303, on 346.

18. Einstein to Michele Besso, 13 May 1911, *CPAE* 5:267, on 295.

19. Einstein to Alfred and Clara Stern, 17 March [1912], *CPAE* 5:374, on 432–433.

20. Einstein to Zangger, [before June 1911], *CPAE* 5/267a, in vol. 10 on 16.

21. Heinrich A. Medicus, "The Friendship among Three Singular Men: Einstein and His Swiss Friends Besso and Zangger," *Isis* 85, no. 3 (September 1994): 456–478, especially on 464.

22. Einstein to Mileva Marić, [4 February 1902], *CPAE* 1:134, on 332.

23. Einstein to Carl Schröter, [1 February 1912], *CPAE* 5:349, on 400.

24. Einstein to Hans Tanner, [24 April 1911], *CPAE* 5:265, on 293. Einstein's biographer Carl Seelig, who knew the physicist personally, recalled hygiene being a constant touchstone in the Zurich–Prague comparisons: "On a later occasion when the conversation turned to this period, Einstein referred in his usual concise way to the hygienic conditions of the East, which could not be compared with the Swiss cult and cleanliness." Seelig, *Albert Einstein: A Documentary Biography*, tr. Mervyn Savill (London: Staples Press Limited, 1956), 129.

25. Einstein to Lucien Chavan, [5–6 July 1911], *CPAE* 5:271, on 304.

26. Ronald W. Clark, *Einstein: The Life and Times* (New York: Thomas Y. Crowell, 1971), 136–137.

27. Ferdinand Seibt, *Deutschland und die Tschechen: Geschichte einer Nachbarschaft in der Mitte Europas* (Munich: Piper, 1993), 217.

28. The Swiss model was discussed for Austria as a whole in September 1900 in the *Ostdeutsche Rundschau*, which ended up rejecting it and any other compromise with the Slavs. In 1906 Karl Renner wrote *Grundlagen und Entwicklungsziele der österreichisch-ungarischen Monarchie*, in which he suggested Swiss methods might be combined with the Moravian personalist system inaugurated the previous year. See the discussion in Elizabeth Wiskemann, *Czechs and Germans: A Study of the Struggle in the Historic Provinces of Bohemia and Moravia*, 2nd. ed. (London: Macmillan, 1967 [1938]), 69.

29. Ernst Denis, *La Bohême depuis la Montagne-Blanche*, 2 vols. (Paris: Ernest Leroux, 1903), 2:227; Jan Havránek, "The Development of Czech Nationalism," *Austrian History Yearbook* 3, pt. 2 (1967): 223–260, on 240–241. Such comparisons and linkages between different national "questions" were a common feature of public political discourse across Europe in the nineteenth century, a phenomenon expertly analyzed in Holly Case, *The Age of Questions: Or, A First Attempt at an Aggregate History of the Eastern, Social, Woman, American, Jewish, Polish, Bullion, Tuberculosis, and Many Other Questions over the Nineteenth Century and Beyond* (Princeton, NJ: Princeton University Press, 2018).

30. Jaroslav Kučera, *Minderheit im Nationalstaat: Die Sprachenfrage in den tschechisch-deutschen Beziehungen 1918–1938* (Munich: Oldenbourg, 1999), 31; Alfred M. De Zayas, *Nemesis at Potsdam: The Anglo-Americans and the Expulsion of the Germans: Background, Execution, Consequences*, rev. ed. (London: Routledge & Kegan Paul, 1979 [1977]), 27. For a thorough discussion on why Switzerland was a nonstarter as a model for the other nations, who considered it little more than an "eye-catcher" (*Blickfang*), see Johann Wolfgang Brügel, *Tschechen und Deutsche, 1918–1938* (Munich: Nymphenberger, 1967), 99–101.

31. Emanuel Rádl, *Der Kampf zwischen Tschechen und Deutschen*, tr. Richard Brandeis (Reichenberg [Liberec]: Gebrüder Stiepel, 1928), 159.

32. "Vortrag Einstein," *Bohemia*, no. 141 (23 May 1911): 10.

33. G. Kowalewski, *Bestand und Wandel: Meine Lebenserinnerungen zugleich ein Beitrag zur neueren Geschichte der Mathematik* (Munich: Oldenbourg, 1950), 237–238.

34. Ibid., 238.

35. Gerhard Kowalewski to Einstein, 14 July 1922, *CPAE* 13:288, on 409–410.

36. Einstein to Gerhard Kowalewski, 25 July 1922, *CPAE* 13:308, on 433; Einstein to Kowalewski, 7 November 1926, *CPAE* 15:408, on 625.

37. Kowalewski, *Bestand und Wandel*, 217.

38. Ibid., 249.

39. Hugo Bergman, "Personal Remembrances of Albert Einstein," in Robert S. Cohen and Marx W. Wartofsky, eds., *Logical and Epistemological Studies in Contemporary Physics* (Dordrecht: D. Reidel, 1974), 388–394, on 389. Perhaps because they were written over half a century after the events in question, there are several inaccuracies in these reminiscences, and they should be used only with caution. For example, Bergmann dated Einstein's time in Prague as beginning in

1910, when it was actually 1911; he thought the physicist was there for four years, instead of only three semesters; and so on.

40. Franz Kafka to Max Brod, 6 February 1914, in Max Brod and Franz Kafka, *Eine Freundschaft*, 2 vols. (Frankfurt am Main: Fischer, 1989), 2:137. See also Reiner Stach, *Kafka: Die frühen Jahre* (Frankfurt am Main: S. Fischer, 2014), 469.

41. Georg Gimpl, ed., *Weil der Boden selbst hier brennt...: Aus dem Prager Salon der Berta Fanta (1865–1918)* (Furth im Wald/Prague: Vitalis, [2000]). For a partial translation, see Wilma Abeles Iggers, *Women of Prague: Ethnic Diversity and Social Change from the Eighteenth Century to the Present* (Providence, RI: Berghahn Books, 1995), ch. 5. Neither Bertha nor Max appears to have been related either to the noted Czech-speaking feminist Julie Fantová-Kusá (1858–1908) or her husband, the celebrated architect Josef Fanta (1856–1954), who built the remarkable façade of the main train station in Prague. On this structure and its architect, see Rostislav Švácha, *The Architecture of New Prague, 1895–1945*, tr. Alexandra Büchler (Cambridge, MA: MIT Press, 1995 [1985]), 33–35.

42. Arnold Heidsieck, *The Intellectual Contexts of Kafka's Fiction: Philosophy, Law, Religion* (Columbia, SC: Camden House, 1994), 6.

43. Max Brod, *Franz Kafka als wegweisende Gestalt* (St. Gallen: Tschudy-Verlag, [1957]), 33–36; idem, *Streitbares Leben 1884–1968* (Munich: F. A. Herbig, 1969), 169–170.

44. Hartmut Binder, "Der Prager Fanta-Kreis: Kafkas Interesse an Rudolf Steiner," *Sudetenland*, no. 2 (1996): 106–150; Amnon Reuveni, "Bertha Fanta," *Das Goetheanum*, no. 50 (12 December 1993): 515–517.

45. Hugo Bergmann, *Das Unendliche und die Zahl* (Halle an der Salle: Max Niemeyer, 1913); Bergmann, diary entry of 20 May 1933, reproduced in Schmuel Hugo Bergman, *Tagebücher & Briefe*, 2 vols., ed. Miriam Sam"`bursky` (Königstein: Athenäum, 1985), 1:343.

46. Max Brod to Franz Kafka, 20 December 1918, in Brod and Kafka, *Eine Freundschaft*, 2:254.

47. John Forrester, "Die Geschichte zweier Ikonen: 'The Jews All over the World Boast of My Name, Pairing Me with Einstein' (Freud, 1926)," tr. Bettina Engels, in Michael Hagner, ed., *Einstein on the Beach: Der Physiker als Phänomen* (Frankfurt am Main: Fischer Taschenbuch, 2005), 96–123, on 100–101.

48. Else Bergmann, "Familiengeschichte," in Gimpl, *Weil der Boden selbst hier brennt*, 198–273, on 201–203.

49. Christian von Ehrenfels, *Sexualethik* (Wiesbaden: J. F. Bergmann, 1907); Edward Ross Dickinson, "Sex, Masculinity, and the 'Yellow Peril': Christian von Ehrenfels' Program for a Revision of the European Sexual Order, 1902–1910," *German Studies Review* 25, no. 2 (May 2002): 255–284; Hans Demetz, "Meine persönlichen Beziehungen und Erinnerungen an den Prager deutschen Dichterkreis," in Eduard Goldstücker, ed., *Weltfreunde: Konferenz über die Prager deutsche Literatur* (Prague: Academia, 1967), 135–145, on 138–139. On von Ehrenfels as the person who may have introduced Einstein, see Binder, "Der Prager Fanta-Kreis," 133.

50. Margarita Pazi, "Franz Kafka, Max Brod und der 'Prager Kreis,'" in Karl Erich Grözinger, Stéphane Mosès, and Hans Dieter Zimmermann, eds., *Franz Kafka und das Judentum* (Frankfurt am Main: Jüdischer Verlag bei Athenäum, 1987), 71–92, on 88. The only surviving Brod diaries run from 4 September 1909 to 20 July 1911, which fortunately includes the beginning of the

Einstein period. The rest were left with Brod's brother in Prague, who destroyed them out of fear of their seizure by the Gestapo. Otto Brod was later murdered in the Holocaust.

51. The classic biographical study of Einsteinová is Desanka Trbuhović-Gjurić, *Im Schatten Albert Einsteins: Das tragische Leben der Mileva Einstein-Marić* (Bern: Paul Haupt, 1985). This book first appeared in Kruševac, Yugoslavia, from the Bagdala Press in 1969 under the title *U senci Alberta Ajnštajna* and was printed in Cyrillic. It argues very strongly for her as a put-upon wife whose own scientific aspirations were snuffed out by her husband's bullying and indifference. This line of reasoning has been replicated in several studies of her life, such as Inge Stephan, *Das Schicksal der begabten Frau: Im Schatten berühmter Männer* (Zurich: Kreuz Verlag, 1989). The evidence for any scientific collaboration between the two is more equivocal than Trbuhović-Gjurić indicates. See the judicious discussion in John Stachel, "Albert Einstein and Mileva Marić: A Collaboration that Failed to Develop," in Stachel, *Einstein from "B" to "Z"* (Boston: Birkhäuser, 2002). See also Alberto A. Martínez, *Science Secrets: The Truth about Darwin's Finches, Einstein's Wife, and Other Myths* (Pittsburgh: University of Pittsburgh Press, 2011).

52. Trbuhović-Gjurić, *Im Schatten Albert Einsteins*, 96.

53. Michele Zackheim, *Einstein's Daughter: The Search for Lieserl* (New York: Riverhead, 1999), 45.

54. Mileva Einstein-Marić to Einstein, [4 October 1911], *CPAE* 5:290, on 331.

55. Georg Winternitz, "Glimpses of the Life of My Father, the Indologist Moriz Winternitz," tr. Debabrata Chakrabarti, *Tagore International* (December 1988): 35–63.

56. Elizabeth Roboz Einstein, *Hans Albert Einstein: Reminiscences of His Life and Our Life Together* (Iowa City: Iowa Institute of Hydraulic Research, University of Iowa, 1991), 18.

57. For extensive details on Einstein's love life, with particular emphasis on his first two wives and this affair, see the gossipy Roger Highfield and Paul Carter, *The Private Lives of Albert Einstein* (New York: St. Martin's Press, 1993).

58. Einstein to Elsa Löwenthal, [30 April 1912], *CPAE* 5:389, on 457.

59. Einstein to Elsa Löwenthal, 21 May [1912], *CPAE* 5:399, on 467.

60. Einstein to Elsa Löwenthal, [3 April 1913], *CPAE* 5:437, on 520.

61. Einstein to Heinrich Zangger, 7 November [1911], *CPAE* 5:303, on 345. See also Zangger to Einstein, [after 28 November 1911], *CPAE* 5:315a, in vol. 13, on 15.

62. Einstein to Marie Curie, 23 November 1911, *CPAE* 5:312a, in vol. 8, on 7.

63. One entirely unreliable biography blames Slav–German tensions in Bohemia as a major contributing factor in the Einsteins' divorce: Antonina Vallentin, *The Drama of Albert Einstein*, tr. Moura Budberg (Garden City, NY: Doubleday & Company, 1954), 52–53.

64. Frank, *Einstein: Sein Leben und seine Zeit*, 145.

65. For example, Clark, *Einstein*, 171; Fölsing, *Albert Einstein*, 71.

66. Einstein, *Hans Albert Einstein*, 20.

67. Einstein to Vladimir Varićak, 24 February 1911, *CPAE* 5:255a, in vol. 10, on 13.

68. Hans Kohn, *Pan-Slavism: Its History and Ideology*, 2nd. ed. (New York: Vintage, 1960 [1953]); Vyšný, *Neo-Slavism and the Czechs*; Bruce M. Garver, *The Young Czech Party, 1874–1901 and the Emergence of a Multi-Party System* (New Haven, CT: Yale University Press, 1978), 12–13.

69. Bergman, "Personal Remembrances of Albert Einstein," 389.

70. Stephan, *Das Schicksal der begabten Frau*, 102.

71. Wiskemann, *Czechs and Germans*, 71. See also Garver, *Young Czech Party*, 271; Robin Okey, "Austria and the South Slavs," in Ritchie Robertson and Edward Timms, eds., *The Habsburg Legacy: National Identity in Historical Perspective* (Edinburgh: Edinburgh University Press, 1994), 46–57.

72. Einstein to Helene Savić, [after 17 December 1912], *CPAE* 5:424, on 508.

73. On the publication context and interpretation, see Maurice Godé, "Un 'petit roman' qui a fait grand bruit: *Une servante tchèque* de Max Brod (1909)," in *Allemands, juifs et tchèques à Prague/ Deutsche, Juden und Tschechen in Prag, 1890–1924: Actes du colloque international de Montpellier, 8–10 décembre 1994* (Montpellier: Université Paul-Valéry-Montpellier III, 1996), 225–240. The book was subsequently translated into Russian by A. Eliasberg and into Czech by J. Osten. Brod would later highlight the significance of this "little novel" as his first major exploration of erotic writing, which later became a staple of the novels of his middle period. Max Brod, *Der Prager Kreis* (Stuttgart: W. Kohlhammer, 1966), 134. See also Gabriela Veselá, "E. E. Kisch und der deutschsprachige Prager erotische Roman," *Philologica Pragensia* 28, no. 4 (1985): 202–215.

74. Max Brod, "Ein tschechisches Dienstmädchen: Kleiner Roman," in Brod, *Arnold Beer: Das Schicksal eines Juden: Roman, und andere Prosa aus den Jahren 1909–1913*, ed. Hans-Gerd Koch and Hans Dieter Zimmermann (Göttingen: Wallstein, 2013 [1909]), 177–258, on 182.

75. On Brod as mediator, see Gaëlle Vassogne, *Max Brod in Prag: Identität und Vermittlung* (Tübingen: Max Niemeyer, 2009); Hans-Gerd Koch, "Max Brod: Ein jüdischer Dichter deutscher Zunge," in Steffen Höhne and Ludger Udolph, eds., *Deutsche–Tschechen–Böhmen: Kulturelle Integration und Desintegration im 20. Jahrhundert* (Köln: Böhlau, 2010), 129–135; Barbora Šrámková, *Max Brod und die tschechische Kultur* (Wuppertal: Arco, 2010). Brod continued to be stung by the rebukes decades later: Max Brod, "Praha a já," *Literární noviny* (10 December 1930): 4; idem, *Streitbares Leben*, 220–222.

76. Jiří Karásek ze Lvovic, "Z německé literatury," *Moderní revue* 21, no. 7 (1909): 361–362, on 361. A very similar point, down to the Strobl comparison, is made in O. Theer, "O knihách," *Česká revue*, no. 12 (1909): 761. See also Šrámková, *Max Brod und die tschechische Kultur*, 309.

77. Božena Benešová, "Krásná Prósa," *Novina* 2, no. 13 (1908–1909): 410–411, on 411. A similar point is made by later literary critics: Agata Zofia Mirecka, *Max Brods Frauenbilder: Im Kontext der Feminitätsdiskurse einiger anderer Prager deutscher Schriftsteller* (Frankfurt am Main: Peter Lang, 2014), 116; Ritchie Robertson, "National Stereotypes in Prague German Fiction," *Colloquia Germanica* 22 (1989): 116–136, on 131–132. On feminism in Bohemia in this period, see Katherine David, "Czech Feminists and Nationalism in the Late Habsburg Monarchy: 'The First in Austria,'" *Journal of Women's History* 3, no. 2 (Fall 1991): 26–45; Melissa Feinberg, *Elusive Equality: Gender, Citizenship and the Limits of Democracy in Czechoslovakia, 1918–1950* (Pittsburgh: University of Pittsburgh Press, 2006); Helena Volet-Jeanneret, *La femme bourgeoise à Prague, 1860–1895: De la philanthropie à l'émancipation* (Geneva: Slatkine, 1988); Jitka Malečková, "The Importance of Being Nationalist," in Iveta Jusová and Jiřina Šiklová, eds., *Czech Feminisms: Perspectives on Gender in East Central Europe* (Bloomington: Indiana University Press, 2016), 46–59.

78. Quoted in Trbuhović-Gjurić, *Im Schatten Albert Einsteins*, 99.

79. Mileva and Albert Einstein to Helene Savić, [December 1912], reproduced in Milan Popović, ed., *In Albert's Shadow: The Life and Letters of Mileva Marić, Einstein's First Wife* (Baltimore: Johns Hopkins University Press, 2003), 107.

80. Einstein to Zangger, [after 27 December 1914], *CPAE* 8:41a, in vol. 10, on 26.

81. Einstein to Mileva Einstein-Marić, 2 April 1914, *CPAE* 8:1, on 11.

82. Einstein to Mileva Einstein-Marić, [15 May 1915], *CPAE* 8:83, on 128–129. See also Einstein to Mileva Einstein-Marić, [27 January 1915], *CPAE* 8:49, on 86; Einstein to Mileva Einstein-Marić, [1 March 1915], *CPAE* 8:58, on 93.

83. Einstein to Mileva Einstein-Marić, 6 February 1916, *CPAE* 8:200, on 270.

84. "Divorce Agreement," 12 June 1918, *CPAE* 8:562, on 795.

85. Thomas de Padova, *Allein gegen die Schwerkraft: Einstein 1914–1918* (Munich: Hanser, 2015), 29–30.

86. Einstein to Paul Ehrenfest, postcard dated 1 January 1912, AEDA 9–318, Box 10, Folder "P. Ehrenfest 1911–1916." The card was sent to Leipzig, where Ehrenfest was visiting G. Herglotz.

87. Einstein to Ehrenfest, [26 January 1912], *CPAE* 5:342, on 393.

88. Einstein to Ehrenfest, 12 February 1912, *CPAE* 5:357, on 408.

89. Martin J. Klein, *Paul Ehrenfest: The Making of a Theoretical Physicist* (Amsterdam: North-Holland, 1970), 175–177.

90. Einstein to Alfred and Clara Stern, 2 February 1912, *CPAE* 5:352, on 402.

91. Einstein to Robert Gnehm, 12 February [19]12, *CPAE* 5:358, on 409.

92. Ludwig Hopf to Einstein, 20 February 1912, *CPAE* 5:363, on 416.

93. "Statement of Reasons for Leaving Prague," [3 August 1912], *CPAE* 5:414, on 499.

94. *Die feierliche Inauguration des Rektors der k.k. Deutschen Karl-Ferdinands-Universität in Prag für das Studienjahr 1912/13* (Prague: K.k. deutschen Karl-Ferdinands-Universität, 1912), 9–10.

CHAPTER 4: EINSTEIN POSITIVE AND EINSTEIN NEGATIVE

1. William James to Carl Stumpf, 26 November 1882, reproduced in Joachim Thiele, *Wissenschaftliche Kommunikation: Die Korrespondenz Ernst Machs* (Kastellaun: Henn, 1978), 169.

2. Philipp Frank, *Einstein: His Life and Times*, tr. George Rosen, ed. and rev. Shuichi Kusaka (Cambridge, MA: Da Capo Press, 2002 [1947]), 170.

3. Einstein to Elsa Einstein, [7 January 1921], *CPAE* 12:11, on 31–32. On Einstein's delight at Frank's makeshift residence and his bride, see Einstein to Paul Ehrenfest, 20 January 1921, *CPAE* 12:24, on 46.

4. Quoted in "Einstein über seine Theorie: Vortrag in der 'Urania,'" *Prager Tagblatt*, no. 6 (8 January 1921): 3. For a complementary (and complimentary) account of Einstein's visit from another local newspaper, see "Prof. Einstein über seine Theorie: Der erste Vortrag; 7. Jänner," *Deutsche Zeitung Bohemia* 94, no. 7 (9 January 1921): 2–3.

5. Einstein to Elsa Einstein, 8 January [1921], *CPAE* 12:12, on 32.

6. Einstein to Ehrenfest, 20 January 1921, *CPAE* 12:24, on 46.

7. Einstein to Elsa Einstein, 8 January [1921], *CPAE* 12:12, on 32.

8. Josef Petráň, *Nástin dějin filozofické fakulty Univerzity Karlovy v Praze (do roku 1948)* (Prague: Univerzita Karlova, 1983), 273.

9. G. Kowalewski, *Bestand und Wandel: Meine Lebenserinnerungen zugleich ein Beitrag zur neueren Geschichte der Mathematik* (Munich: R. Oldenbourg, 1950), 260.

10. Ibid., 260–261.

11. Frank, *Einstein: His Life and Times*, 172.

12. J. C. Nyíri, "The Austrian Element in the Philosophy of Science," in Nyíri, ed., *Von Bolzano zu Wittgenstein: Zur Tradition der österreichischen Philosophie* (Vienna: Hölder-Pichler-Tempsky, 1986), 141–146; Barry Smith, *Austrian Philosophy: The Legacy of Franz Brentano* (Chicago: Open Court, 1994).

13. Frank tends to be sidelined in the philosophical literature about Austrian logical empiricism in favor of Otto Neurath, Rudolf Carnap, and others. For example, he is essentially ignored except for one page in the otherwise excellent Michael Friedman, *Reconsidering Logical Positivism* (Cambridge: Cambridge University Press, 1999).

14. Philipp Frank, "Oral History Transcript," interview with Thomas Kuhn, Cambridge, Massachusetts, 16 July 1962, available at http://www.aip.org/history/ohilist/4610.html, on 16, accessed 30 August 2013.

15. "Report to the Philosophical Faculty of the German University on a Successor to the Chair of Theoretical Physics," [before 23 May 1912], *CPAE* 5:400, on 470.

16. Philipp Frank, "Das Relativitätsprinzip und die Darstellung der physikalischen Erscheinungen im vierdimensionalen Raum," *Annalen der Naturphilosophie* 10 (1911): 129–161; Philipp Frank and Hermann Rothe, "Über die Transformation der Raumzeitkoordinaten von ruhenden auf bewegte Systeme," *Annalen der Physik* 34, no. 5 (1911): 825–855. On p. 827 of the latter, the authors note that the constancy of the speed of light might be an unnecessary principle, thus directly engaging with Einstein's recent pronouncements on the static theory, as described in chapter 2.

17. "Report to the Philosophical Faculty," 472.

18. Frank, "Oral History Transcript," 10. The article in question is Philipp Frank, "Kausalgesetz und Erfahrung," *Annalen der Naturphilosophie* 6 (1907): 443–450.

19. Idem, *The Law of Causality and Its Limits*, ed., Robert S. Cohen, tr. Marie Neurath and Robert S. Cohen (Dordrecht: Kluwer Academic, 1998 [1932]), 196, see also 268. On the Bergson–Einstein debates, see Jimena Canales, *The Physicist and the Philosopher: Einstein, Bergson, and the Debate that Changed Our Understanding of Time* (Princeton, NJ: Princeton University Press, 2015).

20. R. Fürth, "Reminiscences of Philipp Frank at Prague," in Robert S. Cohen and Marx W. Wartofsky, eds., *In Honor of Philipp Frank*, vol. 2 of *Boston Studies in the Philosophy of Science* (New York: Humanities Press, 1965), xiii–xvi, on xiii. On the beard, see also Kurt Sitte, "To Philipp Frank," in Cohen and Wartofsky, *In Honor of Philipp Frank*, xxix–xxx, on xxix.

21. Rudolf Haller and Friedrich Stadler, "The First Vienna Circle," in Thomas E. Uebel, ed., *Rediscovering the Forgotten Vienna Circle: Austrian Studies on Otto Neurath and the Vienna Circle* (Dordrecht: Kluwer Academic, 1991), 95–108. See also the autobiographical introduction in Philipp Frank, *Modern Science and Its Philosophy* (New York: George Braziller, 1955), 1. This volume consists of English translations or originals of Frank essays produced over a period just shy of 40 years.

22. Biographical information is drawn from Josef Sajner, "Ernst Machs Beziehungen zu seinem Heimatort Chirlitz (Chrlice) und zu Mähren," *Bohemia* 24 (1983): 358–368; John T. Blackmore, Ryoichi Itagaki, and Setsuko Tanaka, eds., *Ernst Mach's Science: Its Character and Influence on Einstein and Others* (Kanagawa: Tokai University Press, 2006); and Ivan Úlehla, "One Hundred and Fifty Years since the Birth of Ernst Mach," in Václav Prosser and Jaroslav Folta, eds.,

Ernst Mach and the Development of Physics (Conference Papers) (Prague: Karolinum, 1991), 25–65. This volume was published in 1991 but contains the proceedings from a conference in Prague held on 14–16 September 1988, before the Velvet Revolution. Given Lenin's hostile attitude toward Mach, the conference was a sign of change.

23. Some of his more influential publications were Ernst Mach, "Neue Versuche zur Prüfung der Doppler'schen Theorie der Ton- und Farbenänderung durch Bewegung," *Sitzungsberichte der Kaiserlichen Akademie der Wissenschaften, Philosophisch-Historische Classe* 77 (1878): 299–310; and idem, "Ueber die Controverse zwischen *Doppler* und *Petzval*, bezüglich der Aenderung des Tones und der Farbe durch Bewegung," *Zeitschrift für Mathematik und Physik* 6 (1861): 120–126. On Doppler and especially his time in Prague, see Jan Seidler and Irena Seidlerová, "Zur Entstehungsgeschichte des Dopplerschen Prinzips," tr. Josef Poláček, *Centaurus* 35 (1992): 259–304; *The Phenomenon of Doppler* (Prague: Czech Technical University, Faculty of Nuclear Sciences and Physical Engineering, 1992); Alec Eden, *The Search for Christian Doppler* (Vienna: Springer-Verlag, 1992); Peter Maria Schuster, *Moving the Stars: Christian Doppler, His Life, His Works and Principle, and the World After*, tr. Lily Wilmes (Pöllauberg: Living Edition, 2005); Joachim Thiele, "Zur Wirkungsgeschichte des Dopplerprinzips im Neunzehnten Jahrhundert," *Annals of Science* 27, no. 4 (1971): 393–407; Jan Klepl, "Christian Doppler a Praha," *Dějiny a současnost*, no. 9 (1959): 10–12; and Luboš Nový, ed., *Dějiny exaktních věd v českých zemích do konce 19. století* (Prague: Nakl. ČSAV, 1961), 171–182.

24. Philipp Frank also emphasizes the importance of Prague for Mach's scholarship: Frank, *Modern Science and Its Philosophy*, 99. For secondary accounts, see Dieter Hoffmann, "Ernst Mach in Prag," in Hoffmann and Hubert Laitko, eds., *Ernst Mach: Studien und Dokumente zu Leben und Werk* (Berlin: Deutscher Verlag der Wissenschaften, 1991), 141–178, on 147; idem, "Ernst Mach and the Conflict of Nations," in John Blackmore, ed., *Ernst Mach—A Deeper Look: Documents and New Perspectives* (Dordrecht: Kluwer Academic, 1992), 29–55; Friedrich Stadler, "Ernst Mach—Zu Leben, Werk und Wirkung," in Rudolf Haller and Friedrich Stadler, eds., *Ernst Mach— Werk und Wirkung* (Vienna: Hölder-Pichler-Tempsky, 1988), 11–63, on 20, 24; Blackmore, Itagaki, and Tanaka, *Ernst Mach's Science*, 126.

25. The stature of the book on mechanics is such that it remains in print (in multiple languages) and contains valuable supplementary material framing Mach's philosophy: Ernst Mach, *The Science of Mechanics: A Critical and Historical Account of Its Development*, 6th ed., tr. Thomas J. McCormack (La Salle, IL: Open Court, 1960 [1893]).

26. Ernst Mach, "The Analysis of Sensations: Antimetaphysical," *Monist* 1, no. 1 (October 1890): 48–68, on 58.

27. William James to his wife, 2 November 1882, reproduced in Thiele, *Wissenschaftliche Kommunikation*, 169.

28. For detailed discussions of which parts of Mach's work Einstein accepted and rejected, and when, see Gerald Holton, *Thematic Origins of Scientific Thought: Kepler to Einstein*, rev. ed. (Cambridge, MA: Harvard University Press, 1988 [1973]), chs. 7–8.

29. Einstein to Ernst Mach, [9 August 1909], *CPAE* 5:174, on 204.

30. Einstein to Mach, 17 August 1909, *CPAE* 5:175, on 205.

31. Georg Pick to Ernst Mach, 2 August 1912, reproduced in Thiele, *Wissenschaftliche Kommunikation*, 224.

32. Friedrich Herneck, "Die Beziehungen zwischen Einstein und Mach, dokumentarisch dargestellt," *Wissenschaftliche Zeitschrift der Friedrich-Schiller-Universität Jena* 15 (1966): 1–14.

33. On the undated letter, see Friedrich Herneck, "Zum Briefwechsel Albert Einsteins mit Ernst Mach (Mit zwei unveröffentlichten Einstein-Briefen)," *Forschungen und Fortschritte* 37, no. 8 (August 1963): 239–243, on 242. On the Society for Positivist Philosophy, see "Aufruf der Gesellschaft für positivistische Philosophie," *Physikalische Zeitschrift* 13 (1912): 735–736.

34. "Ernst Mach," reproduced in *CPAE* 6:29. For his praise of Frank's essay, see Einstein to Kathia Adler [wife of Friedrich Adler], 20 February [1917], *CPAE* 8:301, on 394; Einstein to Zanger, 1 February 1917, *CPAE* 8:291a, in vol. 10, on 68.

35. Philipp Frank, "Einsteins Stellung zur Philosophie," *Deutsche Beiträge* 2 (1949): 146–157, on 151. See also Gerald Holton, "Ernst Mach in America," in Prosser and Folta, *Ernst Mach*, 357–380.

36. This is an extremely contentious issue in the scholarly literature, and I have deferred to the consensus view. Gereon Wolters has forcefully argued that Mach never reneged on his support of relativity theory and that the offending preface was actually a forgery by Mach's son Ludwig, acting under the influence of conservative philosopher of science Hugo Dingler: Gereon Wolters, *Mach I, Mach II, Einstein und die Relativitätstheorie: Eine Fälschung und ihre Folgen* (Berlin: Walter de Gruyter, 1987); idem, "Ernst Mach and the Theory of Relativity," tr. Steven Gilles, *Philosophia Naturalis* 21 (1984): 630–641; idem, "Mach and Einstein, or, Clearing Troubled Waters in the History of Science," in Christoph Lehner, Jürgen Renn, and Matthias Schemmel, eds., *Einstein and the Changing Worldviews of Physics* (New York: Springer, 2012), 39–57. Wolters's claims rest very heavily on a controversial ordering of the undated Einstein–Mach correspondence and a sense that it would have been irrational for Mach—who was increasingly incapacitated and irrelevant to mainstream physics—to reject the relativity theory that had brought him back to prominence. The detailed criticisms of this view by other historians and philosophers strike me as convincing: Gerald Holton, "More on Mach and Einstein: Investigation of an Important Phase in the Rise of Logical Positivism/Empiricism," *Methodology and Science* 22, no. 2 (1989): 67–81; John Blackmore, "Ernst Mach Leaves 'the Church of Physics,'" *British Journal for the Philosophy of Science* 40, no. 4 (December 1989): 519–540; Ryoichi Itagaki, "Three Batches of Reasons for Mach's Rejection of Einstein's Theory of Relativity," in Blackmore, *Ernst Mach*, 277–295. Dingler always maintained that Mach rejected relativity in his final years, though this is consistent with both the pro- and anti-Wolters positions: Hugo Dingler, *Die Grundgedanken der Machschen Philosophie* (Leipzig: J. A. Barth, 1924), ch. 5.

37. Anton Lampa, *Ernst Mach* (Prague: Deutsche Arbeit, 1918), 59.

38. Ibid., 33. The credit to Frank is on 34n1.

39. See, for example, the homage he wrote on the occasion of Mach's centenary: Philipp Frank, "Ernst Mach—The Centenary of His Birth," *Erkenntnis* 7 (1937): 247–256.

40. Einstein to Besso, [29 April 1917], *CPAE* 8:331, on 441. For more in this vein, see Besso to Einstein, 5 May 1917, *CPAE* 8:334, on 444; Einstein to Besso, 13 May 1917, *CPAE* 8:339, on 451; and Friedrich Adler to Einstein, 9 March 1917, *CPAE* 8:307, on 403.

41. Einstein to Arnold Sommerfeld, 13 July 1921, *CPAE* 12:175, on 218.

42. "Zur Enthüllung von Ernst Machs Denkmal," *Neue Freie Presse* (12 June 1926): 11, reproduced in *CPAE* 15:303, on 500.

43. Albert Einstein, "Remarks on Bertrand Russell's Theory of Knowledge," in Paul Arthur Schilpp, ed., *The Philosophy of Bertrand Russell* (Evanston, IL: Library of Living Philosophers, 1946), 277–291, on 287.

44. Albert Einstein, "Autobiographical Notes," tr. Paul Arthur Schilpp, in Schilpp, ed., *Albert Einstein: Philosopher-Scientist*, 2 vols. (New York: Harper & Row, 1951 [1949]), 1:1–95, on 1:21.

45. See especially the very rich Einstein to Besso, 6 January 1948, in Albert Einstein and Michele Besso, *Correspondance 1903–1955*, ed. and tr. Pierre Speziali (Paris: Hermann, 1972), 391. On occasion Frank would grant this interpretation some credence—see his "Einsteins Stellung zur Philosophie," 147–148—but in later writings he backtracked to a focus on Mach.

46. This is not to say that Frank was the most doctrinaire or dogmatic defender of a Machian reading of relativity; his own position was subtle and adapted over time. For more rigid contemporary interpretations, see Friedrich Adler to Einstein, 9 March 1917, *CPAE* 8:307 (see note 40); and the works of Joseph Petzoldt: *Die Stellung der Relativitätstheorie in der geistigen Entwicklung der Menschheit*, 2nd. ed. (Leipzig: Johann Ambrosius Barth, 1923); "Das Verhältnis der Machschen Gedankenwelt zur Relativitätstheorie," in Ernst Mach, *Die Mechanik in ihrer Entwickelung historisch-kritisch dargestellt*, 5th ed. (Leipzig: F. A. Brockhaus, 1921), 490–517. Petzoldt's basic position was also articulated in H. E. Hering, "Mach als Vorläufer des physikalischen Relativitätsprinzip," *Kölner Universitäts-Zeitung* 17, no. 1 (1920): 3–4.

47. Frank, "Oral History Transcript," 7. See also idem, "Einstein's Philosophy of Science," *Reviews of Modern Physics* 21, no. 2 (July 1949): 349–355, on 350.

48. Idem, "Die Bedeutung der physikalischen Erkenntnistheorie Machs für das Geistesleben der Gegenwart," *Die Naturwissenschaften* 5, no. 5 (2 February 1917): 65–72, on 68.

49. Idem, "Einstein, Mach, and Logical Positivism," in Schilpp, *Albert Einstein*, 1:269–286, on 1:279. For similar essays in the same volume, see Hans Reichenbach, "The Philosophical Significance of the Theory of Relativity" (287–311); and P. W. Bridgman, "Einstein's Theories and the Operational Point of View" (333–354).

50. Frank, "Einstein, Mach, and Logical Positivism," 1:282.

51. Frank, *Modern Science and Its Philosophy*, 18–19. Philosopher of science Don Howard remarks that Frank's interpretations of relativity were often more astute than the mainstream views of the Vienna Circle logical positivists: Howard, "Einstein and the Development of Twentieth-Century Philosophy of Science," in Michel Janssen and Christoph Lehner, eds., *The Cambridge Companion to Einstein* (Cambridge: Cambridge University Press, 2014), 354–376, on 354.

52. Philipp Frank, *Interpretations and Misinterpretations of Modern Physics*, tr. Olaf Helmer and Milton B. Singer (Paris: Hermann et Cie, 1938), 34–35.

53. Idem, *Relativity: A Richer Truth* (Boston: Beacon, 1950). Two years later, this work appeared in a German translation: idem, *Wahrheit—relativ oder absolut?*, tr. Gertrud Deuel (Zurich: Pan-Verlag, 1952).

54. Josef Winternitz, *Relativitätstheorie und Erkenntnislehre: Eine Untersuchung über die erkenntnistheoretischen Grundlagen der Einsteinschen Theorie und die Bedeutung ihrer Ergebnisse für die allgemeinen Probleme des Naturerkennens* (Wiesbaden: Springer Fachmedien, 1923); Albert Einstein, Review of J. Winternitz, *Relativitätstheorie und Erkenntnislehre*, *Deutsche Literaturzeitung für Kritik der internationalen Wissenschaft* 45 (n.s. 1) (1924): 20–22. It seems likely that Josef

Winternitz was the Prague philosopher whom Leopold Infeld encountered outside Einstein's home in Berlin on his first visit. Infeld, *Quest: An Autobiography* (Providence, RI: AMS Chelsea, 1980 [1941]), 91–92.

55. Moritz Schlick to Einstein, 29 August 1920, *CPAE* 10:116, on 391.

56. Moritz Schlick to Einstein, 27 December 1925, *CPAE* 15:140, on 246; Hans Reichenbach to Einstein, 16 March 1926, *CPAE* 15:224, on 380; Einstein to Reichenbach, 20 March 1926, *CPAE* 15:230, on 387; Reichenbach to Einstein, 24 March 1926, *CPAE* 15:235, on 397.

57. Eckehart Köhler, "Gödel und Carnap in Wien und Prag," in Rudolf Haller and Friedrich Stadler, eds., *Wien-Berlin-Prag: Der Aufstieg der wissenschaftlichen Philosophie* (Vienna: Hölder-Pichler-Tempsky, 1993), 165–174, on 167–168; Gerald Holton, *Science and Anti-Science* (Cambridge, MA: Harvard University Press, 1993), 33.

58. Michael Stöltzner, "Philipp Frank and the German Physical Society," in Werner DePauli-Schimanovich, Eckehart Köhler, and Friedrich Stadler, eds., *The Foundational Debate: Complexity and Constructivity in Mathematics and Physics* (Dordrecht: Kluwer Academic, 1995), 293–302.

59. Max Brod, "Tagung für Erkenntnislehre: Gelten die Naturgesetze unumschränkt?," *Prager Tagblatt* 54, no. 218 (17 September 1929): 4. See also "Physiker- und Mathematikertag: Erkenntnisprobleme der modernen Physik," *Prager Tagblatt* 54, no. 218 (17 September 1929): 3–4.

60. Brod, "Tagung für Erkenntnislehre," 4.

61. The lecture in question was later published as Philipp Frank, "Was bedeuten die gegenwärtigen physikalischen Theorien für die allgemeine Erkenntnislehre?," *Erkenntnis* 1, no. 1 (1930): 126–157. On Sommerfeld's interjection, see Frank, "Einsteins Stellung zur Philosophie," 157.

62. See Friedrich Stadler and Thomas Uebel, eds., *Wissenschaftliche Weltauffassung: Der Wiener Kreis, Hrsg. vom Verein Ernst Mach*, reprint of 1st ed. (Vienna: Springer, 2012), Prague reference on 303. On the enthusiasm for Einstein, see Don Howard, "Einstein, Kant, and the Origins of Logical Empiricism," in Wesley Salmon and Gereon Wolters, eds., *Logic, Language, and the Structure of Scientific Theories* (Pittsburgh: University of Pittsburgh Press, 1994), 45–105.

63. Gerald Holton, "From the Vienna Circle to Harvard Square: The Americanization of a European World Conception," in Friedrich Stadler, ed., *Scientific Philosophy: Origins and Developments* (Dordrecht: Kluwer Academic, 1993), 47–73, on 59; Frank, *Modern Science and Its Philosophy*, 39–40.

64. Hans-Joachim Dahms, "Vertreibung und Emigration des Wiener Kreises zwischen 1931 und 1940," in Dahms, ed., *Philosophie, Wissenschaft, Aufklärung: Beiträge zur Geschichte und Wirkung des Wiener Kreises* (Berlin: Walter de Gruyter, 1985), 307–365, on 347.

65. Oskar Kraus, "Wissenschaftliche Weltauffassung," *Archiv für die gesamte Psychologie* 75 (1930): 287–288, on 288.

66. Biographical details are largely drawn from two self-authored pieces: Oskar Kraus, "Oskar Kraus," in Raymund Schmidt, ed., *Der Philosophie der Gegenwart in Selbstdarstellungen* (Leipzig: Felix Meiner, 1929), 7:161–203; and his entry in Werner Zeigenfuss, ed., *Philosophen-Lexikon: Handwörterbuch der Philosophie nach Personen*, 2 vols. (Berlin: Walter de Gruyter, 1949), 1:681. The secondary literature on Kraus is quite limited and tends to recount his career selectively. See Traugott Konstantin Oesterreich, ed., *Friedrich Ueberwegs Grundriss der Geschichte der Philosophie*, vol. 4, *Die deutsche Philosophie des XIX. Jahrhunderts und der Gegenwart* (Basel: Benno Schwabe, 1951), 502–503; Thomas Binder, "Der Brentano-Schüler Oskar Kraus: Leben, Werk und

Nachlaß," *Nachrichten der Forschungsstelle und Dokumentationszentrum für Österreichische Philosophie* 2 (1991): 15–21; and Vlastimil Hála, "Oskar Kraus—pražský představitel brentanovské školy (K. 60. výročí úmrtí)," *Filosofický časopis* 51, no. 1 (2003): 19–37.

67. Kraus, "Oskar Kraus," 165.

68. Idem, *Die Meyeriade: Humoristisches Epos aus dem Gymnasialleben* (Leipzig: Philipp Reclam, 1925 [1891]). On its popularity, see Josef Körner, "Dichter und Dichtung aus dem deutschen Prag," *Donauland* 1, no. 7 (September 1917): 777–784, on 784; and Reiner Stach, *Kafka: Die frühen Jahre* (Frankfurt am Main: S. Fischer, 2014), 137–138.

69. Oskar Kraus, *Das Bedürfnis: Ein Beitrag zur beschreibenden Psychologie* (Leipzig: Wilhelm Friedrich, 1894).

70. Alfred Kastil, *Die Philosophie Franz Brentanos: Eine Einführung in seine Lehre* (Bern: A. Francke, 1951), 21. See also the colorful and somewhat disparaging descriptions in Herbert Cysarz, "Beiträge der Prager Deutschen Universität zur Philosophie und Grundlagenforschung in der Zwischenkriegszeit," in Karl Bosl and Ferdinand Seibt, eds., *Kultur und Gesellschaft in der Ersten Tschechoslowakischen Republik* (Munich: Oldenbourg, 1982), 255–268, on 264.

71. On editing the *Nachlass*, see Kastil, *Die Philosophie Franz Brentanos*, 7. On the unreliability of the final product, see Karl Schuhmann, "Brentano's Impact on Twentieth-Century Philosophy," in Dale Jacquette, ed., *The Cambridge Companion to Brentano* (Cambridge: Cambridge University Press, 2004), 277–297, on 277.

72. See, for example, Anton Marty, *Raum und Zeit*, ed. Josef Eisenmeier, Alfred Kastil, and Oskar Kraus (Halle an der Salle: Max Niemeyer, 1916), 245; Peter M. Simons, "Marty on Time," in Kevin Mulligan, ed., *Mind, Meaning and Metaphysics: The Philosophy and Theory of Language of Anton Marty* (Dordrecht: Kluwer Academic, 1990), 157–170; and Franz Brentano, "Zur Lehre vom Raum und Zeit," ed. Oskar Kraus, *Kantstudien* 25 (1920): 1–23.

73. Franz Brentano to Ernst Mach, 31 January 1908, reproduced in Franz Brentano, *Über Ernst Machs "Erkenntnis und Irrtum,"* ed. Roderick M. Chisholm and Johann C. Marek (Amsterdam: Rodopi, 1988), 215.

74. A good account of these events is available in Oskar Kraus, "Biographical Sketch of Franz Brentano," tr. Linda L. McAlister, in McAlister, ed., *The Philosophy of Brentano* (London: Duckworth, 1976), 1–9. Oskar Kraus also wrote what is still the most commonly used biography of his idol: *Franz Brentano* (Munich: C. H. Beck, 1919). On Brentano's broader school, including Marty, see the essays in Liliana Albertazzi, Massimo Libardi, and Roberto Poli, eds., *The School of Franz Brentano* (Dordrecht: Kluwer Academic, 1996). Kraus's own history of philosophy in Prague was highly Brentanist in orientation, often consigning other strands of thought to irrelevance: Oskar Kraus, "Besonderheit und Aufgabe der deutschen Philosophie in Böhmen," in *Actes du huitième Congrès international de philosophie à Prague, 2–7 Septembre 1934* (Prague: Comité d'organisation du Congrès, 1936), 766–771.

75. Arnold Heidsieck, *The Intellectual Contexts of Kafka's Fiction: Philosophy, Law, Religion* (Columbia, SC: Camden House, 1994), 11.

76. Appointment letter, 6 October 1911, AUK, Collection of the Faculty of Arts of the German University in Prague, inv. numb. 510, sig. Pl/1, box 49, personal file of Oskar Kraus.

77. Oskar Kraus, ed., *Jeremy Benthams Grundsätze für ein künftiges Völkerrecht und einen dauernden Frieden (Principles of international law)*, tr. Camill Klatscher (Halle an der Salle: Max

Niemeyer, 1915). See also idem, *Der Machtgedanke und die Friedensidee in der Philosophie der Engländer: Bacon und Bentham* (Leipzig: C. L. Hirschfeld, 1926); idem, "Zur Philosophie des Krieges," *Schweizerische Juristen-Zeitung* 13, no. 23/24 (1917): 376–380; and idem, "Der Krieg, die Friedensfrage und die Philosophen," *Sammlung gemeinnütziger Vorträge* 24, no. 470/471 (January–February 1918).

78. Idem, *Das Recht zu strafen: Eine rechtsphilosophische Untersuchung* (Stuttgart: Ferdinand Enke, 1911). On the connection to Kafka, see Heidsieck, *Intellectual Contexts*; and Barry Smith, "Kafka and Brentano: A Study in Descriptive Psychology," in Smith, ed., *Structure and Gestalt: Philosophy and Literature in Austria-Hungary and Her Successor States* (Amsterdam: John Benjamins, 1981), 113–159.

79. Luigi Dappiano, "Theories of Values," in Albertazzi, Libardi, and Poli, *School of Franz Brentano*, 377–422; Reinhard Fabian and Peter M. Simons, "The Second Austrian School of Value Theory," in Wolfgang Grassl and Barry Smith, eds., *Austrian Economics: Historical and Philosophical Background* (London: Croom Helm, 1986), 37–101. A 1930 American study of Austrian thought gratefully thanked Kraus for his assistance: Howard O. Eaton, *The Austrian Philosophy of Values* (Norman: University of Oklahoma Press, 1930).

80. Oskar Kraus, *Die Werttheorien: Geschichte und Kritik* (Brno: Rudolf M. Rohrer, 1937), 5n1, 433n1. His citation of Philipp Frank's work on causality (130n1), however, was respectful.

81. Quoted in Milena Wazeck, *Einstein's Opponents: The Public Controversy about the Theory of Relativity in the 1920s*, tr. Geoffrey S. Koby (Cambridge: Cambridge University Press, 2014 [2009]), 1.

82. E. Gehrcke, *Kritik der Relativitätstheorie: Gesammelte Schriften über absolute und relative Bewegung* (Berlin: Hermann Meusser, 1924); idem, *Die Massensuggestion der Relativitätstheorie: Kulturhistorisch-psychologische Dokumente* (Berlin: Hermann Meusser, 1924).

83. On the anti-Einstein movement, see especially Wazeck, *Einstein's Opponents*; and David E. Rowe, "Einstein's Allies and Enemies: Debating Relativity in Germany, 1916–1920," in Vincent F. Hendricks, Klaus Frovin Jørgensen, Jesper Lützen, and Stig Andur Pedersen, eds., *Interactions: Mathematics, Physics and Philosophy, 1860–1930* (Berlin: Springer, 2006), 231–280.

84. Hoffmann, "Ernst Mach," 34. On the history of Lotos, see Emilie Těšínská, "Fyzikální vědy v pražském německém přírodovědném spolku 'Lotos,'" *Pokroky matematiky, fyziky a astronomie* 42, no. 1 (1997): 35–47; and Irena Seidlerová, "Science in a Bilingual Country," in Mikuláš Teich, ed., *Bohemia in History* (Cambridge: Cambridge University Press, 1998), 229–243, on 235.

85. "Die Relativitätstheorie Einsteins: Diskussion im 'Lotos,'" *Prager Tagblatt* 45, no. 49 (27 February 1920): 1. For a conceptual analysis of the Frank–Kraus dispute, see Klaus Hentschel, *Interpretationen und Fehlinterpretationen der speziellen und der allgemeinen Relativitätstheorie durch Zeitgenossen Albert Einsteins* (Basel: Birkhäuser, 1990).

86. Oskar Kraus, "Die Wunder der Wissenschaft: Ein Fastnachtsscherz von Prof. Neinstein," [probably 1921], D.4.3, NOK, 8–9.

87. Ibid., 9.

88. Oskar Kraus, "Ueber die Deutung der Relativitäts-Theorie Einsteins," *Lotos* 67/68 (1919/1920): 146–152, on 149.

89. Ibid., 150.

90. Philipp Frank, "Zur Relativitäts-Theorie Einsteins," *Lotos* 67/68 (1919/1920): 152–156, on 153.

91. Oskar Kraus, "Zur Abwehr der vorstehenden Erwiderung Professor Franks," *Lotos* 67/68 (1919/1920): 156–161, on 161. Despite the public fracas, Frank and Kraus continued to cooperate professionally in the years to come, including on the Eighth International Congress of Philosophy, which took place in Prague on 2–7 September 1936. Kraus proofed and edited the German entries, and Frank served on the organizing committee. See *Actes du huitième Congrès international de philosophie*.

92. Hans Reichenbach, "Die Einsteinsche Bewegungslehre," *Die Umschau* 25, no. 35 (27 August 1921): 501–505; Oskar Kraus, "Die Unmöglichkeit der Einsteinschen Bewegungslehre: Eine Erwiderung an Herrn Hans Reichenbach," *Die Umschau* 46 (1921): 681–684; Reichenbach, "Entgegnung," *Die Umschau* 25 (1921): 684–685; Benno Urbach, "Kritische Bemerkungen zur philosophischen Bekämpfung der Einsteinschen Relativitätstheorie durch Prof. Dr. O. Kraus," *Lotos* 70 (1922): 309–332.

93. Editor L. Freund's statement appended to Oskar Kraus, "Eine neue Verteidigung der Relativitätstheorie," *Lotos* 70 (1922): 333–342, on 342.

94. H. Vaihinger to J. Petzoldt, 24 May 1920, as quoted in Hentschel, *Interpretationen und Fehlinterpretationen*, 173.

95. For the refusal, see Einstein to Hans Vaihinger, 3 May 1919, *CPAE* 9:33, on 51.

96. Max Wertheimer to Einstein, 15 May 1920, *CPAE* 10:16, on 260–261. Emphasis in original. On the Wertheimer–Einstein relationship, see Mitchell G. Ash, *Gestalt Psychology in German Culture, 1890–1967: Holism and the Quest for Objectivity* (Cambridge: Cambridge University Press, 1995), 291.

97. Einstein to Elsa Einstein, [19 May 1920], *CPAE* 10:19, on 265. This letter was written from Leiden.

98. Joseph Petzoldt to Einstein, 6 July 1920, *CPAE* 10:72, on 332.

99. Oskar Kraus, "Fiktion und Hypothese in der Einsteinschen Relativitätstheorie: Erkenntnistheoretische Betrachtungen," *Annalen der Philosophie* 2, no. 3 (1921): 335–396, on 338.

100. Ibid., 337n1. On the meeting and the reception of Kraus's speech, see Hubert F. Goenner, "The Reaction to Relativity Theory I: The Anti-Einstein Campaign in Germany in 1920," *Science in Context* 6 (1993): 107–133, on 111.

101. Albert Einstein, "Dialog über Einwände gegen die Relativitätstheorie," *Die Naturwissenschaften* 16, no. 48 (29 November 1918): 697–702, on 700.

102. Oskar Kraus, "Einwendungen gegen Einstein: Philosophische Betrachtungen gegen die Relativitätstheorie," *Neue Freie Presse*, no. 20130 (11 September 1920): 2–3. At that notorious meeting, Philipp Lenard specifically invoked Kraus in his comments (reprinted in *CPAE* 7:46, on 356).

103. Wazeck, *Einstein's Opponents*, 241; Goenner, "Reaction to Relativity Theory I," 118; Siegfried Grundmann, *Einsteins Akte: Einsteins Jahre in Deutschland aus der Sicht der deutschen Politik* (Berlin: Springer, 1998), on 152. For the claims that Kraus's visa was denied, see Johannes Riem, "Gegen den Einsteinrummel!," *Die Umschau* 24, no. 39 (2 October 1920): 583–584.

104. For example, Oskar Kraus, "Zum Kampfe gegen Einstein und die Relativitätstheorie," *Deutsche Zeitung Bohemia* 93, no. 208 (3 September 1920): 5; idem, "Die Verwechselungen von

'Beschreibungsmittel' und 'Beschreibungsobjekt' in der Einsteinschen speziellen und allgemeinen Relativitätstheorie," *Kant-Studien* 26 (1921): 454–486; idem, "Der gegenwärtige Stand der Relativitätstheorie," part 1, *Neue Freie Presse*, no. 21238 (25 October 1923): 16–17; part 2, *Neue Freie Presse*, no. 21240 (27 October 1923): 15–16; idem, "Millers Versuch und die Relativitätstheorie," *Frankfurter Zeitung und Handelsblatt* 71, no. 163 (3 March 1927): 1–2; Kraus's introduction to Franz Brentano, *Über die Zukunft der Philosophie*, ed. Oskar Kraus (Leipzig: Felix Meiner, 1929); and Kraus, "Die Grenzen der Relativität," *Hochschulwissen: Monatsschrift für das deutsche Volk und seine Schule* 6 (1929): 647–655, 748–752.

105. Idem, *Offene Briefe an Albert Einstein u. Max v. Laue über die gedanklichen Grundlagen der speziellen und allgemeinen Relativitätstheorie* (Vienna: Wilhelm Braumüller, 1925), 31–32.

106. Hans Israel, Erich Ruckhaber, and Rudolf Weinmann, eds., *Hundert Autoren gegen Einstein* (Leipzig: R. Voigtländer, 1931), with Kraus's contribution on 17–19. On this volume, see Hubert Goenner, "The Reaction to Relativity Theory in Germany, III: 'A Hundred Authors Against Einstein,'" in John Earman, Michel Janssen, and John D. Norton, eds., *The Attraction of Gravitation: New Studies in the History of General Relativity* (Boston: Birkhäuser, 1993), 248–273.

107. Oskar Kraus, *Albert Schweitzer: Sein Werk und seine Weltanschauung* (Bern: Paul Haupt, 1926).

108. On the Circle, see especially Jindřich Toman, *The Magic of a Common Language: Jakobson, Trubetzkoy, and the Prague Linguistic Circle* (Cambridge, MA: MIT Press, 1995). For the Carnap and Kraus lectures, see Bruce Kochis, comp., "List of Lectures Given in the Prague Linguistic Circle (1926–1948)," in Ladislav Matejka, ed., *Sound, Sign and Meaning: Quinquagenary of the Prague Linguistic Circle* (Ann Arbor: Department of Slavic Languages and Literatures, University of Michigan, 1978), 607–622, on 613.

109. Diary entry entitled "Zar Ferdinand," dated 2 October 1935, reproduced in Schmuel Hugo Bergman, *Tagebücher & Briefe*, 2 vols., ed. Miriam Sambursky (Königstein: Athenäum, 1985), 1:410. See also the curious description of Kraus from a few years later in Elias Canetti, *Das Augenspiel: Lebensgeschichte, 1931–1937* (Munich: Carl Hanser, 1994 [1985]), 288–289.

110. Oskar Kraus, "Über die Missdeutungen der Relativitätstheorie," in *Naturwissenschaft und Metaphysik: Abhandlungen zum Gedächtnis des 100. Geburtstages von Franz Brentano* (Brünn: Rudolf M. Rohrer, 1938): 31–77, on 37. Emphasis in original.

111. See Fürth's own contribution to *Naturwissenschaft und Metaphysik*, "Der Streit um die Deutung der Relativitätstheorie," on 1–29.

112. Midia Kraus, "[Kraus gesteht einen Irrtum betreffs der Relativitätstheorie ein]," undated, L.5.6, NOK.

113. Oskar Kraus to Albert Einstein, 10 July 1940, NOK. The copy from Einstein's archive is dated 15 July 1940, 55–490, AEDA C0701, Box 81, Folder "1940–44 K."

114. Philipp Frank, *Philosophy of Science: The Link Between Science and Philosophy* (Englewood Cliffs, NJ: Prentice-Hall, 1962), 33. Interestingly, Kraus was cited in a single footnote in the famous 1947 volume edited by Paul Arthur Schilpp: Aloys Wenzl, "Einstein's Theory of Relativity, Viewed from the Standpoint of Critical Realism, and Its Significance for Philosophy," tr. Paul Arthur Schilpp, in Schilpp, *Albert Einstein*, 2:581–606, on 2:586 and 2:587n1.

115. For Frank's American career, see Gerald Holton, "Philipp Frank at Harvard University: His Work and Influence," *Synthese* 153, no. 2 (November 2006): 297–311; idem, "From the Vienna

Circle"; and George A. Reisch, *How the Cold War Transformed the Philosophy of Science: To the Icy Slopes of Logic* (Cambridge: Cambridge University Press, 2005), passim.

116. Gerald Holton, "On the Vienna Circle in Exile: An Eyewitness Report," in DePauli-Schimanovich, Köhler, and Stadler, *Foundational Debate*, 269–292.

117. Frank, *Einstein: His Life and Times*, xiii–xiv.

118. Holton, *Science and Anti-Science*, 38–39.

119. Frank, *Einstein: Sein Leben und seine Zeit*, 5.

120. For example, see Jiří Bičák's defense of the relation of his own account to Frank's: "It is without question influenced by this—the author did not have at his disposal manuscript or other materials, the gradual publication of which expands the constantly developing Einsteinian historiography (which does not only touch the natural sciences), but none of the newer biographies carries in relation to Einstein's Prague stay the imprimatur of immediacy and contains so much insight and humor as Frank's book." Jiří Bičák, ed., *Einstein a Praha: K stému výročí narození Alberta Einsteina* (Prague: Jednota československých matematiků a fyziků, 1979), 7. Most other authors are not so explicit or generous about their debts to Frank. In a 2014 article on the same topic, Bičák is much more critical of the accuracy of Frank's account: "Einstein in Prague: Relativity Then and Now," in Jiří Bičák and Tomáš Ledvinka, eds., *General Relativity, Cosmology and Astrophysics: Perspectives 100 Years after Einstein's Stay in Prague* (Cham: Springer, 2014), 33–63.

121. Besso to Einstein, 27 April 1949, in Einstein and Besso, *Correspondance*, 398.

122. Rowe, "Einstein's Allies and Enemies," 234.

123. Max Jammer, *Einstein and Religion: Physics and Theology* (Princeton, NJ: Princeton University Press, 1999), 21.

124. Frank, "Oral History Transcript," 11.

125. Emil Utitz, "Einstein und die Prager Juden," [undated, perhaps 1952], MÚA AV ČR, Emil Utitz Personal Papers, inventory number 78, sig. III.g)2, box 2.

126. For an example of apparent projection, Frank included an aside about how he had attended faculty meetings with Einstein and enjoyed watching Einstein's entertainment at the academic vanity of his colleagues. Frank, *Einstein: Sein Leben und seine Zeit*, 145. Of course, Frank could not have done this, as he was never in Prague with Einstein aside from the visit in 1921. Frank's student Reinhold Fürth recorded of his mentor regarding the Academic Senate: "In contrast to many of my other colleagues he did not like to be involved in academic affairs, and when I happened to sit next to him at Board meetings, he used to make amusing and sarcastic asides to me about some of the proceedings." Fürth, "Reminiscences of Philipp Frank at Prague," xvi. No doubt Einstein would have agreed with these asides, but here Frank may have been placing his own thoughts in Einstein's head. On the contentious politicking within the philosophy faculty in Einstein's and Frank's days, see Kowalewski, *Bestand und Wandel*, 242.

CHAPTER 5: THE HIDDEN KEPLER

1. Edward Rosen, tr. and commentary, *Kepler's* Somnium: *The Dream, or Posthumous Work on Lunar Astronomy* (Madison: University of Wisconsin Press, 1967), 11.

2. This was not Kepler's only fanciful work that drew directly from his Prague surroundings. His classic description of the snowflake, replete with humanist embellishments, also drew from

the courtly culture in Bohemia. See Anthony Grafton, "Humanism and Science in Rudolphine Prague: Kepler in Context," in Grafton, *Defenders of the Text: The Traditions of Scholarship in an Age of Science, 1450–1800* (Cambridge, MA: Harvard University Press, 1991), 178–203.

3. Philipp Frank, *Einstein: His Life and Times*, tr. George Rosen, ed. and rev. Shuichi Kusaka (Cambridge, MA: Da Capo Press, 2002 [1947]), 85. In the German original, "consciously or unconsciously" is actually "bewußt oder halb unbewußt," and I have made the emendation in square brackets. Frank, *Einstein: Sein Leben und seine Zeit* (Munich: Paul List, 1949), 151–152. This is, if anything, even a stronger claim than appears in the English version. Frank's ventriloquized Nernst has been rendered in multiple ways, such as "The image of Kepler reminds me of you!" ("Die Gestalt von Kepler erinnert mich an Sie!"; Carl Seelig, *Albert Einstein: Leben und Werk eines Genies unserer Zeit* [Zurich: Bertelsmann Lesering, 1960], 209) and "Brod's Kepler, that's you!" ("Kepler bei Brod, das sind Sie!"; Hans-Jürgen Treder, "Albert Einstein an der Berliner Akademie der Wissenschaften," in Christa Kirsten and Hans-Jürgen Treder, eds., *Albert Einstein in Berlin, 1913–1933*, 2 vols. [Berlin: Akademie-Verlag, 1979], 1:7–78, on 11n). None of these works offers any source other than Frank or explains why it quotes the passage differently.

4. Cornelius Lanczos, *The Einstein Decade (1905–1915)* (New York: Academic Press, 1974), 96.

5. Albert Einstein, *Out of My Later Years*, rev. reprint ed. (New York: Citadel Press, 1995 [1956]), 233–235. On Frank's repeated errors of interpretation when it comes to Nernst (and the Nernst–Einstein relationship), see Diana Kormos Barkan, *Walther Nernst and the Transition to Modern Physical Science* (Cambridge: Cambridge University Press, 1999).

6. Einstein to Hedwig Born, 8 September 1916, *CPAE* 8:257, on 336.

7. For some of many examples, see Seelig, *Albert Einstein*, 122; Jürgen Neffe, *Einstein: A Biography*, tr. Shelley Frisch (New York: Farrar, Straus and Giroux, 2007 [2005]), 34–35; Isaacson, *Einstein*, 166–167; B. G. Kuznetsov, *Einshtein* (Moscow: Izd. AN SSSR, 1962), 193; Thomas Levenson, *Einstein in Berlin* (New York: Bantam Books, 2003), 99–100; Armin Hermann, *Einstein: Der Weltweise und sein Jahrhundert* (Munich: Piper, 1994), 181, 218. The exceptions are Banesh Hoffmann with Helen Dukas, *Albert Einstein: Creator and Rebel* (New York: Viking, 1972); and a slim popular volume in Czech, Jan Horský, *Albert Einstein: Genius lidstva* (Prague: Prometheus, 1998), although this latter work does mention Brod's novels in passing on 10.

8. See David Reichinstein, *Albert Einstein: A Picture of His Life and His Conception of the World* (Prague: Stella Publishing House, 1934); H. Gordon Garbedian, *Albert Einstein: Maker of Universes* (New York: Funk & Wagnalls, 1939); Anton Reiser, *Albert Einstein: A Biographical Portrait* (New York: Albert & Charles Boni, 1930).

9. Max Brod and Felix Weltsch, *Anschauung und Begriff: Grundzüge eines Systems der Begriffsbildung* (Leipzig: Kurt Wolff, 1913). On Frank's lectures, see Max Brod, *Streitbares Leben 1884–1968* (Munich: F. A. Herbig, 1969), 199.

10. Max Brod, "Vom Sinn und Würde des historischen Romans [1956]," in Brod, *Über die Schönheit häßlicher Bilder: Essays zu Kunst und Ästhetik* (Göttingen: Wallstein Verlag, 2014), 344.

11. For biographical details, see Gaëlle Vassogne, *Max Brod in Prag: Identität und Vermittlung* (Tübingen: Max Niemeyer, 2009); Berndt W. Wessling, *Max Brod: Ein Portrait* (Stuttgart: W. Kohlhammer, 1969); Hugo Gold, ed., *Max Brod: Ein Gedenkbuch, 1884–1968* (Tel Aviv: Olamenu, 1969); and Margarita Pazi, *Max Brod: Werk und Persönlichkeit* (Bonn: H. Bouvier u. Co., 1970).

12. Hugo Bergmann to Martin Buber, 19 September 1919, reproduced in Schmuel Hugo Bergman, *Tagebücher & Briefe*, 2 vols., ed. Miriam Sambursky (Königstein: Athenäum, 1985), 1:128. On the Czech language, see Barbora Šrámková, *Max Brod und die tschechische Kultur* (Wuppertal: Arco, 2010), 95–97.

13. Karel Jezdinský, "Presse und Rundfunk in der Tschechoslowakei 1918–1938," in Karl Bosl and Ferdinand Seibt, eds., *Kultur und Gesellschaft in der Ersten Tschechoslowakischen Republik* (Munich: Oldenbourg, 1982), 135–149, on 145.

14. Werner Kayser and Horst Gronemeyer, *Max Brod* (Hamburg: Hans Christian Verlag, 1972); Vassogne, *Max Brod in Prag*, 243–331.

15. Felix Weltsch, ed., *Dichter, Denker, Helfer: Max Brod zum 50. Geburtstag* (Mähr.-Ostrau: Julius Kittls Nachfolger, 1934); Ernst F. Taussig, ed., *Ein Kampf um Wahrheit: Max Brod zum 65. Geburtstag* (Tel Aviv: ABC Verlag, [1950]).

16. Max Brod, *Der Prager Kreis* (Stuttgart: W. Kohlhammer, 1966); Eduard Goldstücker, ed., *Weltfreunde: Konferenz über die Prager deutsche Literatur* (Prague: Academia, 1967). On Germanophone writers in Prague in the generation before Brod, see Peter Demetz, *René Rilkes Prager Jahre* (Düsseldorf: Eugen Diedrichs, 1953).

17. Josef Körner, "Dichter und Dichtung aus dem deutschen Prag," *Donauland* 1, no. 7 (September 1917): 777–784, on 782.

18. Paul Raabe, "Der junge Max Brod und der Indifferentismus," in Goldstücker, *Weltfreunde*, 253–269.

19. Idem, "Die frühen Werke Max Brods," in Gold, *Max Brod*, 137–152. See also Camill Hoffmann, "Tycho Brahes Weg zu Gott," *Die Zukunft* 24, no. 38 (24 June 1916): 332–334, on 333; Arno Gassmann, *Lieber Vater, Lieber Gott?: Der Vater-Sohn-Konflikt bei den Autoren des engeren Prager Kreises (Max Brod—Franz Kafka—Oskar Baum—Ludwig Winder)* (Oldenburg: Igel Verlag Wissenschaft, 2002), 34.

20. Max Brod, "Distanzliebe," *Europäische Begegnung* 4 (1964): 149–153; Hans Dieter Zimmermann, "'Distanzliebe': Max Brod zwischen Deutschen und Tschechen," in Marek Nekula and Walter Koschmal, eds., *Juden zwischen Deutschen und Tschechen: Sprachliche und kulturelle Identitäten in Böhmen 1800–1845* (Munich: R. Oldenbourg, 2006), 233–248. On Brod as an intermediary for Czech culture, especially among German-identified audiences in and out of Bohemia, see Šrámková, *Max Brod und die tschechische Kultur*. On Brod and Jewishness, see Claus-Ekkehard Bärsch, *Max Brod im Kampf um das Judentum: Zum Leben und Werk eines deutsch-jüdischen Dichters aus Prag* (Vienna: Passagen, 1992).

21. Robert Weltsch, "Max Brod and His Age," *Leo Baeck Memorial Lecture* 13 (1970), on 15.

22. Max Brod, *Franz Kafka: A Biography*, tr. G. Humphreys Roberts and Richard Winston (New York: Da Capo Press, 1995 [1937]), 81; idem, *Streitbares Leben*, 82.

23. Max Brod to Kurt Wolff, 30 June 1914, in Kurt Wolff, *Briefwechsel eines Verlegers, 1911–1963*, ed. Bernhard Zeller and Ellen Otten (Frankfurt am Main: Verlag Heinrich Scheffler, 1966), 176.

24. Max Brod, *Tycho Brahes Weg zu Gott: Roman* (Göttingen: Wallstein Verlag, 2013 [1915]), 17.

25. Ibid., 18.

26. Ibid., 23. On the following page, John Dee and Edward Kelley, the well-known alchemists who served in the employ of Rudolf II, are rejected as "English swindlers" ("englischen

Schwindler"). On Dee and Kelley in Prague, see Jennifer Rampling, "John Dee and the Alchemists: Practising and Promoting English Alchemy in the Holy Roman Empire," *Studies in History and Philosophy of Science* 43 (2012): 498–508.

27. Brod, *Tycho Brahes Weg zu Gott*, 28.

28. Ibid., 111.

29. Ibid., 175–176.

30. Ibid., 252.

31. Tengnagel's story would make for a gripping novel in its own right: after marrying Tycho's daughter and gaining control of his estate (much to Kepler's annoyance), Tengnagel and his children served as diplomatic fixers across the Habsburg lands. See Jan Bedřich Novák, *Rudolf II. a jeho pád* (Prague: Nákl. Českého zemského výboru, 1935), on 25, 68n1, 239–240; John Robert Christianson, *On Tycho's Island: Tycho Brahe, Science, and Culture in the Sixteenth Century* (Cambridge: Cambridge University Press, 2003 [2000]), 237, 301–305.

32. On the court dynamics that resulted in Tycho's appointment, see R.J.W. Evans, *Rudolf II and His World: A Study in Intellectual History 1576–1612* (Oxford: Clarendon, 1973), 135–137.

33. The most detailed account of these historical events is Edward Rosen, *Three Imperial Mathematicians: Kepler Trapped Between Tycho Brahe and Ursus* (New York: Abaris, 1986). Rosen repeatedly censures Brod's novelistic account as doing violence to the historical material and as based on inadequate research in local archives, points that he also published as Edward Rosen, "Brod's Brahe: Fact vs Fiction," *Sudhoffs Archiv* 66, no. 1 (1982): 70–78. Rosen never fully acknowledges the gap between historical scholarship and historical fiction.

34. Brod, *Tycho Brahes Weg zu Gott*, 129.

35. Ibid., 145. Ellipses in original. For the full text of Kepler's letter to Ursus as published by the latter, see Nicholas Jardine, *The Birth of History and Philosophy of Science: Kepler's A Defence of Tycho Against Ursus with Essays on Its Provenance and Significance* (Cambridge: Cambridge University Press, 1984), 10.

36. For a complete edition, translation, and commentary, see Jardine, *Birth of History and Philosophy*. On the philological mastery exhibited by Kepler in the *Apologia*, see Grafton, "Humanism and Science," 199.

37. As noted in André Neher, *Jewish Thought and the Scientific Revolution of the Sixteenth Century: David Gans (1541–1613) and His Times*, tr. David Maisel (Oxford: Oxford University Press, 1986), this conversation could never have happened at this place and time, because the Maharal's only audience with the emperor was in 1592, and Tycho did not arrive in Prague until 1599. Nonetheless, Tycho also reportedly met with Löw on another occasion, and he had a pedagogical relationship with David Gans, one of the few Jewish astronomers of this period, who had also studied with Löw. On Gans, see George Alter, *Two Renaissance Astronomers: David Gans, Joseph Delmedigo* (Prague: Československá Akademie věd, 1958); Mordechai Breuer, "Modernism and Traditionalism in Sixteenth-Century Jewish Historiography: A Study of David Gans' *Tzemaḥ David*," in Bernard Dov Cooperman, ed., *Jewish Thought in the Sixteenth Century* (Cambridge, MA: Harvard University Press, 1983), 49–88; André Neher, "L'exégèse biblique juive face à Copernic au XVIème et au XVIIème siècles," in M.S.H.G. Heerma van Voss, Ph. H. J. Houwink ten Cate, and N. A. van Uchelen, eds., *Travels in the World of the Old Testament: Studies Presented to Professor M. A. Beek on the Occasion of His 65th Birthday* (Assen: Van Gorcum & Comp., 1974),

190–196. On Löw and the (much) later ascription of the Golem story, see Moshe Idel, *Golem: Jewish Magical and Mystical Traditions on the Artificial Anthropoid* (Albany: State University of New York Press, 1990); Frederic Thieberger, *The Great Rabbi Loew of Prague: His Life and Work and the Legend of the Golem* (London: East and West Library, 1954); Vladimír Sadek, "Stories of the Golem and Their Relation to the Work of Rabbi Löw of Prague," *Judica Bohemiae* 23, no. 2 (1987): 85–91; idem, "Rabbi Loew—sa vie, héritage pédagogique et sa légende (à l'occasion de la 370e anniversaire de sa mort)," *Judica Bohemiae* 15 (1979): 27–41; and Hillel J. Kieval, "Pursuing the Golem of Prague: Jewish Culture and the Invention of a Tradition," *Modern Judaism* 17, no. 1 (February 1997): 1–23.

38. Brod, *Tycho Brahes Weg zu Gott*, 288.

39. Ibid., 289.

40. On the link to Buber, see Vassogne, *Max Brod in Prag*, 57; Pazi, *Max Brod*, 87; Peter Fenves, "The Kafka-Werfel-Einstein Effect," in Max Brod, *Tycho Brahe's Path to God: A Novel*, tr. Felix Warren Crosse (Evanston, IL: Northwestern University Press, 2007), vii–lviii, on xxxv; and Felix Weltsch, "Philosophie eines Dichters," in Weltsch, *Dichter, Denker, Helfer*, 8–26, on 14.

41. Max Caspar, *Kepler*, tr. and ed. C. Doris Hellman (New York: Dover Publications, 1993 [1959]); Arthur Koestler, *The Sleepwalkers: A History of Man's Changing Vision of the Universe* (London: Arkana, 1989 [1959]).

42. J.L.E. Dreyer, *Tycho Brahe: A Picture of Scientific Life and Work in the Sixteenth Century* (Edinburgh: Adam and Charles Black, 1890); idem, *Tycho Brahe: Ein Bild wissenschaftlichen Lebens und Arbeitens im sechzehnten Jahrhundert*, tr. M. Bruhns (Karlsruhe: G. Braun, 1894). For a more recent scholarly biography, which places significantly less emphasis on the Ursus affair than do Dreyer or Brod, see Victor E. Thoren, *The Lord of Uraniborg: A Biography of Tycho Brahe* (Cambridge: Cambridge University Press, 1990).

43. Johannes Kepler, *Opera Omnia*, 8 vols., ed. Ch. Frisch (Frankfurt am Main: Heyder & Zimmer, 1858–1871), vol. 8, pt. 2. On Brod and Latin, see Scott Spector, *Prague Territories: National Conflict and Cultural Innovation in Franz Kafka's Fin de Siècle* (Berkeley: University of California Press, 2000), 212.

44. G. Kowalewski, *Bestand und Wandel: Meine Lebenserinnerungen zugleich ein Beitrag zur neueren Geschichte der Mathematik* (Munich: R. Oldenbourg, 1950), 249. On Kowalewski's search for new Kepler manuscripts from the Strahov monastery, see 234.

45. Johannes Heinrich von Mädler, *Geschichte der Himmelskunde von der ältesten bis auf die neueste Zeit*, 2 vols. (Braunschweig: George Westermann, 1873); F. J. Studnička, *Bis an's Ende der Welt!: Astronomische Causerien* (Prague: Šimáček, 1896); idem, *Bericht über die astrologischen Studien des Reformators der beobachtenden Astronomie Tycho Brahe: Weitere Beiträge zur bevorstehenden Saecularfeier der Erinnerung an sein vor 300 Jahren erfolgtes Ableben* (Prague: Kön. Böhm. Gesellschaft der Wissenschaften, 1901). On František Josef Studnička, the Jesuit and mathematician at the Czech University, see Josef Petráň, "The Czech Philosophical Faculty 1882–1918," in František Kavka and Josef Petráň, eds., *A History of Charles University*, 2 vols. (Prague: Karolinium, 2001), 2:147–161, on 156.

46. Josef von Hasner, *Tycho Brahe und J. Kepler in Prag: Eine Studie* (Prague: J. G. Calve, 1872), 4.

47. This point is illustrated in the discussions of both Kepler and Ursus as "Germans" in Ferdinand B. Mikowec, *Tycho Brahe: Žiwotopisný nástin* (Prague: Jar. Puspišil, 1847).

48. Ernst Lemke, Review of Max Brod's *Tycho Brahes Weg zu Gott*, *Neuphilologische Blätter* 23 (1915–1916): 415–416.

49. On Julius Kraus, see Michel Vanoosthuyse, "L'espace partagé: Considérations sur quelques romans pragois au tournant de siècle," in *Allemands, juifs et tchèques à Prague/Deutsche, Juden und Tschechen in Prag, 1890–1924: Actes du colloque international de Montpellier, 8–10 décembre 1994* (Montpellier: Université Paul-Valéry-Montpellier III, 1996), 311–319, on 313. See also Auguste Hauschner, *Der Tod des Löwen* (Berlin: Egon Fleischel, 1916); and the 1924 Tycho fictionalization by Karl Hans Strobl, *Die Vaclavbude: Eine Prager Studentengeschichte* (Leipzig: L. Stackmann, 1924). Brod's views of Hauschner's work are outlined in Brod, *Der Prager Kreis*, 45.

50. Figures from Roland Reuß's afterword to Brod, *Tycho Brahes Weg zu Gott*, 319–320; and Kayser and Gronemeyer, *Max Brod*, 67. The best English-language secondary source on the novel is Fenves, "Kafka-Werfel-Einstein Effect."

51. Max Brod to Kurt Wolff, 22 February 1916, in Wolff, *Briefwechsel eines Verlegers*, 179–180. Emphasis in original.

52. Max Brod to Kurt Wolff, 10 July 1916, in Wolff, *Briefwechsel eines Verlegers*, 181.

53. Max Brod, "Umgang mit Verlegern [1968]," in Brod, *Über die Schönheit häßlicher Bilder*, 353.

54. Hugo Bergmann, "Max Brods neuer Roman," *Der Jude* 1/2 (May 1916): 134–136, on 135.

55. Felix Weltsch, "Tycho Brahes Weg zu Gott," *Die Schaubühne* 13, no. 1 (1917): 474–478, on 478. Decades later, Weltsch returned to this point about mirror characters, extending it to both *Rëubeni* and *Galilei*: Weltsch, "Aus Zweiheit zur Einheit: Max Brods Weg als Dichter und Denker," in Taussig, *Ein Kampf um Wahrheit*, 8–17.

56. Otto Pick, "Ein Weg zu Gott," *Die Neue Rundschau* 27 (1918): 862–864, on 864.

57. Hoffmann, "Tycho Brahes Weg zu Gott," 332; Rudolf Fuchs, "Literarische Neuerscheinungen," *Die Aktion* 6 (1916): 656–657; Elise Bergmann, Review of Max Brod's *Tycho Brahes Weg zu Gott*, *Preußische Jahrbücher* 166 (1916): 302–304; Karl Münzer, Review of Max Brod's *Tycho Brahes Weg zu Gott*, *Das literarische Echo* 18 (1915/1916): 765–766; Max Herrmann, Review of Brod's *Tycho Brahes Weg zu Gott*, *Zeit-Echo* 2, no. 8 (May 1916): 125–126; A. Blau, "Max Brod's Roman 'Tycho Brahes Weg zu Gott,'" *Jeschurun* 3 (1916): 210–217; Kurt Pinthus, Review of Max Brod's *Tycho Brahes Weg zu Gott*, *Zeitschrift für Bücherfreunde* 8, no. 2 (December 1916): 458–459; Ella Seligmann, Review of Max Brod's *Tycho Brahes Weg zu Gott*, *Liberales Judentum* 8 (1916): 96. A common theme of these reviews was their comparison of the book with Gustav Meyrink's *Der Golem*, which came out a year before *Tycho*, with some reviewing the two works alongside each other: see, for example, Hans Schorn, "Geschichtliche Romane und Erzählungen," *Die schöne Literatur* 18 (6 January 1917): 6.

58. J. d'Ouckh, "*Tycho Brahes Weg zu Gott*: Ein Roman von Max Brod," *Das Reich* 1 (1916/1917): 299–300.

59. Wolfgang Schumann, "Geschichtliche Romane," *Deutscher Wille des Kunstwarts* 29, no. 4 (1915/1916): 231–236, on 234.

60. Brod quoted in Wenig, "O knize a autorovi," in Brod, *Tychona Brahe cesta k Bohu*, tr. Adolf Wenig (Prague: F. Topič, [1917]), 6.

61. Max Brod quoted in ibid., 7.

62. Brod quoted in ibid., 7.

63. K. M., Review of Brod's *Tychona Brahe cesta k Bohu*, *Zvon* 18 (1918): 264–265, on 265. On Hájek/Hagecius's scientific accomplishments, see the brief discussions in Dreyer, *Tycho Brahe*, 83; Evans, *Rudolf II and His World*, 152; Luboš Nový, ed., *Dějiny exaktních věd v českých zemích do konce 19. století* (Prague: Nakl. ČSAV, 1961), 39.

64. K. M., Review of Brod's *Tychona Brahe cesta k Bohu*, 265.

65. Quido Vetter, "Čestí hvězdáři v Brodově knize 'Tychona Brahe cesta k Bohu,'" *Nová síla* 21 (2 October 1920): 2.

66. František Xaver Šalda, "Židovský román staropražský," in Šalda, *Kritické projevy X: 1917– 1918* (Prague: Československý spisovatel, 1957), 284–287, on 286. Emphasis in original. For another review that juxtaposed Brod and Meyrink, see Antonín Veselý, "Dva německé romány," *Česká revue* (1916–1917): 359–366.

67. K., "Dva německé romány ze staré Prahy," *Právo lidu* 25, no. 58 (27 February 1916): 2.

68. Šalda, "Židovský roman staropražský," 287.

69. F. Marek, "Knihy přeložené," *Cesta* 1 (1919): 48.

70. Paul Adler, "Tycho Brahes Weg zu Gott," *Die Rheinlande* 26 (1916): 376.

71. For details on the conflict, see Donald G. Daviau, "Max Brod and Karl Kraus," in Margarita Pazi, ed., *Max Brod 1884–1984: Untersuchungen zu Max Brods literarischen und philosophischen Schriften* (New York: Peter Lang, 1987), 207–231.

72. Karl Kraus, "Selbstanzeige," *Die Fackel* 13, nos. 326/327/328 (8 July 1911): 34–36, on 35–36.

73. Max Brod to Richard Dehmel, 21 December 1913, in Kayser and Gronemeyer, *Max Brod*, 29.

74. Ibid., 29. On the Jewish dimensions of the antagonism, see also Pazi, *Max Brod*, 16.

75. Max Brod to Richard Dehmel, 27 December 1913, in Kayser and Gronemeyer, *Max Brod*, 30. Emphasis in original.

76. Karl Kraus to Kurt Wolff, 9 December 1913, in Wolff, *Briefwechsel eines Verlegers*, 123.

77. Max Brod, *Diesseits und Jenseits*, 2 vols. (Winterthur: Mondial, 1947), 1:331. See also 1:49.

78. Kurt Krolop, "Prager Autoren im Lichte der 'Fackel,'" in *Prager deutschsprachige Literatur zur Zeit Kafkas* (Vienna: Braumüller, 1989), 92–117.

79. Max Brod, "Wie ich Franz Werfel entdeckte," *Prager Montagsblatt* 60, no. 52 (27 December 1937): 58.

80. Anselm Ruest, "Der Max Brod-Abend," *Die Aktion* 1, no. 45 (12 December 1911): 1425– 1426. For a contemporary description of the sensation Werfel's verses made in Prague, see Otto Pick, "Erinnerungen an den Winter 1911/12," *Die Aktion* 6 (1916): 605.

81. Michel Reffet, "Pro Werfel contra Kraus: Eine alte Polemik aus neuer Sicht," in Klaas-Hinrich Ehlers, Steffen Höhne, Václav Maidl, and Marek Nekula, eds., *Brücken nach Prag: Deutschsprachige Literatur im kulturellen Kontext der Donaumonarchie und der Tschechoslowakei* (Frankfurt am Main: Peter Lang, 2000), 201–219; Josef Čermák, "Junge Jahre in Prag: Ein Beitrag zum Freundeskreis Franz Werfels," in Ehlers et al., *Brücken nach Prag*, 125–162.

82. Frank Field, *The Last Days of Mankind: Karl Kraus and His Vienna* (London: Macmillan, 1967), 128–129.

83. Brod, *Streitbares Leben*, 13.

84. Idem, "Wie ich Franz Werfel entdeckte," 58.

85. Adler, "Tycho Brahes Weg zu Gott," 376.

86. Brod, *Streitbares Leben*, 203.

87. "Meinem Freunde *Franz Kafka*." Brod, *Tycho Brahes Weg zu Gott*, 15.

88. Idem, *Franz Kafka*, 131.

89. Franz Kafka to Max Brod, 6 February 1914, in Max Brod and Franz Kafka, *Eine Freundschaft*, 2 vols. (Frankfurt am Main: Fischer, 1989), 2:137.

90. Franz Kafka to Felice Bauer, 20 April 1914, in Franz Kafka, *Briefe an Felice: und andere Korrespondenz aus der Verlobungszeit*, ed. Erich Heller and Jürgen Born (New York: Schocken, 1967), 559.

91. See Vivian Liska, "Neighbors, Foes, and Other Communities: Kafka and Zionism," *Yale Journal of Criticism* 13, no. 2 (Fall 2000): 343–360, on 349.

92. Franz Kafka to Felice Bauer, [5 December 1915], in Kafka, *Briefe an Felice*, 645.

93. Ritchie Robertson, "The Creative Dialogue Between Brod and Kafka," in Mark H. Gelber, ed., *Kafka, Zionism, and Beyond* (Tübingen: Max Niemeyer Verlag, 2004), 283–296, on 289.

94. Brod, *Streitbares Leben*, 171.

95. Ibid., 201–202.

96. Ibid., 202.

97. Peter Stephan Jungk, *Franz Werfel: A Life in Prague, Vienna, and Hollywood*, tr. Anselm Hollo (New York: Grove Weidenfeld, 1990 [1987]), 154–155.

98. Brod to Einstein, 30 November 1938, AEA 34–063.

99. Ibid.

100. Max Brod, *Franz Kafka als wegweisende Gestalt* (St. Gallen: Tschudy-Verlag, [1957]), 48.

101. Ibid., 47.

102. Brod to Einstein, 28 June 1940, AEA 34–064.

103. Gassmann, *Lieber Vater, Lieber Gott?*, 43.

104. Brod, *Diesseits und Jenseits*, 1:175. For references to Frank, see 1:125, 137, 162, 186, 198, 219; 2:30.

105. That said, there are a number of references in the novel to Kepler and Tycho (Max Brod, *Galilei in Gefangenschaft* [Winterthur: Mondial, 1948], 65, 76, 87–88, 249, 572), the Bohemian defeat at White Mountain (212–213), and Jan Hus (342, 763), and one of the characters—Simon Delmedigo—travels to and from Prague (280, 776). Outside of Rome, Prague is probably the most frequently invoked city in the novel.

106. Max Brod, "Praha a já," *Literární noviny* (10 December 1930): 4. Brod did return to Prague 25 years after his exile, but it was to a very non-Germanophone city that was in the throes of a fascination with Kafka's writings. "Max Brod v Praze po 25 letech," *Literární noviny* 13, no. 28 (1964): 3.

107. On the Czechoslovak State Prize, see Jürgen Serke, *Böhmische Dörfer: Wanderungen durch eine verlassene literarische Landschaft* (Vienna: Paul Zsolnay, 1987), 56; Ines Koeltzsch, *Geteilte Kulturen: Eine Geschichte der tschechisch-jüdisch-deutschen Beziehungen in Prag (1918–1938)* (Munich: Oldenbourg, 2012), 193.

108. Brod to Einstein, 23 June 1949, AEA 34–067.

109. On the events in question, and for translations of the central documents, see Maurice A. Finnochiario, ed., *The Galileo Affair: A Documentary History* (Berkeley: University of California, 1989).

110. Brod, *Galilei in Gefangenschaft*, 686. For more information on the historical Delmedigo, see Alter, *Two Renaissance Astronomers*.

111. Einstein to Brod, 4 July 1949, AEA 34–068. For a contrasting, and much more flattering, interpretation, see Martin Buber's reading of the Galileo figure in the novel in Buber, "Der Galilei Roman."

112. Quoted in Sebastian Schirrmeister, "On Not Writing Hebrew: Max Brod and the 'Jewish Poet of the German Tongue' Between Prague and Tel Aviv," *Leo Baeck Institute Year Book* 60 (2015): 25–42, on 39.

113. Brod to Einstein, 22 July 1949, AEA 34–069.

114. Brod to Einstein, 15 September 1949, AEA 34–070.

115. Brod to Einstein, 15 December 1951, AEA 34–073.

116. This detail is cited in lots of places, such as Peter Demetz, *Prague in Black and Gold: Scenes in the Life of a European City* (New York: Hill and Wang, 1997), 33.

117. Einstein, foreword to Carola Baumgardt, *Johannes Kepler: Life and Letters* (New York: Philosophical Library, 1951), 12–13.

118. On Kepler's religious views, see especially Aviva Rothman, *The Pursuit of Harmony: Kepler on Cosmos, Confession, and Community* (Chicago: University of Chicago Press, 2017).

119. Einstein to Maurice Solovine, 1 January 1951, in Albert Einstein, *Letters to Solovine* (New York: Philosophical Library, 1987), 102.

120. C. Lanczos, "The Greatness of Albert Einstein," in Gerald E. Tauber, ed., *Albert Einstein's Theory of General Relativity* (New York: Crown, 1979), 16–26, on 18.

121. Hitler quoted in Dietrich Eckart, *Der Bolschewismus von Moses bis Lenin: Zwiegespräch zwischen Adolf Hitler und mir* (Munich: Hoheneichen-Verlag, 1924), 12. My thanks to Alex Csiszar in helping me obtain a copy of this work.

CHAPTER 6: OUT OF JOSEFOV

1. Martin Buber, *Drei Reden über das Judentum* (Frankfurt am Main: Rütten & Loening, 1911), 17.

2. See, for example, the essays in Karl Erich Grözinger, Stéphane Mosès, and Hans Dieter Zimmermann, eds., *Franz Kafka und das Judentum* (Frankfurt am Main: Jüdischer Verlag bei Athenäum, 1987).

3. Philipp Frank, *Einstein: His Life and Times*, tr. George Rosen, ed. and rev. Shuichi Kusaka (Cambridge, MA: Da Capo Press, 2002 [1947]), 85.

4. Albert Einstein, "Autobiographical Notes," tr. Paul Arthur Schilpp, in Schilpp, ed., *Albert Einstein: Philosopher-Scientist*, 2 vols. (New York: Harper & Row, 1951 [1949]), 1:1–95, on 3–5. For a survey of Einstein's negative views of organized religion and a personal god, see Matthew Stanley, "Myth 21: That Einstein Believed in a Personal God," in Ronald L. Numbers, ed., *Galileo Goes to Jail: And Other Myths about Science and Religion* (Cambridge, MA: Harvard University Press, 2009), 187–195.

5. In his memoirs, Einstein's son-in-law would provide a somewhat repellent characterization of Einstein's distance from Jewishness: "His prayer-shawl ancestors had made no introduction into his bearing, his acts, or his life. He was also entirely free from that racial element which

undisputably confines Jewry—a cold, money-loving trait, which dusts her soul, and has for centuries." Dimitri Marianoff with Palma Wayne, *Einstein: An Intimate Study of a Great Man* (Garden City, NY: Doubleday, Doran and Co., 1944), 169.

6. On his registration in Zurich, see the editor's commentary, "Einstein and the Jewish Question," *CPAE* 7, on 222; and Albrecht Fölsing, *Albert Einstein: A Biography*, tr. and abridged Ewald Osers (New York: Penguin Books, 1997 [1993]), 41.

7. Ze'ev Rosenkranz, *Einstein Before Israel: Zionist Icon or Iconoclast?* (Princeton, NJ: Princeton University Press, 2011), 32–33.

8. Einstein to Heinrich Zangger, 24 August [1911], *CPAE* 5:279, on 314. Despite his ostensible firsthand credentials, Frank claimed in his biography that Einstein "did not go through any formal ceremony," and that his registration of himself as Jewish only amounted to changing a line on a form. Frank, *Einstein: His Life and Times*, 79.

9. Antonina Vallentin, *The Drama of Albert Einstein*, tr. Moura Budberg (Garden City, NY: Doubleday & Company, 1954), 52.

10. Einstein to Ehrenfest, 25 April 1912, AEDA 9–324, Box 10, Folder "P. Ehrenfest 1911–1916." Emphasis in original.

11. Einstein to Alfred Kleiner, 3 April 1912, *CPAE* 5:381, on 446.

12. Einstein to Zangger, [before 29 February 1912], *CPAE* 5:366, on 421.

13. Ehrenfest's biographer speculated that the insistence might have stemmed from the conditions of his marriage as imposed by Russian authorities. Tsarist law did not permit interfaith marriage, and therefore it was essential that both Ehrenfest and Afanas'eva declare themselves to be without religion. Martin J. Klein, *Paul Ehrenfest: The Making of a Theoretical Physicist* (Amsterdam: North-Holland, 1970), 178, 180.

14. On Orthodoxy, see Michele Zackheim, *Einstein's Daughter: The Search for Lieserl* (New York: Riverhead, 1999), 68. On Lutherans, see Einstein to Mileva Einstein-Marić, Hans Albert, and Eduard, [10 April 1914], *CPAE* 8:3, on 14.

15. "Deposition in Divorce Proceedings," 23 December 1918, *CPAE* 8:676, on 974; "Divorce Decree," 14 February 1919, *CPAE* 9:6, on 8.

16. Ilse Einstein to the Protestant Synod of Berlin, 9 March 1920, *CPAE* 9:346, on 467.

17. Einstein to Jewish Community of Berlin, 22 December 1920, *CPAE* 10:238, on 534.

18. Jewish Community of Berlin to Einstein, 30 December 1920, *CPAE* 10:253, on 550.

19. Einstein to Jewish Community of Berlin, 5 January 1921, *CPAE* 12:8, on 29.

20. Einstein to Malwin Warschauer, 8 March 1921, *CPAE* 12:86, on 124. This was in response to Malwin Warschauer to Einstein, 2 March 1921, *CPAE* 12:74, on 112–113. He made fun of the claim that his membership would help Zionism in a letter to a friend: Einstein to Alfred Kerr, 7 March 1921, *CPAE* 12:81, on 120.

21. Thomas Levenson, *Einstein in Berlin* (New York: Bantam Books, 2003), 304.

22. R. K., "Einstein in Prag," *Prager Tagblatt* 37, no. 144 (26 May 1912): 2.

23. Ibid., 2.

24. Ibid.

25. Ibid.

26. "A. Einstein," *Montags-Revue* 43, no. 31 (29 July 1912): 1–2, on 2.

27. Ibid., 1.

28. Einstein to Emil Starkenstein, 14 July 1921, *CPAE* 12:181, on 223.

29. The most comprehensive study of this issue is Rosenkranz, *Einstein Before Israel.*

30. Hillel J. Kieval, *Languages of Community: The Jewish Experience in the Czech Lands* (Berkeley: University of California Press, 2000), 11; Hans Kohn, "Before 1918 in the Historic Lands," in *The Jews of Czechoslovakia: Historical Studies and Surveys*, 3 vols. (Philadelphia: Jewish Publication Society of America, 1968–1984), 1:12–20; Rachel L. Greenblatt, *To Tell Their Children: Jewish Communal Memory in Early Modern Prague* (Stanford, CA: Stanford University Press, 2014), 14.

31. Hillel J. Kieval, "Pursuing the Golem of Prague: Jewish Culture and the Invention of a Tradition," *Modern Judaism* 17, no. 1 (February 1997): 1–23, on 9; idem, *Languages of Community*, 15, 18.

32. Idem, *The Making of Czech Jewry: National Conflict and Jewish Society in Bohemia, 1870–1918* (New York: Oxford University Press, 1988), 12.

33. On *Mauscheldeutsch*, see Gary B. Cohen, *The Politics of Ethnic Survival: Germans in Prague, 1861–1914*, 2nd. rev. ed. (West Lafayette, IN: Purdue University Press, 2006 [1981]), 61.

34. For these figures, see Jiří Pešek, "Jüdische Studenten an den Prager Universitäten, 1882–1939," in Marek Nekula, Ingrid Fleischmann, and Albrecht Greule, eds., *Franz Kafka im sprachnationalen Kontext seiner Zeit: Sprache und nationale Identität in öffentlichen Institutionen der böhmischen Länder* (Köln: Böhlau, 2007), 213–227, on 214; Kieval, *Languages of Community*, 145; and idem, *Making of Czech Jewry*, 57. On the earlier history of Jews at the university in Prague, see Guido Kisch, "Die Prager Universität und die Juden: Mit Beiträgen zur Geschichte des Medizinstudiums," *Jahrbuch der Gesellschaft für Geschichte der Juden in der Čechoslovakischen Republik* 6 (1934): 1–144.

35. Institutionally, Jews occupied high positions at the most significant German-focused establishments, such as the German Casino. Gerhard Kurz, "Kafka zwischen Juden, Deutschen und Tschechen," in Karl Bosl and Ferdinand Seibt, eds., *Kultur und Gesellschaft in der Ersten Tschechoslowakischen Republik* (Munich: Oldenbourg, 1982), 37–50, on 39.

36. Hillel J. Kieval, "Death and the Nation: Ritual Murder as Political Discourse in the Czech Lands," in *Allemands, juifs et tchèques à Prague/Deutsche, Juden und Tschechen in Prag, 1890–1924: Actes du colloque international de Montpellier, 8–10 décembre 1994* (Montpellier: Université Paul-Valéry-Montpellier III, 1996), 83–99; Michal Frankl, "The Background of the Hilsner Case: Political Antisemitism and Allegations of Ritual Murder 1896–1900," *Judaica Bohemiae* 36 (2000): 34–118. On Czech political anti-Semitism, see idem, "'Sonderweg' of Czech Antisemitism?: Nationalism, National Conflict and Antisemitism in Czech Society in the Late 19th Century," *Bohemia* 46, no. 1 (2005): 120–134. On the broader climate of pan-Germanism in Austria, see Andrew G. Whiteside, *The Socialism of Fools: Georg Ritter von Schönerer and Austrian Pan-Germanism* (Berkeley: University of California Press, 1975).

37. Wilfried Brosche, "Das Ghetto von Prag," in *Die Juden in den böhmischen Ländern: Vorträge der Tagung des Collegium Carolinum in Bad Wiessee vom 27. bis 29. November 1981* (Munich: Oldenbourg, 1983), 87–122, on 114; Richard Burton, *Prague: A Cultural History* (Northampton, MA: Interlink, 2009 [2003]), 64–65. On contemporary protests against the *asanace*, see Reiner Stach, *Kafka: Die frühen Jahre* (Frankfurt am Main: S. Fischer, 2014), 125–126.

38. Kieval, *Making of Czech Jewry*, 135–136, 202–203; Hagit Lavsky, *Before Catastrophe: The Distinctive Path of German Zionism* (Detroit: Wayne State University Press, 1996), 147–148.

39. Margarita Pazi, "Franz Kafka, Max Brod und der 'Prager Kreis,'" in Grözinger, Mosès, and Zimmermann, *Franz Kafka und das Judentum*, 71–92, on 77–78.

40. On Bar Kochba, see Kieval, *Making of Czech Jewry*, ch. 4; Hannelore Rodlauer, "Ein anderer 'Prager Frühling': Das Verein 'Bar Kochba' in Prag," *Das jüdische Echo* 49 (2000): 181–188. On the publication history of *Selbstwehr*, see Avigdor Dagan, "The Press," in *Jews of Czechoslovakia*, 1:523–531, on 525. On the later intellectual history of the Prague Zionists, see Dimitry Shumsky, *Zweisprachigkeit und binationale Idee: Der Prager Zionismus 1900–1930*, tr. Dafna Mach (Göttingen: Vandenhoeck & Ruprecht, 2013 [2010]). On Max Brod and Bar Kochba, see Gaëlle Vassogne, "Prague Zionism, the Czechoslovak State, and the Rise of German National Socialism: The Figure of Max Brod, 1914–1933," in Rebecka Lettevall, Geert Somsen, and Sven Widmalm, eds., *Neutrality in Twentieth-Century Europe* (New York: Routledge, 2012), 207–225. On numbers, see Cohen, *Politics of Ethnic Survival*, 166.

41. Max Brod, *Franz Kafka: A Biography*, tr. G. Humphreys Roberts and Richard Winston (New York: Da Capo Press, 1995 [1960, 1937]), 14. On Bergmann's biography, see Miriam Sambursky, "Zionist und Philosoph: Das Habilitierungsproblem des jungen Hugo Bergmann," *Bulletin des Leo Baeck Instituts*, no. 58 (1981): 17–40.

42. Hugo S. Bergman, "Erinnerungen an Franz Kafka," in Reuben Klingsberg, ed., *Exhibition Franz Kafka, 1883–1924: Catalogue* (Jerusalem: Berman Hall, Jewish National and University Library, 1969), 5–12, on 5.

43. See Hugo Bergmann to Else Bergmann, 20 July 1911, from Göttingen, reproduced in Schmuel Hugo Bergman, *Tagebücher & Briefe*, 2 vols., ed. Miriam Sambursky (Königstein: Athenäum, 1985), 1:41.

44. See the letter from Franz Brentano to Hugo Bergmann, 16 December 1907, reproduced in Hugo Bergmann and Franz Brentano, "Briefe Franz Brentanos an Hugo Bergmann," *Philosophy and Phenomenological Research* 7, no. 1 (September 1946): 83–158, on 103.

45. Hugo Bergmann, *Das philosophische Werk Bernard Bolzanos* (Halle an der Salle: Max Niemeyer, 1909), 145 and 176. See also his earlier perceptive analysis in idem, "Das philosophische Bedürfnis in der modernen Physik," *Philosophische Wochenschrift und Literatur-Zeitung* 1 (January–March 1906): 332–338.

46. Oskar Kraus, "Bolzano," *Bohemia* 83, no. 308 (8 November 1910): 1–2.

47. Hugo Bergmann to Carl Stumpf, [1914], quoted in Sambursky, "Zionist und Philosoph," 36.

48. Hugo Bergmann, "Bemerkungen zur arabischen Frage," *Palästina* 7 (1911): 190–195, on 190. On Bar Kochba's sponsorship of discussion about Palestine, see Kateřina Čapková, "'Ich akzeptiere den Komplex, der ich bin': Zionisten um Franz Kafka," in Peter Becher, Steffen Höhne, and Marek Nekula, eds., *Kafka und Prag: Literatur-, kultur-, sozial- und sprachhistorische Kontexte* (Köln: Böhlau, 2012), 81–95, on 86.

49. Bergmann, "Bemerkungen zur arabischen Frage," 195.

50. G. Kowalewski, *Bestand und Wandel: Meine Lebenserinnerungen zugleich ein Beitrag zur neueren Geschichte der Mathematik* (Munich: R. Oldenbourg, 1950), 245–246.

51. See especially Rosenkranz, *Einstein Before Israel*, and references therein.

52. Einstein, "Wie ich Zionist wurde," *Jüdische Rundschau* (21 June 1921): 351–352, reproduced in *CPAE* 7:57, on 428.

53. Kurt Blumenfeld, *Erlebte Judenfrage: Ein Vierteljahrhundert deutscher Zionismus* (Stuttgart: Deutsche Verlags-Anstalt, 1962), 126.

54. H. Bergman, "Personal Remembrances of Albert Einstein," in Robert S. Cohen and Marx W. Wartofsky, eds., *Logical and Epistemological Studies in Contemporary Physics* (Dordrecht: D. Reidel, 1974), 388–394, on 390–391.

55. On the symbolic opposition of *Ostjuden*—romanticized and demonized by so-called West-ernized Jews—see the excellent account in Steven E. Aschheim, *Brothers and Strangers: The East European Jew in German and German Jewish Consciousness, 1800–1923* (Madison: University of Wisconsin Press, 1982). Thomas Levenson (*Einstein in Berlin*, 172) claims that Einstein first en-countered *Ostjuden* in Prague, but to my mind this mischaracterizes the Jewish community with which he interacted in Bohemia, who were just as Westernized as the Berliners. As discussed in earlier chapters, the understanding of Prague as part of "Eastern Europe" is often a retrospective projection by scholars who emphasize the Slavic aspects of the city, which were not the parts Einstein most noticed.

56. Einstein to Central Association of German Citizens of the Jewish Faith, 5 April 1920, *CPAE* 9:368, on 495. See also his comments in Einstein, "Assimilation und Antisemitismus," [3 April 1920], reproduced in *CPAE* 7:34, on 290; and Einstein to Ehrenfest, 22 March 1919, *CPAE* 9:10, on 16.

57. Ofer Ashkenazi, "Reframing the Interwar Peace Movement: The Curious Case of Albert Einstein," *Journal of Contemporary History* 46, no. 4 (2011): 741–766, on 750, 760.

58. Hugo Bergmann to Einstein, 22 October 1919, *CPAE* 9:147, on 211. See also Rosenkranz, *Einstein Before Israel*, 61–62.

59. Einstein to Bergmann, 5 November 1919, *CPAE* 9:155, on 222.

60. Einstein to Ehrenfest, 12 January 1920, *CPAE* 9:254, on 352.

61. Quoted in Blumenfeld, *Erlebte Judenfrage*, 127–128.

62. Maja Winteler-Einstein, "Albert Einstein—Beitrag für sein Lebensbild," *CPAE* 1:lx.

63. Einstein to Born, 9 November 1919, in Albert Einstein, Hedwig Born, and Max Born, *Brief-wechsel, 1916–1955* (Frankfurt am Main: Edition Erbrich, 1982 [1969]), 36.

64. Michael Berkowitz, *Western Jewry and the Zionist Project, 1914–1933* (Cambridge: Cam-bridge University Press, 1997), 48–54. Einstein commented in astonishment on his global appeal to Jews in Einstein to Besso, 5 June 1925, *CPAE* 15:2, on 50.

65. Yfaat Weiss, "Central European Ethnonationalism and Zionist Binationalism," *Jewish So-cial Studies* 11, no. 1 (Fall 2004): 93–117; Lavsky, *Before Catastrophe*, ch. 9; Shumsky, *Zweisprachig-keit und binationale Idee*; and Ofer Ashkenazi, "'Tsioni aval lo yehudi leumi': Albert Ainshtayin u'Brit Shalom' nokhah meoraot tarpat," in Adi Gordon, ed., *"Berit Shalom" veha-Tsiyonut ha-du-le'umit: "Ha-She'elah ha-'Arvit" ki-shee'lah Yehudit* (Jerusalem: Merkaz Minervah, 2008), 123–148.

66. Viktor G. Ehrenberg to Einstein, 23 November 1919, *CPAE* 9:173, on 243.

67. Georg Schlesinger to Einstein, 20 December 1921, *CPAE* 12:334, on 389.

68. In the phone conversation logs of Hanna Fantova (whom we will meet properly in the conclusion to this book) from January 1954, Einstein is noted as saying: "The Israelis ought to have selected the English language instead of Hebrew; that would have been much better, but the Jews were too fanatical." Entry for 2 January 1954 in Hanna Fantova, "Gespräche mit Einstein," undated, HFC, Box 1, Folder 6, p. 14.

69. Einstein, "Travel Diary Japan, Palestine, Spain," 3 February 1923, *CPAE* 13:379, on 558.

70. Hugo Bergmann, *Der Kampf um das Kausalgesetz in der jüngsten Physik* (Braunschweig: Friedr. Vieweg & Sohn, 1929). Interestingly, on page 3 Bergmann disputed the notion that quantum theory was reasoning about "fictions," thus implicitly critiquing Oskar Kraus's contemporary attacks on relativity. For a discussion of Einstein's views on causality in quantum physics, see Arthur Fine, *The Shaky Game: Einstein, Realism, and the Quantum Theory* (Chicago: University of Chicago Press, 1986).

71. Einstein to Hugo Bergmann, 27 September 1929, AEDA 45–553, Box 65, Folder "1927–31 B Folder II."

72. Hugo Bergmann to Einstein, 8 October 1929, AEDA 45–555, Box 65, Folder "1927–31 B Folder II."

73. Ofer Ashkenazi, "Zionism and Violence in Albert Einstein's Political Outlook," *Journal of Jewish Studies* 63, no. 2 (Autumn 2012): 331–355.

74. Einstein to Hugo Bergmann, 19 June 1930, AEDA 45–571, Box 65, Folder "1927–31 B Folder II." Emphasis in original.

75. Hugo Bergmann to Einstein, 14 January 1947, AEDA 59–211, Box 84, Folder "1947–49 B"; and Einstein to Bergmann, 25 January 1947, AEDA 59–211, Box 84, Folder "1947–49 B."

76. Bergmann to Einstein, 6 March 1950, reproduced in Bergman, *Tagebücher & Briefe*, 2:44. On the offer of the presidency, see Yitzhak Navon, "Einstein and the Presidency of Israel," in Gerald Holton and Yehuda Elkana, eds., *Albert Einstein: Historical and Cultural Perspectives* (Princeton, NJ: Princeton University Press, 1982), 293–296.

77. On the recognition of Jewish nationality in Czechoslovakia, see Tatjana Lichtenstein, "'Making' Jews at Home: Zionism and the Construction of Jewish Nationality in Inter-war Czechoslovakia," *East European Jewish Affairs* 36, no. 1 (June 2006): 49–71; Elizabeth Wiskemann, *Czechs and Germans: A Study of the Struggle in the Historic Provinces of Bohemia and Moravia*, 2nd. ed. (London: Macmillan, 1967 [1938]), 226. On Masaryk and Jewish nationalism, see Felix Weltsch, "Masaryk and Zionism," in Ernst Rychnovsky, ed., *Thomas G. Masaryk and the Jews: A Collection of Essays*, tr. Benjamin R. Epstein (New York: B. Pollak, 1941), 74–114; and Hugo Bergmann, "Masaryk in Palestine," in Rychnovsky, *Thomas G. Masaryk and the Jews*, 259–271. As noted in Andrea Orzoff, *Battle for the Castle: The Myth of Czechoslovakia in Europe, 1914–1948* (New York: Oxford University Press, 2009), 83, Masaryk's support of Jewish causes was double-edged: his belief that Jews were a separate nation also indicated limitations on their capacity to assimilate into a Czechoslovak *ethnos*. On Masaryk's political thought leading up to his presidency, see H. Gordon Skilling, *T. G. Masaryk: Against the Current, 1882–1914* (University Park: Pennsylvania State University Press, 1994).

78. Einstein to the Nobel Committee of the Norwegian Parliament, 19 January 1921, *CPAE* 12:22, on 44. See also Otto Nathan and Heinz Norden, *Einstein on Peace* (New York: Schocken, 1968 [1960]), 41.

79. Reproduced in Nathan and Norden, *Einstein on Peace*, 130.

80. Reproduced in ibid., 130–131. See discussion in Roman Kotecký, "Korespondence Einsteina s Masarykem: Domněle ztracený dopis a pozapomenutá historie jedné intervence," *Vesmír* 72 (1993): 566–568.

81. Reproduced in Kotecký, "Korespondence Einsteina s Masarykem," 567. The italicized words represent English terms in the original; the underlining is Masaryk's.

82. Reproduced in Nathan and Norden, *Einstein on Peace*, 131.

83. The complex saga is described in detail in ibid., 265–267.

84. Einstein to Besso, 9 June 1937, in Albert Einstein and Michele Besso, *Correspondance 1903–1955*, ed. and tr. Pierre Speziali (Paris: Hermann, 1972), 313.

85. Emil Nohel's birth certificate, 3 January 1886, AEA 91–747.

86. See Emil Nohel's curriculum vitae, [undated but around 1913], AEA 91–748; and his transcript from 1904–1905 to 1908–1909, AEA 91–752. For Pick's report on Nohel's dissertation, dated 11 December 1913, see AEA 91–749. The doctoral committee consisted of Pick, Gerhard Kowalewski, and Philipp Frank. Interestingly, Nohel did take a one-hour rigorous examination in philosophy on 19 February 1914 from Oskar Kraus and Christian von Ehrenfels, receiving an "excellent" mark from both of them. Jan Havránek, "Materiály k Einsteinovu Pražskému působení z Archivu Univerzity Karlovy," *Acta universitatis Carolinae—Historia universitatis Carolinae Pragensis* 22, no. 1 (1980): 109–134, on 132.

87. There is very little secondary literature on Nohel; much of what exists consists of asides in Einstein biographies. For useful accounts, see Emilie Těšínská, "Profilování teoretické fyziky na pražské univerzitě a vazby s pražským působením A. Einsteina před 100 lety," *Pokroky matematiky, fyziky a astronomie* 57, no. 2 (2012): 146–168, on 152; Havránek, "Materiály k Einsteinovu Pražskému působení z Archivu Univerzity Karlovy," 110–111, 128ff; Abraham Pais, *"Subtle Is the Lord ...": The Science and Life of Albert Einstein* (New York: Oxford University Press, 1982), 485–486; Armin Hermann, *Einstein: Der Weltweise und sein Jahrhundert* (Munich: Piper, 1994), 170; and Frank, *Einstein: His Life and Times*, 82–83. Frank personally knew Nohel, and it is difficult to determine—as always with his book—how much of the account was derived from conversations with Einstein and how much from Frank's own experiences of the man.

88. Quoted in Pais, *"Subtle Is the Lord ..."*, 486.

89. Yeshayahu Nohel, "My Family," undated typescript, AEA 85–569.

90. On Stern as Einstein's assistant, see Emilio Segrè, *Otto Stern, 1888–1969: A Biographical Memoir* (Washington, DC: National Academy of Sciences, 1973), 216. On Hopf, see Lewis Pyenson, "Einstein's Early Scientific Collaboration," *Historical Studies in the Physical Sciences* 7 (1976): 83–123. Fölsing's assessment of Nohel is characteristically blunt, but it is hard to see it as entirely unjustified: "However, Nohel does not seem to have been much of a replacement for Hopf. Einstein never published anything with Nohel and never mentioned Nohel in his letters. Nor is anything known about any contributions Nohel might have made to science." Fölsing, *Albert Einstein*, 280.

91. Yeshayahu Nohel, "My Family," undated typescript, AEA 85–569.

92. Walter Kohn, "Mein verehrter Wiener Lehrer, Professor Doktor Emil Nohel," in Friedrich Stadler, ed., *Österreichs Umgang mit dem Nationalsozialismus: Die Folgen für die naturwissenschaftliche und humanistische Lehre* (Vienna: Springer, 2004), 43–50. On Kohn's time in Vienna with Nohel, see Andrew Zangwill, "The Education of Walter Kohn and the Creation of Density Functional Theory," *Archive for History of Exact Sciences* 68 (2014): 775–848, on 778–779.

93. Einstein to Emil Nohel, 9 May 1939, AEDA 54–127, Box 78, Folder "1937–39 N."

94. Einstein to Arthur E. Ruark, 16 July 1939, AEDA 54–129, Box 78, Folder "1937–39 N." This was in response to Ruark to Einstein, 6 July 1939, AEDA 54–128, Box 78, Folder "1937–39 N."

95. Einstein to Isidor Rabinovitz, 16 July 1939, AEDA 54–131; this was in response to Rabinovitz to Einstein, 8 July 1939, AEDA 54–130, Box 78, Folder "1937–39 N."

96. H. H. Higbie to Einstein, 1 August 1939, AEDA 54–132; Einstein to Higbie, August 1939, AEDA 54–133; Louis Margolis to Einstein, 31 August 1939, AEDA 54–135, Box 78, Folder "1937–39 N."

97. Emil Nohel to Yeshayahu Nohel, 20 June 1941 and 12 October 1941, YVS, Record Group O.7.cz, Collection on the Protectorate of Bohemia and Moravia, File 352, p. 1.

98. Emil Nohel to Yeshayahu Nohel, 20 June 1941 and 12 October 1941, YVS, Record Group O.7.cz, Collection on the Protectorate of Bohemia and Moravia, File 352, p. 3.

99. Emil Nohel to Yeshayahu Nohel, 26 December 1941, YVS, Record Group O.7.cz, Collection on the Protectorate of Bohemia and Moravia, File 352, p. 10. The most important secondary source on this camp remains H. G. Adler, *Theresienstadt 1941–1945: Das Antlitz einer Zwangsgemeinschaft*, 2nd. ed. (Göttingen: Wallstein, 2005 [1960]). Adler was himself a member of the generation of Germanophone Prague writers surrounding Max Brod.

100. Emil Nohel to Yeshayahu Nohel, 26 December 1941, YVS, Record Group O.7.cz, Collection on the Protectorate of Bohemia and Moravia, File 352, p. 14.

101. Emil Nohel to Yeshayahu Nohel, 2 January 1942, YVS, Record Group O.7.cz, Collection on the Protectorate of Bohemia and Moravia, File 352, p. 18. The last letter was Emil Nohel to Yeshayahu Nohel, 23–27 November 1942, LBI, Yehoshua Nohel Collection, AR 10231, p. 59.

102. "Ústřední kartoteka—transporty" for Emil Nohel, YVS.

103. Einstein to Fred R. Schwarz, 7 February 1946, AEA 89–295.

104. Einstein to Planck, 6 April 1933, AEDA 19–391, Box 24, Folder "M. Planck 1929–1934."

105. A parallel conflation can be seen in the case of Hitler, who repeatedly blended Czechs and Jews as plagues on the German *Volk*, an interpretation based on his reaction to local Viennese politics as a young man. See Brigitte Hamann, *Hitler's Vienna: A Portrait of the Tyrant as a Young Man* (New York: Tauris Parke, 2010 [1999]), 324.

106. Entry for 12 February 1954 in Hanna Fantova, "Gespräche mit Einstein," undated, HFC, Box 1, Folder 6, p. 21.

CHAPTER 7: FROM REVOLUTION TO NORMALIZATION

1. Emanuel Rádl, *Moderní věda: Její podstata, methody, výsledky* (Prague: Čin, 1926), 64.

2. Albert Einstein, *Theorie relativity speciální i obecná* (Prague: Fr. Borový, 1923), 7.

3. Heinrich Teweles, "Einstein," *Prager Tagblatt* 44 (18 December 1919).

4. Einstein to Heinrich Teweles, 23 December [1919], *CPAE* 9:231, on 323.

5. Message for the 10th anniversary of the Prague "Urania," 26 February 1927, published in *Urania* 4, no. 3/4 (March–April 1927): 9, *CPAE* 15:491, on 774.

6. This dynamic is discussed in greater detail in Michael D. Gordin, *Scientific Babel: How Science Was Done Before and After Global English* (Chicago: University of Chicago Press, 2015).

7. Joseph Kalousek, *Geschichte der kön. böhmischen Gesellschaft der Wissenschaften: Sammt einer kritischen Übersicht ihrer Publicationen aus dem Bereiche der Philosophie, Geschichte und Philologie* (Prague: Kön. böhm. Gesellschaft der Wissenschaften, 1884), 23.

8. "Deutsche Geselleschaft der Wissenschaften und Künste für die Tschechoslowakische Republik in Prag," in Otto Kletzl, ed., *Von deutscher Kultur in der Tschechoslowakei* (Eger: Johannes Stauda, [1928]), 126–135, on 126. See also Joseph Frederick Zacek, "The *Virtuosi* of Bohemia: The Royal Bohemian Society of Sciences," *East European Quarterly* 2, no. 2 (June 1968): 147–159, on

157; Stanley Z. Pech, *The Czech Revolution of 1848* (Chapel Hill: University of North Carolina Press, 1969), 30–31; Soňa Štrbáňová, "Patriotism, Nationalism and Internationalism in Czech Science: Chemists in the Czech Revival," in Mitchell G. Ash and Jan Surman, eds., *The Nationalization of Scientific Knowledge in the Habsburg Empire, 1848–1918* (Basingstoke: Palgrave Macmillan, 2012), 138–156, esp. 141. On Czech scientific societies in this period, see idem, "Congresses of the Czech Naturalists and Physicians in the Years 1880–1914 and the Czech-Polish Scientific Collaboration," in *Studies of Czechoslovak Historians for the 18th International Congress of the History of Science* (Prague: Institute of Czechoslovak and General History CSAS, 1989), 79–122.

9. Irena Seidlerová, "Science in a Bilingual Country," in Mikuláš Teich, ed., *Bohemia in History* (Cambridge: Cambridge University Press, 1998), 229–243, on 236–237.

10. Ivan Úlehla, "One Hundred and Fifty Years since the Birth of Ernst Mach," in Václav Prosser and Jaroslav Folta, eds., *Ernst Mach and the Development of Physics (Conference Papers)* (Prague: Karolinum, 1991), 25–65, on 48–59; Dieter Hoffmann, "Ernst Mach und die Teilung der Prager Universität," in Hans Lemberg, ed., *Universitäten in nationaler Konkurrenz: Zur Geschichte der Prager Universitäten im 19. und 20. Jahrhundert* (Munich: R. Oldenbourg, 2003), 33–61, on 45. On his fluency in Czech and willingness to speak it with his students after class, see Vladimír Novák, *Vzpomínky a paměti (Životopis)* (Brno: Tiskl Pokorný, 1939), 82.

11. Emilie Těšínská, "Profilování teoretické fyziky na pražské univerzitě a vazby s pražským působením A. Einsteina před 100 lety," *Pokroky matematiky, fyziky a astronomie* 57, no. 2 (2012): 146–168; Ladislav Zachoval, "Česká fysika před rokem 1918," *Sborník pro dějiny přírodních věd a techniky* 1 (1954): 37–47; J. Folta, M. Rotter, and E. Těšínská, *Fyzika na Karlově Univerzitě* (Prague: Univerzita Karlova, 1988); and Břetislav Fajkus, "The Faculty of Natural Sciences 1918–1945," in František Kavka and Josef Petráň, eds., *A History of Charles University*, 2 vols. (Prague: Karolinium, 2001), 2:235–241, on 236.

12. Jan Surman, "Science and Its Publics: Internationality and National Languages in Central Europe," in Ash and Surman, *Nationalization of Scientific Knowledge*, 30–56; Štrbáňová, "Patriotism, Nationalism and Internationalism"; Seidlerová, "Science in a Bilingual Country."

13. Jiří Kořalka, *Tschechen im Habsburgerreich und in Europa 1815–1914: Sozialgeschichtliche Zusammenhänge der neuzeitlichen Nationsbildung und der Nationalitätenfrage in den böhmischen Ländern* (Vienna: Verlag für Geschichte und Politik, 1991), 283–284.

14. Vilém Mathesius, *Kulturní aktivismus: Anglické paralely k českému životu* (Prague: Gustav Voleský, 1925), 88.

15. Jaroslav Batušek, "Zur Problematik der deutsch-tschechischen Beziehungen im Bereich der Geschichte der tschechischen physikalischen Terminologie," in B. Havránek and R. Fischer, eds., *Deutsch-tschechische Beziehungen im Bereich der Sprache und Kultur: Aufsätze und Studien, II, Abhandlungen der Sächsischen Akademie der Wissenschaften zu Leipzig, Philologisch-historische Klasse*, Band 59, Heft 2 (Berlin: Akademie Verlag, 1968), 85–95; Luboš Nový, ed., *Dějiny exaktních věd v českých zemích do konce 19. století* (Prague: Nakl. ČSAV, 1961); idem, "Entwicklungslinien der tschechischen Mathematik 1850–1918," tr. H. Laitko, in Horst Kant, ed., *Fixpunkte—Wissenschaft in der Stadt und der Region: Festschrift für Hubert Laitko anläßlich seines 60. Geburtstages* (Berlin: Verlag für Wissenschafts- und Regionalgeschichte Dr. Michael Engel, 1996), 117–134; Blanka Ondráčková, "Zaměření českých chemických časopisů v 2. polovine 19. století," *Dějiny věd a techniky* 6 (1973): 215–228.

16. Gerald Druce, *Two Czech Chemists: Bohuslav Brauner (1855–1935), Frantisek Wald (1861–1930)* (London: New Europe Publishing Company, 1944), 11, 40; American Institute in Czechoslovakia, *Czechoslovak Literature and Science* (Prague: Orbis, 1935); Franco Calascibetta, "Chemistry in Czechoslovakia Between 1919 and 1939: J. Heyrovský and the Prague Polarographic School," *Centaurus* 39 (1997): 368–381.

17. Andrea Orzoff, *Battle for the Castle: The Myth of Czechoslovakia in Europe, 1914–1948* (New York: Oxford University Press, 2009); Ines Koeltzsch, *Geteilte Kulturen: Eine Geschichte der tschechisch-jüdisch-deutschen Beziehungen in Prag (1918–1938)* (Munich: Oldenbourg, 2012). For population numbers, see *Ottův slovník naučný nové doby*, vol. 1/2 (Prague: Jan Otto, 1931), 1082–1083.

18. For contemporary justifications of the Lex Mareš, see *Universitas Carolina* (Prague: [Orbis], 1934), 10–12; and *L'Université Charles IV dans le passé et dans le présent* (Prague: Pražska Akciová Tiskárna, 1923). On the implications of the German University being declared new, see Petr Lozoviuk, *Interethnik im Wissenschaftsprozess: Deutschsprachige Volkskunde in Böhmen und ihre gesellschaftlichen Auswirkungen* (Leipzig: Leipziger Universitätsverlag, 2008), 94.

19. Hans-Joachim Härtel, "Die beiden Philosophischen Fakultäten in Prag im Spiegel ihrer Dissertationen 1882–1939/45," in *Die Teilung der Prager Universität 1882 und die intellektuelle Desintegration in den böhmischen Ländern: Vorträge der Tagung des Collegium Carolinum in Bad Wiessee vom 26. bis 28. November 1982* (Munich: R. Oldenbourg, 1984), 81–94; *Přírodovědecká fakulta Univerzity Karlovy 1920–1980: Dějiny, současnost, perspektivy* (Prague: Univerzita Karlova, 1981). The latter book does not even mention the German University, although it does refer to Einstein.

20. *Universitas Carolina*, 7, 9.

21. Herbert Cysarz, *Prag im deutschen Geistesleben: Blicke durch ein Jahrtausend* (Mannheim-Sandhofen: Kessler Verlag, 1961), 52; Ferdinand Seibt, *Mittelalter und Gegenwart: Ausgewählte Aufsätze: Festgabe zu seinem 60. Geburtstag*, ed. Winfried Eberhard and Heinz-Dieter Heimann (Sigmaringen: Jan Thorbecke Verlag, 1987), 346; Jiří Pešek, "Die Prager Universitäten im ersten Drittel des 20. Jahrhunderts: Versuch eines Vergleichs," in Lemberg, *Universitäten in nationaler Konkurrenz*, 145–166, on 156; Ota Konrád, "Die Deutsche Universität Prag in der Ersten Tschechoslowakischen Republik—Zwischen Kooperation und Konfrontation," in Elmar Schübl and Harald Heppner, eds., *Universitäten in Zeiten des Umbruchs: Fallstudien über das mittlere und östliche Europa im 20. Jahrhundert* (Vienna: Lit Verlag, 2011), 29–42.

22. Alena Míšková, "Die Lage der Juden an der Prager Deutschen Universität," in Jörg K. Hoensch, Stanislav Biman, and L'ubomír Lipták, eds., *Judenemanzipation–Antisemitismus–Verfolgung in Deutschland, Österreich-Ungarn, den Böhmischen Ländern und in der Slowakei* (Essen: Klartext, 1998), 117–129; Bruno Blau, "Nationality among Czechoslovak Jewry," *Historia Judaica* 10 (1948): 147–154, on 153–154; Pešek, "Die Prager Universitäten im ersten Drittel des 20. Jahrhunderts," 156; Dana Kasperová, "Integrationsfrage der jüdischen Bevölkerung und das Schulwesen am Anfang des 20. Jahrhunderts," in Steffen Höhne and Ludger Udolph, eds., *Deutsche–Tschechen–Böhmen: Kulturelle Integration und Desintegration im 20. Jahrhundert* (Köln: Böhlau, 2010), 63–71, on 65–66. On Steinherz, see Jiří Pešek, Alena Míšková, Petr Svobodný, and Jan Janko, "The German University of Prague 1918–1939," in Kavka and Petráň, *History of Charles University*, 2:245–256, on 246–247.

23. Philipp Frank to Einstein, 30 May [1919], *CPAE* 9:49, on 77.

24. Philipp Frank, *Einstein: His Life and Times*, tr. George Rosen, ed. and rev. Shuichi Kusaka (Cambridge, MA: Da Capo Press, 2002 [1947]), 77. Frank also attacked Lampa for being hyper-nationalistic and anti-Czech, despite the fact that he himself was of Czech heritage (81).

25. On Lampa's biography, see Andreas Kleinert, "Anton Lampa and Albert Einstein: Die Neubesetzung der physikalischen Lehrstühle an der deutschen Universität Prag 1909 und 1910," *Gesnerus* 32, no. 3/4 (1975): 285–292; idem, *Anton Lampa, 1868–1938: Eine Biographie und eine Bibliographie seiner Veröffentlichungen* (Mannheim: Bionomica, 1985); and Walter Hofmann, "Anton Lampa zum Gedächtnis," *Werk und Wille* 5, no. 2/3 (1938): 33–36. It is worth bearing in mind that the last appeared in a right-wing journal and especially celebrated Lampa's pro-German patriotism.

26. Anton Lampa to Einstein, [19 January 1920], *CPAE* 9:267, on 365. Einstein complied in a letter to Rudolf Wegscheider, 20 January 1920, *CPAE* 9:269, on 367.

27. Lampa to Einstein, 3 March 1920, *CPAE* 9:338, on 461.

28. Lampa to Einstein, [19 May 1920], *CPAE* 10:39, on 286.

29. See the discussion about Bergmann's bona fides in Peter Gabriel Bergmann to Einstein, 14 March 1936, AEDA 6–222, Box 6, Folder "P. Bergmann (1936–1938)"; Philipp Frank to Einstein, [Spring 1936?], AEDA 11–082; Einstein to Frank, 4 July 1936, AEDA 11–080; and Frank to Einstein, 4 August 1936, AEDA 11–084, Box 12, Folder "Ph. Frank F-Misc (II)."

30. Philipp Frank, *Modern Science and Its Philosophy* (New York: George Braziller, 1955), 45.

31. On the reception of relativity theory in Habsburg Bohemia and Czechoslovakia, see Rudolf Kolomý, "Ohlas vzniku teorie relativity ve fyzice v Československu," *Dějiny věd a techniki* 12 (1979): 209–225. Záviška was the first Czech theorist to take relativity seriously. The equivalent for quantum theory was Viktor Trkal. See Břetislav Fajkus, "Přírodovědecká fakulta 1920–1945," in Jan Havránek and Zdeněk Pousta, eds., *1918–1990*, vol. 4 of *Dějiny Univerzity Karlovy* (Prague: Univerzita Karlova, 1998), 163–180, on 167.

32. Augustin Žáček, "O principu relativity ve fysice," *Živa* 21 (1911): 135–137, on 136. On Žáček, see Těšínská, "Profilování teoretické fyziky na pražské univerzitě a vazby s pražským působením A. Einsteina před 100 lety," 163.

33. Arnošt Dittrich, "Princip relativnosti," *Časopis pro pěstování matematiky a fysiky* 43 (1914): 43–53, 200–211; Jul. Suchý, "Princip relativity," *Přehled* 9 (1911): 401–403, 419–421; Augustin Žáček, "Odvození Einsteinova addičního theoremu pro skládání rychlosti v případě rychlostí parallelních," *Časopis pro pěstování matematiky a fysiky* 41 (1912): 538–541.

34. František Nachtikal, *Princip relativity: Názorný výklad* (Brno: A. Píša, 1921), 4. See also the reference on 50. On anti-Einstein critiques, see 58–59 and 113.

35. Arnošt Dittrich, *O principu relativnosti, nové teorii světa, 4-rozměrna* (Třeboň: Karl Brandeis, 1922). See also idem, "Problém gravitace," *Časopis pro pěstování matematiky a fysiky* 52, no. 4 (1923): 387–402. On Dittrich's work in general, see A. Dratvová, "K osmdesátým narozeninám Arnošta Dittricha," *Pokroky matematiky, fyziky a astronomie* 3, no. 3 (1958): 366–368.

36. Rádl, *Moderní věda*, 226. On the context of philosophy in Czechoslovakia, including Rádl, see Jean Patocka, "La philosophie en Tschécoslovaquie et son orientation actuelle," *Études philosophiques* (n.s.) 3, no. 1 (January/March 1948): 63–74.

37. František Nachtikal, "Einsteinův posuv spektrálních car slunečních," *Časopis pro pěstování matematiky a fysiky* 53, no. 4 (1924): 414–416, on 416.

38. On Zavíška's biography, see the excellent study by Emilie Těšínská, "František Záviška (1879–1945) Physiker: Ein großer Verlust für die tschechische Physik," in Monika Glettler and Alena Míšková, eds., *Prager Professoren 1938–1948: Zwischen Wissenschaft und Politik* (Essen: Klartext, 2001), 483–511; as well as idem, "Profilování teoretické fyziky na pražské univerzitě a vazby s pražským působením A. Einsteina před 100 lety"; and Miroslav Brdička, "Dílo profesora dr. Františka Závišky," *Československý časopis pro fyziku* A20 (1970): 673–680.

39. Miroslav Brdička, "Einstein a Praha: Česká einsteinovská pohlednice," *Československý časopis fyziky* A29 (1979): 269–275, on 274. The lack of contact with any Czech colleagues is confirmed by Philipp Frank, *Einstein: Sein Leben und seine Zeit* (Munich: Paul List, 1949), 139.

40. Těšínská, "Profilování teoretické fyziky na pražské univerzitě a vazby s pražským působením A. Einsteina před 100 lety," 160–161.

41. Idem, "František Záviška (1879–1945) Physiker," 492. See also Reinhold Fürth, "Reminiscences of Philipp Frank at Prague," in Robert S. Cohen and Marx W. Wartofsky, eds., *In Honor of Philipp Frank*, vol. 2 of *Boston Studies in the Philosophy of Science* (New York: Humanities Press, 1965), xiii–xvi, on xv.

42. This is especially true of František Záviška, "O principu relativnosti," part 1, *Časopis pro pěstování mathematiky a fysiky* 43, no. 3–4 (1914): 363–395; part 2, *Časopis pro pěstování mathematiky a fysiky* 43, no. 5 (1914): 564–593; and idem, "K Einsteinově teorii gravitační," *Časopis pro pěstování matematiky a fysiky* 51, no. 1 (1922): 58–60.

43. Idem, Review of Jindřich Skokan, *Einsteinův princip relativity* (1920), *Časopis pro pěstování mathematiky a fysiky* 50, no. 2–3 (1921): 160–167, on 160.

44. Ibid., 164.

45. *Nové Atheneum* 2, no. 2 (1920): 214–215, on 215.

46. Bohuslav Hostinský, "Einsteinovy přednášky o theorii relativity z r. 1921," *Časopis pro pěstování matematiky a fysiky* 53, no. 3 (1924): 308–319, on 318.

47. František Záviška, "K předešlému článku prof. Hostinského," *Časopis pro pěstování matematiky a fysiky* 53, no. 3 (1924): 319–324. On the Mount Wilson findings, see Jeffrey Crelinstein, *Einstein's Jury: The Race to Test Relativity* (Princeton, NJ: Princeton University Press, 2006).

48. Bohuslav Hostinský, "Doslov k článku o Einsteinových přednáškách," *Časopis pro pěstování matematiky a fysiky* 53, no. 4 (1924): 401–403; František Záviška, "K Doslovu prof. Hostinského," *Časopis pro pěstování matematiky a fysiky* 53, no. 4 (1924): 403–404.

49. František Záviška, *Einsteinův princip relativnosti a teorie gravitační* (Prague: Jednoty Čs. matematiku a fysiků, 1925).

50. Těšínská, "František Záviška (1879–1945) Physiker," 507.

51. Jaroslav Cesar and Bohumil Černý, "Die deutsche antifaschistische Emigration in der Tschechoslowakei (1933–1934)," tr. Jiří Wehle, *Historica* 12 (1966): 147–184; Bohumil Černý, *Most k novému životu: Německá emigrace v ČSR v letech 1933–1939* (Prague: Lidová demokracie, 1967); S. P. Postnikov, ed., *Russkie v Prage, 1918–1928 g.g.* (Prague: Legografie, 1928).

52. Claudio Magris, "Prag als Oxymoron," *Neohelicon* 7, no. 2 (1979/1980): 11–65, on 60–61.

53. Folta, Rotter, and Těšínská, *Fyzika na Karlově Univerzitě*, 37; Peter G. Bergmann, "Homage to Professor Philipp G. Frank," in Cohen and Wartofsky, *In Honor of Philipp Frank*, ix–x, on ix.

54. Frank to Einstein, [Spring 1936?], AEDA 11–082, Box 12, Folder "Ph. Frank F-Misc (II)."

55. For example, he asked her to solicit manuscripts from Czech colleagues for the new logical empiricist journal *Erkenntnis*, provided they were in French, German, or English. P. Frank to A. Dratvová, 19 August 1938, reproduced in Václav Podaný, "Philipp Frank, Albína Dratvová, Jaroslav Heyrovský (Mnichov 1938 a poválečné osudy)," *Dějiny věd a techniky* 28, no. 3 (1995): 129–143, on 133.

56. Frank to Dratvová, 1 September 1938, reproduced in Podaný, "Philipp Frank, Albína Dratvová, Jaroslav Heyrovský," 136.

57. The controversy is expertly detailed in Chad Bryant, *Prague in Black: Nazi Rule and Czech Nationalism* (Cambridge, MA: Harvard University Press, 2007).

58. Theodore Procházka, Sr., *The Second Republic: The Disintegration of Post-Munich Czechoslovakia (October 1938–March 1939)* (Boulder, CO: East European Monographs, 1981), 53.

59. Einstein to Besso, 10 October 1938, in Albert Einstein and Michele Besso, *Correspondance 1903–1955*, ed. Pierre Speziali (Paris: Hermann, 1972), 330. Einstein was almost certainly unaware that the Latin phrase he interjected—*O sancta simplicitas*, meaning "O holy simplicity!" and used to chastise the naïve—is attributed to the Bohemian religious reformer Jan Hus, who uttered it to a peasant woman who was adding wood to the fire consuming him at the stake.

60. Einstein to Solovine, 23 December 1938, in Albert Einstein, *Letters to Solovine* (New York: Philosophical Library, 1987), 76.

61. Frank to Dratvová, 25 October [1938], reproduced in Podaný, "Philipp Frank, Albína Dratvová, Jaroslav Heyrovský," 136.

62. Ibid., 137.

63. P. Frank to J. Heyrovský, 31 October 1938, reproduced in Podaný, "Philipp Frank, Albína Dratvová, Jaroslav Heyrovský," 138–139.

64. Elizabeth Wiskemann, *Czechs and Germans: A Study of the Struggle in the Historic Provinces of Bohemia and Moravia*, 2nd. ed. (London: Macmillan, 1967 [1938]), 214.

65. Jan Krčmář, *The Prague Universities: Compiled According to the Sources and Records* (Prague: Orbis, 1934), 12.

66. Konrad Bittner, *Deutsche und Tschechen: Zur Geistesgeschichte des böhmischen Raumes* (Brno: Rudolf M. Rohrer, 1936), 1:103–104; Gray C. Boyce and William H. Dawson, *The University of Prague: Modern Problems of the German University in Czechoslovakia* (London: Robert Hale, 1938).

67. Diary entry of 16 March 1939, reproduced in Schmuel Hugo Bergman, *Tagebücher & Briefe*, 2 vols., ed. Miriam Sambursky (Königstein: Athenäum, 1985), 1:499.

68. Mark Mazower, *Hitler's Empire: How the Nazis Ruled Europe* (New York: Penguin, 2008), 185.

69. Shiela Grant Duff, *A German Protectorate: The Czechs under Nazi Rule* (London: Frank Cass and Co., 1970 [1942]), 48.

70. Karl Hans Strobl, *Prag: Schicksal, Gestalt und Seele einer Stadt* (Prague: Buchhandlung Czerny, 1943), 16. The last three sentences were added to the 1939 edition, in celebration of Nazi occupation. The general point was reinforced that same year in Wolfgang Wolfram von Wolmar, *Prag und das Reich: 600 Jahre Kampf deutscher Studenten* (Dresden: Franz Müller Verlag, 1943), 12, 32, and passim.

71. Peter Demetz, *Prague in Danger: The Years of German Occupation, 1939–45* (New York: Farrar, Straus and Giroux, 2008), 35. On the limbo into which Jews fell during the Second Republic,

see Heinrich Bodensieck, "Das Dritte Reich und die Lage der Juden in der Tschecho-Slowakei nach München," *Vierteljahrshefte für Zeitgeschichte* 9, no. 3 (July 1961): 249–261.

72. Hans Lemberg, "Universität oder Universitäten in Prag—und der Wandel der Lehrsprache," in Lemberg, *Universitäten in nationaler Konkurrenz*, 19–32, on 32.

73. V. Patzak, "The Caroline University of Prague," *Slavonic and East European Review* 19, no. 53/54 (1939–1940): 83–95, on 95.

74. Bryant, *Prague in Black*, 49; Duff, *German Protectorate*, 190–191; Monika Glettler, "Tschechische, jüdische und deutsche Professoren in Prag: Möglichkeiten und Grenzen biographischer Zugänge," in Glettler and Míšková, *Prager Professoren 1938–1948*, 13–26, on 15. On the pressing of the university into wartime service and other scientific projects under the Protectorate, see Alena Míšková, "Německá univerzita za druhé světové války," in Havránek and Pousta, *Dějiny Univerzity Karlovy*, 4:213–231; and Miloš Hořejš and Ivana Lorencová, eds., *Věda a technika v českých zemích v období 2. světové války* (Prague: Národní technické muzeum, 2009).

75. Bryant, *Prague in Black*, 3; idem, "Either German or Czech: Fixing Nationality in Bohemia and Moravia, 1939–1946," *Slavic Review* 61, no. 4 (Winter 2002): 683–706. On Jews in postwar Czechoslovakia, see Petr Brod, "Die Juden in der Nachkriegstschechoslowakei," in Hoensch, Biman, and Lipták, *Judenemanzipation–Antisemitismus–Verfolgung*, 211–228. Many of those who survived soon emigrated to Palestine and then (after 1948) Israel.

76. The literature on the expulsions is vast. For a survey of the major debates, and for statistics, see Benjamin Frommer, *National Cleansing: Retribution Against Nazi Collaborators in Postwar Czechoslovakia* (Cambridge: Cambridge University Press, 2005); Wolfgang Benz, ed., *Die Vertreibung der Deutschen aus dem Osten: Ursachen, Ereignisse, Folgen* (Frankfurt am Main: Fischer, 1985); Tomáš Staněk, *Tábory v českých zemích, 1945–1948* (Ostrava: Nakl. Tilia, 1996); idem, *Německá menšina v českých zemích, 1948–1989* (Prague: Institut pro středoevropskou kulturu a politiku, 1993); Eagle Glassheim, *Cleansing the Czechoslovak Borderlands: Migration, Environment, and Health in the Former Sudetenland* (Pittsburgh: University of Pittsburgh Press, 2016); Alfred-Maurice de Zayas, *The German Expellees: Victims in War and Peace*, tr. John A. Koehler (New York: St. Martin's Press, 1993 [1986]); and idem, *Nemesis at Potsdam: The Anglo-Americans and the Expulsion of the Germans: Background, Execution, Consequences*, rev. ed. (London: Routledge & Kegan Paul, 1979 [1977]). On the terminological issues, see Ronald M. Smelser, "The Expulsion of the Sudeten Germans: 1945–1952," *Nationalities Papers* 24, no. 1 (1996): 79–92, on 79; and Bradley F. Abrams, "Morality, Wisdom and Revision: The Czech Opposition of the 1970s and the Expulsion of the Sudeten Germans," *East European Politics and Societies* 9, no. 2 (Spring 1995): 234–255, on 244–245.

77. For a contemporary justification of the closure of the German University, see Otakar Odložilík, *The Caroline University: 1348–1948* (Prague: [Orbis], 1948).

78. H. Gordon Skilling, "The Formation of a Communist Party in Czechoslovakia," *American Slavic and East European Review* 14, no. 3 (October 1955): 346–358.

79. Paul E. Zinner, "Problems of Communist Rule in Czechoslovakia," *World Politics* 4, no. 1 (October 1951): 112–129, on 119; Bradley F. Abrams, *The Struggle for the Soul of the Nation: Czech Culture and the Rise of Communism* (Lanham, MD: Rowman & Littlefield, 2004), 57. See also Paul E. Zinner, *Communist Strategy and Tactics in Czechoslovakia, 1918–48* (New York: Praeger, 1963).

80. Bryant, *Prague in Black*, 259.

81. Arnošt Kolman, *Die verirrte Generation: So hätten wir nicht leben sollen. Eine Autobiographie*, 2nd. exp. ed., ed. Hanswilhelm Haefs and František Janouch, tr. Elisabeth Mahler-Berger (Frankfurt am Main: S. Fischer, 1982 [1979]), 265.

82. Martin Buber, *Tři řeči o židovství*, tr. A. Kollmann (*sic*) (Prague: Nákl. Spolku židovských akademiků Theodor Herzl, 1912). On the milieu of Czech-speaking Zionist leaders of this era, including Kolman, see S. Goshen, "Zionist Students' Organizations," in *The Jews of Czechoslovakia: Historical Studies and Surveys*, 3 vols. (Philadelphia: Jewish Publication Society of America, 1968–1984), 2:173–184, on 178.

83. On various editions of the memoirs, see Arnošt Kolman, *Die verirrte Generation: So hätten wir nicht leben sollen. Eine Biographie*, tr. Elisabeth Mahler-Berger (Frankfurt am Main: S. Fischer, 1979); Kolman, *Die verirrte Generation* (1982 ed.); E. Kol'man, *My ne dolzhny byli tak zhit'* (New York: Chalidze, 1982); and Arnošt Kolman, *Zaslepená generace: Paměti starého bolševika* (Brno: Host, 2005). Also useful are the recollections of Kolman's daughter—which are principally about her mother—though they rely extensively on the same memoirs for important events. Ada Kol'man, "Pamiat' ne stynet . . . (Vospominaniia docheri o pisatel'nitse Ekaterine Kontsevoi)," *Metsenat i mir*, no. 49–52 (2011), http://www.mecenat-and-world.ru/49-52/kolman.htm.

84. Kolman, *Die verirrte Generation* (1979 ed.), 34–36.

85. Kurt Marko, "No Juvenal of Bolshevism," tr. T. J. Blakeley, *Studies in Soviet Thought* 22, no. 2 (May 1981): 147–149, on 147.

86. For secondary accounts of Kolman's career, see Michael D. Gordin, "The Trials of Arnošt K.: The Dark Angel of Dialectical Materialism," *Historical Studies in the Natural Sciences* 47, no. 3 (2017): 320–348; Pavel Kovaly, "Arnošt Kolman: Portrait of a Marxist-Leninist Philosopher," *Studies in Soviet Thought* 12, no. 4 (December 1972): 337–366; idem, *Rehumanization or Dehumanization?: Philosophical Essays on Current Issues of Marxist Humanism in Arnost Kolman, György Lukács, Adam Schaff, Alexander Solzhenitsyn* (Boston: Branden Press, 1974), ch. 2; Yakov M. Rabkin, "On the Origins of Political Control over the Content of Science in the Soviet Union," *Canadian Slavonic Papers* 21, no. 2 (June 1979): 225–237; and Eugene Seneta, "Mathematics, Religion, and Marxism in the Soviet Union in the 1930s," *Historia Mathematica* 31 (2004): 337–367.

87. Kolman, *Die verirrte Generation* (1979 ed.), 74. This account is confirmed by the memoirs of Dutch-American mathematician Dirk Struik, who had heard it when he met Kolman in the 1920s when both were doing underground work in Germany. Struik, "[Personal Recollections]," [1973?], Struik Papers, MC 418, Box 1, Folder "Personal Recollections, 1973?," pp. 24–25.

88. Kolman, *Die verirrte Generation* (1979 ed.), 117.

89. Michal V. Simunek and Uwe Hossfeld, "Trofim D. Lysenko in Prague 1960: A Historical Note," *Istoriko-biologicheskie issledovaniia* 5, no. 2 (2013): 84–88, on 85.

90. Einstein, "Opinion on Engels' 'Dialectics of Nature,'" 30 June 1924, *CPAE* 14:277, on 414.

91. On dialectical materialism in this period, see Helena Sheehan, *Marxism and the Philosophy of Science: A Critical History: The First Hundred Years* (London: Verso, 2017 [1985]); and David Joravsky, *Soviet Marxism and Natural Science, 1917–1932* (London: Routledge and Kegan Paul, 1961).

92. Alexander Vucinich, *Einstein and Soviet Ideology* (Stanford, CA: Stanford University Press, 2001), 62–63, 80–81, 120; A. S. Sonin, *"Fizicheskii idealizm": Istoriia odnoi ideologicheskoi kampanii* (Moscow: Fiziko-Matematicheskaia Literatura, 1994), 41.

93. E. Kol'man, "K vystupleniiu Einshteina po voprosu o sovremennoi fizike," *Pod znamenem marksizma*, no. 12 (1940): 100–105, on 100. This article was a direct response to a Soviet reprint and translation of the physicist's lecture at the 8th Panamerican Congress in Washington, DC, on 5 May 1940: A. Einstein, "Soobrazheniia k obosnovaniiu teoreticheskoi fiziki," *Pod znamenem marksizma*, no. 12 (1940): 106–113. For similar statements, see E. Kol'man, "Vozrozhdenie pifagoreizma v sovremennoi fizike," *Pod znamenem marksizma*, no. 8 (1938): 138–160, on 151; and idem, "Teoriia kvant i dialekticheskii materializm," *Pod znamenem marksizma*, no. 10 (1939): 129–145, on 138.

94. Idem, "Khod zadom filosofii Einshteina," *Nauchnoe slovo* (1931): 11–15, on 13.

95. Reproduced in idem, *Die verirrte Generation* (1982 ed.), 355. The fullest exposition of this view is idem, "Teoriia otnositel'nosti i dialekticheskii materializm," *Pod znamenem marksizma*, no. 6 (1939): 106–120.

96. On the situation in mathematics, see Loren Graham and Jean-Michel Kantor, *Naming Infinity: A True Story of Religious Mysticism and Mathematical Creativity* (Cambridge, MA: Belknap Press, 2009); and Aleksey E. Levin, "Anatomy of a Public Campaign: 'Academician Luzin's Case' in Soviet Political History," *Slavic Review* 49, no. 1 (Spring 1990): 90–108.

97. E. Kol'man, "Vreditel'stvo v nauke," *Bol'shevik*, no. 2 (1931): 73–81; and idem, "Pis'mo tov. Stalina i zadachi fronta estestvoznaniia i meditsiny," *Pod znamenem marksizma*, no. 9–10 (1931): 163–172.

98. Zdeněk Dittrich, "Die Prager Tschechische Universität 1945–1948 in meiner Erinnerung," in Glettler and Míšková, *Prager Professoren 1938–1948*, 657–661, on 658; Antonín Kostlán, "Die Prager Professoren in den Jahren 1945 bis 1950: Versuch einer prosopographischen Analyse," in Glettler and Míšková, *Prager Professoren 1938–1948*, 605–655, on 612; N. Lobkowicz, *Marxismus-Leninismus in der ČSR: Die tschechoslowakische Philosophie seit 1945* (Dordrecht: D. Reidel, 1961), 19; John Connelly, *Captive University: The Sovietization of East German, Czech, and Polish Higher Education, 1945–1956* (Chapel Hill: University of North Carolina Press, 2000), 33; František Janouch, "Portrét Arnošta Kolmana," in Janouch and Arnošt Kolman, *Jak jste tak mohli žít?: Dialog generací* (Prague: Novela bohemica, 2011), 17; Abrams, *Struggle for the Soul*, 59, 192; Peter Hruby, *Fools and Heroes: The Changing Role of Communist Intellectuals in Czechoslovakia* (Oxford: Pergamon, 1980), 207 (see also 187); Zinner, *Communist Strategy and Tactics*, 128.

99. Arnošt Kolman, *Diskuse s univ. prof. dr. Arnoštem Kolmanem* (Prague: Orbis, 1946); idem, *Ideologie německého fašismu* (Prague: Svoboda, 1946).

100. Kolman, *Die verirrte Generation* (1979 ed.), 191–192.

101. Zinner, "Problems of Communist Rule," 122; D. E. Viney, "Czech Culture and the 'New Spirit,' 1948–52," *Slavonic and East European Review* 31, no. 77 (June 1953): 466–494, on 467n2, 480n40.

102. S. S. Ilizarov, "Ernest Kol'man, Nikita Khrushchev i IIET," in *80 let Institutu istorii nauki i tekhniki: 1932–2012* (Moscow: Izd. RTSoft, 2012), 198–205; I. R. Grinina and S. S. Ilizarov, eds., "Dokumenty o prebyvanii Ernesta Kol'mana v Institute istorii estestvoznaniia i tekhniki AN SSSR," *Voprosy istorii estestvoznaniia i tekhniki*, no. 1 (1998): 156–161.

103. E. Kol'man, *Velikii russkii myslitel' N. I. Lobachevskii*, 2nd. ed. (Moscow: Gos. izd. politicheskoi literatury, 1956); idem, *Bernard Bolzano*, ed. Alfred Händel and Günther Höpfner (Berlin: Akademie-Verlag, 1963); idem, "Chto takoe kibernetika?," *Voprosy filosofii*, no. 4 (1955): 148–159;

idem, *Kibernetika (O mashinakh, vypolniaiushchikh nekotorye psikhicheskie funktsii cheloveka)* (Moscow: Znanie, 1956); idem, "The Adventures of Cybernetics in the Soviet Union," *Minerva* 16, no. 3 (Autumn 1978): 416–424.

104. Idem, "Kuda vedet fizikov sub"ektivizm," *Voprosy filosofii*, no. 6 (1953): 173–189; idem, "K sporam o teorii otnositel'nosti," *Voprosy filosofii*, no. 5 (1954): 178–189; idem, *Lenin i noveishaia fizika* (Moscow: Gos. izd. politicheskoi literatury, 1959), 23, 44–48, 117; idem, *Filosofskie problemy sovremennoi fiziki* (Moscow: Znanie, 1957), 9.

105. See Josef Zeman, "Ještě k problémům rozpornosti pohybu," *Filosofický časopis* 8 (1960): 240–243, on 242; and Zdeněk Mlynář, "Filosofie a praxe," *Nová mysl* 3 (March 1960): 287–299.

106. Arnošt Kolman, speech at the plenary session of Czechoslovak writers, 10 December 1962, MÚA AV ČR, fond Drobné fondy, karton 5, složka Kolman Arnošt, p. 8.

107. H. Gordon Skilling, *Czechoslovakia's Interrupted Revolution* (Princeton, NJ: Princeton University Press, 1976), 577n51.

108. Kolman, *Die verirrte Generation* (1979 ed.), 247.

109. Arnošt Kolman, "Proti dogmatismu v naší filosofii," *Filosofický časopis* 11, no. 2 (1963): 222–227, on 223.

110. Milan Šimečka, *The Restoration of Order: The Normalization of Czechoslovakia, 1969–1976* (London: Verso, 1984).

111. E. Kolman to Struik, postcard of 28 May 1971, Struik Papers, MC 418, Box 7, Folder "K–M"; Kol'man, *My ne dolzhny byli tak zhit'*, 357–358.

112. Ada Kolman-Janouch to Struik, 11 December 1974, Struik Papers, MC 418, Box 7, Folder "K–M."

113. E. Kolman, "The Philosophical Interpretation of Contemporary Physics," tr. E. M. Swiderski, *Studies in Soviet Thought* 21, no. 1 (February 1980): 1–14, on 14.

114. Arnosht Kolman, "Is Marching On," *New York Times* (13 October 1976): 38.

115. "I also do not belong to those who, from the fact that socialism has never so far been implemented anywhere except through being foisted by an oligarchy, draw the false conclusion that it could never be implemented. I very much want to hope that it will work out with the Eurocommunists." Kolman to Struik, 31 May 1977, Struik Papers, MC 418, Box 7, Folder "K–M."

116. See the letters from Kolman to Struik dated 30 October 1974 and 4 September 1975, Struik Papers, MC 418, Box 7, Folder "K–M"; and that from 4 September [1978] in Folder "K." For Cohen, see Kolman to Cohen, letters of 5 September and 18 October 1977, Cohen Papers, #1666, Box 29, Folder 18, "Kolman, Arnost." There are many other such letters in Struik's and Cohen's archives.

117. Robert S. Cohen to Arnost Kolman, 17 November 1976, Cohen Papers, #1666, Box 29, Folder 18, "Kolman, Arnost." See also Cohen to Kolman, 12 December 1975, Cohen Papers, #1666, Box 29, Folder 18, "Kolman, Arnost."

118. Katya Coleman to Cohen, 1 August 1978, Cohen Papers, #1666, Box 29, Folder 18, "Kolman, Arnost."

119. "Prof. Ernst [*sic*] Kolman, a Confidant of Lenin, Dies in Sweden at 85," *New York Times* (26 January 1979): A23. See also the biographical description in Kolman's posthumous "Gelehrter und Freiheitskämpfer," *Europäische Ideen* 48 (1980): 6–8. Earlier Western news stories, by contrast, did not mention any connection with Einstein: "Soviet Communist, Disillusioned at 84, Resigns from Party," *New York Times* (7 October 1976): 4.

120. Vladimir V. Kusin, *From Dubček to Charter 77: A Study of "Normalisation" in Czechoslovakia, 1968–1978* (Edinburgh: Q Press, 1978), 202.

121. František Lehar, "Praha zapomíná na Einsteina," *Mladá fronta* 25 (18 April 1969): 1. The obituaries were delayed—as in the Soviet Union—by the difficulty of categorizing Einstein ideologically. See, for example, Jiří Vrána, "Albert Einstein," *Pokroky matematiky, fysiky a astronomie* 2 (1957): 320–333.

122. Zdeněk Guth, "Einsteinovi a Smíchov," *Neděle s Lidová demokracie* 31, no. 15 (12 April 1975): 9–10. See also Rudolf Kolomý, "Albert Einstein a jeho vztah v Praze," *Pokroky matematiky, fysiky a astronomie* 17 (1972): 265–272.

123. Hugo Bergman, "Personal Remembrances of Albert Einstein," in Robert S. Cohen and Marx W. Wartofsky, eds., *Logical and Epistemological Studies in Contemporary Physics* (Dordrecht: D. Reidel, 1974), 388–394, on 394n1; Bergmann to E. Kolmann, 3 November 1974, reproduced in Bergman, *Tagebücher & Briefe*, 2:702.

124. Libor Pátý, "Pamětní deska Albertu Einsteinovi v Praze v Lesnické ulici," *Československý časopis pro fyziku* A29 (1979): 311.

125. Martin Černohorský and Jiří Komrska, "Pocta Einsteinovi—Praha 1979," *Československý časopis pro fyziku* A29 (1979): 309–311.

126. Jiří Bičák, ed., *Einstein a Praha: K stému výročí narození Alberta Einsteina* (Prague: Jednota československých matematiků a fyziků, 1979). The main Czech physics journal produced a special issue to memorialize the occasion, which included several historical articles: Brdička, "Einstein a Praha"; Jiří Bičák, "Einsteinova cesta k obecné teorii relativity," *Československý časopis pro fyziku* A29 (1979): 222–243; and Jan Fischer, Review of Jiří Bičák, *Einstein a Praha*, *Československý časopis pro fyziku* A29 (1979): 296–297.

127. Nancy M. Wingfield, *Flag Wars and Stone Saints: How the Bohemian Lands Became Czech* (Cambridge, MA: Harvard University Press, 2007); Michaela Marek, "Baudenkmäler im tschechoslowakischen Grenzland nach dem Zweiten Weltkrieg: Strategien der (Wieder-) Aneignung," in Höhne and Udolph, *Deutsche-Tschechen-Böhmen*, 193–229.

CONCLUSION: PRINCETON, TEL AVIV, PRAGUE

1. Max Brod, "Unmodernes Prag," *Die Aktion* 2 (1912): 944–949, on 944.

2. On Charter 77, see H. Gordon Skilling, *Charter 77 and Human Rights in Czechoslovakia* (London: George Allen & Unwin, 1981). On the events of 1989, see Timothy Garton Ash, *The Magic Lantern: The Revolution of '89 Witnessed in Warsaw, Budapest, Berlin and Prague* (New York: Vintage, 1999 [1990]). On the challenges of understanding what happened, especially in Czechoslovakia, as "dissidence," see Jonathan Bolton, *Worlds of Dissent: Charter 77, The Plastic People of the Universe, and Czech Culture under Communism* (Cambridge, MA: Harvard University Press, 2012).

3. Annotation in Karel Plicka, *Prag: Ein fotografisches Bilderbuch* (Prague: Artia, 1953), Princeton Rare Books and Special Collections (Ex) DB2620.P56 1953q.

4. Carl Seelig, *Albert Einstein und die Schweiz* (Zurich: Europa-Verlag, 1952), Princeton Rare Books and Special Collections (Ex) QC16.E5S33 1952.

5. On Einstein's postwar antipathy to speaking German with Germans, though he gladly did so with Russians, Hungarians, Americans, or others, see Michael D. Gordin, *Scientific Babel: How*

Science Was Done Before and After Global English (Chicago: University of Chicago Press, 2015), ch. 7.

6. Biographical information is drawn from assorted documents preserved in her personal papers, including an affidavit to the West German government requesting compensation for forced emigration and the back of a library catalogue card where she jotted down additional information. The details do not always agree. The relevant documents are a certificate of citizenship in the Czechoslovak Republic for Otto Fanta and his wife Jana Bobaschová; Mr. Kalies of the Federal Ministry of the Interior, Wiedergutmachungsbescheid to the application of Frau Hanna Fanta, Bonn, 8 January 1958; and information written on the back of a discarded library catalogue card; all in HFC C0703, Box 1, Folder 4. Almost nothing has been said about the Einstein–Fantova relationship in the scholarly literature; a rare Czech newspaper article has the general outlines correct but includes numerous errors of fact: Marie Homolová, "Einstein očima Johanny z Čech," *Lidové noviny* (15 May 2004): 20.

7. Jan Havránek, "Materiály k Einsteinovu Pražskému působení z Archivu Univerzity Karlovy," *Acta universitatis Carolinae—Historia universitatis Carolinae Pragensis* 22, no. 1 (1980): 109–134, on 114.

8. G. Kowalewski, *Bestand und Wandel: Meine Lebenserinnerungen zugleich ein Beitrag zur neueren Geschichte der Mathematik* (Munich: R. Oldenbourg, 1950), 249.

9. Translation of a document by Dr. Zámiš, head of the Criminal Department at the Police Headquarters in Prague, 9 December 1938, HFC C0703, Box 1, Folder 4. Fanta's passion for handwriting analysis prompted him to do one for Max Brod in a *Festschrift* and a "controlled experiment" using a sample from his mother. See, respectively, Otto Fanta, "Die Handschrift," in Felix Weltsch, ed., *Dichter, Denker, Helfer: Max Brod zum 50. Geburtstag* (Mähr.-Ostrau: Julius Kittls Nachfolger, 1934), 102–107; and idem, "Die Kontrollanalyse: Ein Beitrag zur Verifizierung graphologischer Gutachten," *Die Schrift* 2 (1936): 91–121.

10. Einstein to Elsa Einstein, [7 January 1921], *CPAE* 12:11, on 31.

11. Einstein to Humboldt-Film-Gesellschaft, 1 June 1922, *CPAE* 13:212, on 325. Einstein clarified in an interview in the *Berliner Tageblatt* that he had nothing to do with the film: "On the 'Einstein Film,'" *CPAE* 13:213, on 327. For details on its release at the Frankfurt Trade Fair in April 1922, see the editorial comments at *CPAE* 13 on 325n1. I would like to thank Devin Fore for sharing with me a copy of an extract of the film containing the animations.

12. E. Gehrcke, *Die Massensuggestion der Relativitätstheorie: Kulturhistorisch-psychologische Dokumente* (Berlin: Hermann Meusser, 1924), 45.

13. "Der Einsteinfilm," *Die Umschau* 26, no. 16 (16 April 1922): 247–249, on 248.

14. Hanna Fantova, Introduction to "Gespräche mit Einstein," undated, HFC C0703, Box 1, Folder 6, p. 3. The complete phone logs are still unpublished, but a summary of them can be accessed in Alice Calaprice, "Einstein's Last Musings," *Princeton University Library Chronicle* 65, no. 1 (Autumn 2003): 51–64.

15. Einstein to Fantova, 31 October 1938, HFC C0703, Box 1, Folder 5.

16. Einstein testimonial, 15 December 1938, HFC C0703, Box 1, Folder 5.

17. Hugo Bergmann to Luise Herrmann, 14 January 1941, reproduced in Schmuel Hugo Bergman, *Tagebücher & Briefe*, 2 vols., ed. Miriam Sambursky (Königstein: Athenäum, 1985), 1:550; Georg Gimpl, "Wäre dieser Krieg nicht gekommen! Hätten wir unser glückliches Leben vor 1914

weitergeführt!," in Gimpl, ed., *Weil der Boden selbst hier brennt . . . : Aus dem Prager Salon der Berta Fanta (1865–1918)* (Furth im Wald/Prague: Vitalis, [2000]), 275–371, on 312.

18. Hanna Fantova, Introduction to "Gespräche mit Einstein," 6.

19. Robert F. Goheen to Johanna Fantova, 27 June 1966, HFC C0703, Box 1, Folder 4. See also William Dix (Librarian of Princeton), September 1954, in HFC C0703, Box 1, Folder 4.

20. Hanna Fantova, Introduction to "Gespräche mit Einstein," 11.

21. Einstein to Besso, 8 August 1938, in Albert Einstein and Michele Besso, *Correspondance 1903–1955*, ed. and tr. Pierre Speziali (Paris: Hermann, 1972), 321.

22. Other Czechs also found him in Princeton. For the story of Rudi W. Mandl, an amateur scientist who had emigrated from Czechoslovakia, and the idea for gravitational lensing that he discussed with Einstein, see Jürgen Renn and Tilman Sauer, "Eclipses of the Stars: Mandl, Einstein, and the Early History of Gravitational Lensing," in Jürgen Renn, Lindy Divarci, Petra Schröter, Abhay Ashtekar, Robert S. Cohen, Don Howard, Sahotra Sarkar, and Abner Shimony, eds., *Revisiting the Foundations of Relativistic Physics: Festschrift in Honor of John Stachel* (Dordrecht: Kluwer Academic, 2003), 69–92; and Jürgen Renn, Tilman Sauer, and John Stachel, "The Origin of Gravitational Lensing: A Postscript to Einstein's 1936 *Science* Paper," *Science* 275, no. 5297 (10 January 1997): 184–186.

23. Diary entry of 9 November 1946, reproduced in Bergman, *Tagebücher & Briefe*, 1:705.

24. Bergmann to Luise Herrmann, 7 January 1949, reproduced in Bergman, *Tagebücher & Briefe*, 2:10.

25. Bergmann to Paul Amann, 11 August 1956, reproduced in Bergman, *Tagebücher & Briefe*, 2:222.

26. Diary entry of 29 November 1966/67, reproduced in Bergman, *Tagebücher & Briefe*, 2:515. On his learning that Frank was dead, see the diary entry of 21 January 1967, reproduced in Bergman, *Tagebücher & Briefe*, 2:522.

27. Diary entry of 23 October 1969, reproduced in Bergman, *Tagebücher & Briefe*, 2:588. He only learned of Kraus's death a few years later: diary entry of 11 March 1972, reproduced in Bergman, *Tagebücher & Briefe*, 2:657–658.

28. Bergmann to Robert Weltsch, 14 January 1967, reproduced in Bergman, *Tagebücher & Briefe*, 2:521.

29. Diary entry of 4 May 1967, reproduced in Bergman, *Tagebücher & Briefe*, 2:530.

30. Carolyn Abraham, *Possessing Genius: The Bizarre Odyssey of Einstein's Brain* (New York: St. Martin's Press, 2002).

31. Josef Nesvadba, *Einsteinův mozek* (Prague: Mladá Fronta, 1960), 5; Jakub Arbes, *Newtonův mozek* (Prague: J. Otta, 1877).

32. A final, rather peculiar recent story can serve as a bookend for the fading—and occasional resurgence—of Einstein's Prague moment. Ludek Zakel was born in the city on 14 April 1932. As he told a reporter for the *New York Times* in 1995, he was Albert Einstein's biological son—at least, so he had been told. The woman who raised him, Eva Zakel, claimed that when she was about to give birth at St. Apollinarus Hospital in Prague, she came across Einstein's second wife, Elsa Einstein, who was ostensibly there having a tumor examined. According to the story, she was secretly pregnant, and neither she nor Albert wanted to risk having a child in Germany given the way politics were moving. When Eva Zakel's own newborn died a day after the Einstein child

was born, Elsa gave hers to Eva. Ludek Zakel became a physicist and bore some vague physical resemblance, at least judging by photographs, to Albert Einstein, but the story does not seem plausible. Setting aside any speculations about conjugal relations between the Einsteins at this point in their marriage, the fact remains that Elsa would have been 54 years old in 1932 (she was three years older than Albert), an unlikely age for a pregnancy. There is also no evidence that at this time Elsa was anywhere other than at the cottage in Caputh, outside Berlin, with her husband. Michael Specter, "Einstein's Son?: It's a Question of Relativity," *New York Times* (22 July 1995): 1, 5. One writer has speculated that while the Elsa part of the story is not true, maybe this was a love child of Einstein's with the Viennese woman Margarete Lebach, with whom he was known to be having an affair. Michele Zackheim, *Einstein's Daughter: The Search for Lieserl* (New York: Riverhead, 1999), 208–209.

Index

Page numbers in *italics* refer to illustrations.

Curie, Marie, 47, 49, 78, 98, 104–5

Curie, Pierre, 52, 98

Czechoslovak Academy of Sciences, 217–18, 246

Czechoslovakia: dismemberment of, 66, 171, 217, 232–33; dissolution of, 254; formation of, 84, 150, 171, 202, 213–14, 217; Germans in, 123, 125, 171–72, 236–37; prewar encirclement of, 230–31; romantic notions of, 220–21

Czecho-Slovakia, 141, 171, 232, 260

Czech Polytechnic, Brno, 225

A Czech Serving Girl (Brod), 101–3, 151, 161

Czech Technical University in Prague, 34, 218

Czech University, Prague, 21, 37–38, 53, 218, 221–22, 225

De astronomicis hypothesibus (Ursus), 154–55

Debye, Peter, 45, 125

Dehmel, Richard, 166

Delmedigo, Simon, 174, 175

Denmark, 28

Deutsche Arbeit, 45

dialectical materialism, 242–44, 246

The Dialectics of Nature (Engels), 242

Dialogue Concerning Two Chief World Systems (Galileo Galilei), 175

Diesseits und Jenseits (Brod), 173

Dingler, Hugo, 114, 297n36

Dittrich, Arnošt, 226

Dollfuss, Engelbert, 230

Doppler, Christian, 18, 115

Doppler effect, 115

Dratvová, Albína, 231, 232, 233

Dreyer, John Lewis Emil, 157, 162

Dreyfus Affair, 191

Dubček, Alexander, 247

Dublin, 1–2

Duhem, Pierre, 114

Dukas, Helen, 15

Eckart, Dietrich, 178

Eckstein, Josef, 109–10

Eddington, Arthur Stanley, 64, 66

Edict of Toleration (1781), 32, 191

Ehrenberg, Viktor, 200

Ehrenfels, Christian, Freiherr von, 94, 192–93

Ehrenfest, Paul, 67, 105, 110, 113, 184, 198

Ehrenhaft, Felix, 224

Einstein (Frank), 142–44, 146–49

Einstein, Albert, 257; Abraham viewed by, 69–72, 73; anti-Semitism suspected and denounced by, 42, 210–11; apocryphal accounts of, 9, 43, 144, 238–41; arrival in Prague of, 1, 3, 18–19, 44–46, 216; atomic bomb and, 143; in Berlin, 12, 184–85, 186, 195–96; in Bermuda, 14; Brod's correspondence with, 171–77; Brod viewed by, 147; as candidate for professorship, 21, 22–26, 38–44; citizenship of, 11–12, 13, 14, 28, 44, 84, 252; communist-era invocation of, 238; death of, 202; departure from Prague of, 1, 19, 75–78; dishevelment of, 5, 180, 239; divorce of, 104, 185; Engels viewed by, 242–43; equivalence principle of, 59–60, 67, 69–70, 72, 75–76, 118; ethnicity of, 16; extramarital affairs of, 97–98, 103, 331–32n32; Hanna Fantova and, 256–61; film project viewed by, 258–59; Frank encouraged by, 124; Frank's biography of, 43, 85, 99, 112, 142–44, 146–49, 157, 164, 170, 178, 182, 223, 251; general relativity attempted by, 8, 50–52, 61–62, 67–68, 72, 118; German Jews assailed by, 211–12; Halle conference shunned by, 135–36; Hitler's rant about, 178–79; homesickness declaimed by, 261–62; inaugural lecture of, 91–92; internationalism of, 11–14; international recognition of, 7–8, 64, 109, 197, 210; isolation of, 55, 56, 89, 224; Israeli presidency offered to, 202; Jewish identification of, 13, 15, 16, 42, 180–212, 240, 252; Kafka's meeting with, 7, 8; Kepler viewed by, 177–78; Kolman's view of, 243–44; Kraus's apology to, 140–41; Kraus's attacks on, 110, 128–29, 132–40; as